41 Springer Series in Solid-State Sciences

Edited by Manuel Cardona and Peter Fulde

Springer Series in Solid-State Sciences

Editors: M. Cardona   P. Fulde   H.-J. Queisser

Volumes 1 – 39 are listed on the back inside cover

Hans L. Skriver

The LMTO Method

Muffin-Tin Orbitals and Electronic Structure

With 34 Figures

Springer-Verlag
Berlin Heidelberg New York Tokyo 1984

Dr. *Hans L. Skriver*

Fysikafdelingen, Forsögsanlag Risø
DK-4000 Roskilde, Denmark

Series Editors:

Professor Dr. Manuel Cardona
Professor Dr. Peter Fulde
Professor Dr. Hans-Joachim Queisser

Max-Planck-Institut für Festkörperforschung, Heisenbergstrasse 1
D-7000 Stuttgart 80, Fed. Rep. of Germany

ISBN 3-540-11519-6 Springer-Verlag Berlin Heidelberg New York Tokyo
ISBN 0-387-11519-6 Springer-Verlag New York Heidelberg Berlin Tokyo

Library of Congress Cataloging in Publication Data. Skriver, Hans L., 1944- The LMTO method. (Springer series in solid-state sciences ; 41) Bibliography: p. Includes index. 1. Energy band theory of solids. 2. Linear free energy relationship. I. Title. II. Series. QC176.8.E4S54 1983 530.4'1 83-14735

Offset printing: Beltz Offsetdruck, 6944 Hemsbach/Bergstr. Bookbinding: J. Schäffer OHG, 6718 Grünstadt
2153/3130-5 4 3 2 1 0

Preface

The simplifications of band-structure calculations which are now referred to as *linear methods* were introduced by Ole K. Andersen almost ten years ago. Since then these ideas have been taken up by several workers in the field and translated into computer programmes that generate the band structure of almost any material. As a result, running times on computers have been cut by orders of magnitude.

One of the strong motivations behind the original proposal was a desire to give the conventional methods a physically meaningful content which could be understood even by the non-specialist. Unfortunately, this aspect of linear methods seems to have been less well appreciated, and most workers are content to use the latter as efficient computational schemes.

The present book is intended to give a reasonably complete description of one particular linear method, the Linear Muffin-Tin Orbital (LMTO) method, without losing sight of the physical content of the technique. It is also meant as a guide to the non-specialist who wants to perform band-structure calculations of his own, for example, to interpret experimental results. For this purpose the book contains a set of computer programmes which allow the user to perform full-scale self-consistent band-structure calculations by means of the LMTO method. In addition, it contains a listing of self-consistent potential parameters which, for instance, may be used to generate the energy bands of metallic elements. A major part of this listing is by Dieter Glötzel, and I wish to thank him for permission to include his results.

Although I have been involved with linear methods for a long time, I was by no means the first to make working LMTO programmes. I have in this respect benefitted greatly from early access to the computer codes generated by my predecessors Ove Jepsen and Uffe Kim Poulsen. In this context, I should also mention that Ove Jepsen performed the first calculations of the canonical bands for the fcc, bcc, and hcp crystal structures which are included in this book.

Part of my work has been performed at the laboratory of the National Research Council (NRC) of Canada in Ottawa and at the Kamerlingh Onnes Laboratory in Leiden, and I owe my sincere thanks to my colleagues at these institutions as well as to those at my home base Risø. I am especially grateful to J.-P. Jan at NRC who, from a desire to interpret his pioneering de Haas-van Alphen measurements on intermetallic compounds, helped perform the first LMTO calculations on alloys in 1975, and who continued to be a true collaborator until his untimely death in 1981.

The present book would not have existed without the work of my colleague and friend, Ole Krogh Andersen. In fact, large parts of the present monograph are based directly on his notes, and I am extremely grateful to him for allowing me to use these unpublished results.

An important role in the genesis of this book has also been played by Allan Mackintosh, who was largely responsible for the initiation of band-structure calculations in Denmark. He has furthermore kindly read the manuscript at various stages of completion and made numerous useful suggestions. Thanks are also due to Niels Egede Christensen who read the manuscript in its penultimate version, and who spotted several weak points in the text. The remaining flaws are my responsibility.

It is a pleasure to thank the Danish Natural Science Foundation for financial support during the writing of the book, and Risø National Laboratory for providing office and writing facilities. In this context, I wish to thank Agnete Gjerløv for her expert typing of the manuscript and her everlasting patience in the preparation of the half-dozen versions in which each chapter has appeared over the writing period.

Finally, to quote a fellow countryman, I should like to thank my parents for making the book possible, and my wife and children for making it necessary.

Roskilde, July 1983   *Hans L. Skriver*

Contents

1. Introduction

This volume proposes to describe one particular method by which the self-consistent electronic-structure problem may be solved in a highly efficient manner. Although the technique under consideration, the Linear Muffin-Tin Orbital (LMTO) method, is quite general, we shall restrict ourselves to the case of crystalline solids. That is, it will be shown how one may perform self-consistent band-structure calculations for infinite crystals, and apply the results to estimate ground-state properties of real materials.

It is obvious from the current literature that the LMTO method is only one of many techniques which may be used to solve the one-electron problem in crystalline solids. However, the method does combine a certain number of very convenient features which, to the author's mind, makes it one of the most desirable techniques currently available. This is so for several reasons.

Firstly, the LMTO method is cast in the form of a standard algebraic eigenvalue problem and therefore has the *speed* needed in self-consistent calculations. Secondly, although the method is an approximately one, it still has the *accuracy* required. Thirdly, it employs the same type of basis functions for all the elements, thus leading to a *conceptually consistent description* of physical trends throughout the periodic table. Fourthly, the method is *physically transparent* and may be used at several levels of approximation. Hence, the LMTO technique may be applied to problems ranging from complicated self-consistent calculations in crystals with many atoms per unit cell to the art of constructing the *simple analytic models* which make the computed results intelligible.

There is, however, a price for this versatility. The LMTO method is one of several *linear methods*, and like all the other linear techniques it is accurate only in a certain energy range. The present technique in particular should not be used for states too far above the Fermi level. If such states are required one may still solve the self-consistency problem by the LMTO technique and then turn to the Linear Augmented Plane Wave (LAPW) method for accurate calculation of the unoccupied high-lying levels. Furthermore, in the form in which it is presented here, the LMTO method is not well suited for open structures with low point symmetry. Again, one may turn to the LAPW method, which on the other hand is more complicated and less physically transparent. For that reason we prefer the LMTO techniques whenever applicable.

The introduction continues with a brief account of the approximations usually made to arrive at a solvable electronic-structure problem, and we shall discuss several of the methods applied to calculate band structures in solids. Then there is a brief summary of the history and development of the linear methods of band-structure calculations, followed by an outline of the remaining chapters.

1.1 The One-Electron Approximation

The theory of electronic states in infinite crystals has for many years been of great value in the quest for a better understanding of the chemical and physical properties of solid-state materials. We have learnt that electrons at the microscopic level govern the behaviour of these materials, and that one may obtain a surprisingly good description of many macroscopic properties in terms of the stationary states of the electronic system. One reason for this state of affairs is that the electrons are light particles which in their motion immediately follow the much heavier nuclei. In theoretical terms this means that the nuclei and the electrons to a good approximation may be treated separately, a procedure known as the Born-Oppenheimer approximation [1.1]. According to this, one may first solve for the electronic structure and then, at a later stage, use the energy of the electronic ground state obtained as a function of nuclear positions as a potential energy for the motion of the nuclei.

The electronic-structure problem consists in finding eigenstates for an infinite number of interacting fermions, and immediately calls for further approximations. The most important of these is the one-electron approximation, which describes each electron as an independent particle moving in the mean field of the other electrons plus the field of the nuclei. In this picture one is led to solve the one-electron Schrödinger equation

$$(-\nabla^2 + v)\Psi_j = E_j\Psi_j \quad , \tag{1.1}$$

where v is the total mean field, in order to find the one-electron energies E_j, and wave functions Ψ_j. Here and throughout we use atomic Rydberg units ($\hbar = 2m = e^2/2 = 1$).

2

The field v is usually referred to as the effective one-electron potential, and consists of the electrostatic field from the nuclei and the charge clouds of all the electrons plus corrections for exchange and correlation. It is often determined in a self-consistent manner in the spirit of the classical self-consistent field method of *Hartree* [1.2] and *Fock* [1.3]. For a supposed total field v one solves the Schrödinger equation (1.1), and populates the one-electron eigenstates according to the Pauli principle to give an electron density of

$$n(\mathbf{r}) = \sum_{j}^{occ.} |\Psi_j|^2 \; . \qquad (1.2)$$

A new field is constructed by solving Poisson's equation

$$\nabla^2 u(\mathbf{r}) = -8\pi n(\mathbf{r}) \qquad (1.3)$$

for the electronic contribution $u(\mathbf{r})$, to which is added the field from the nuclear point charges and the exchange-correlation corrections. With a weighted average of the new and old fields, the calculation is repeated and the cycle iterated until the two fields are consistent. When a self-consistent solution is obtained, the total electronic energy may be estimated from the one-electron energies and wave functions, the one-electron energies themselves being approximations to the one-electron excitation energies.

There is, in principle, nothing which limits the self-consistent field method to any particular form of the exchange-correlation potential, and the procedure outlined above has been used in connection with several approximations for exchange and correlation. Most notable in this respect is SLATER's Xα method [1.4] which has been applied to all atoms in the periodic table, to some molecules, and in the majority of the existing electronic-structure calculations for crystalline solids.

At present the most satisfactory foundation of the one-electron picture for metals is provided by the *local approximation* to the *density-functional formalism* of *Hohenberg* and *Kohn* [1.5] and *Kohn* and *Sham* [1.6]. This one-electron theory is presented in Chap.7 when ground-state properties of crystalline solids are discussed. Here we note that the *local-density* (LD) formalism, like the Xα method, leads to an effective one-electron potential which is a function of the local electron density as expressed by (1.2). Since the density in turn depends on the solutions of the effective one-electron Schrödinger equation (1.1), we are forced to perform self-consistent electronic-structure calculations. This proves to be a formidable task, and practicable, accurate solutions are possible only on large-scale digital computers.

1.2 The Energy-Band Problem

An energy-band structure consists of the eigenvalues of the one-electron Schrödinger equation

$$[- \nabla^2 + v(\mathbf{r})]\Psi_j(\mathbf{k},\mathbf{r}) = E_j(\mathbf{k})\Psi_j(\mathbf{k},\mathbf{r}) \qquad (1.4)$$

for a single electron moving in the local potential $v(\mathbf{r})$. The form of the equation and its solution hinge on the symmetries of the Hamiltonian $[- \nabla^2 + v(\mathbf{r})]$, which in turn are governed by $v(\mathbf{r})$. With a view to introducing the concepts usually encountered in band theory, we shall immediately discuss the implications of the symmetries of the potential.

In an infinite crystal the potential is invariant under lattice translations. One may therefore apply Bloch's theorem, and take the wave functions in the form of eigenfunctions of the translational operators, i.e.,

$$T_{\mathbf{R}}\Psi(\mathbf{r}) \equiv \Psi(\mathbf{r} + \mathbf{R}) = e^{i\mathbf{k}\cdot\mathbf{R}}\Psi(\mathbf{r}) \quad . \qquad (1.5)$$

Hereby, the electronic-structure problem is reduced to that of finding eigensolutions inside a single unit cell.

The lattice vectors $\mathbf{R}$ of real space are constructed as the integer linear combinations

$$\mathbf{R} = n_1\mathbf{a}_1 + n_2\mathbf{a}_2 + n_3\mathbf{a}_3 \qquad (1.6)$$

of the set of translational vectors $\{\mathbf{a}_i\}$, which defines the primitive lattice. An example of such a set of vectors may be found in Fig.1.1.

The Wigner-Seitz primitive cell of the lattice is the smallest region, of volume

$$\Omega = \mathbf{a}_1 \cdot (\mathbf{a}_2 \times \mathbf{a}_3) \quad , \qquad (1.7)$$

enclosed by planes which bisect lattice vectors. For crystals with only one kind of atom, the Wigner-Seitz atomic cell, or atomic polyhedron, is defined as the smallest volume enclosed by planes bisecting the interatomic distances. For structures containing only one atom per primitive cell the two types of Wigner-Seitz cells are identical, and an example of such cells is shown in Fig.1.2. The atomic Wigner-Seitz sphere is centred at the atom and has a volume equal to that of the atomic polyhedron. Its radius S is determined by

$$(4\pi/3)S^3 = \Omega \qquad (1.8)$$

for crystals with one atom per primitive cell.

The Bloch vector $\mathbf{k}$ which may be used to label the one-electron states is conveniently viewed as a vector in reciprocal space. A lattice vector $\mathbf{G}$ in

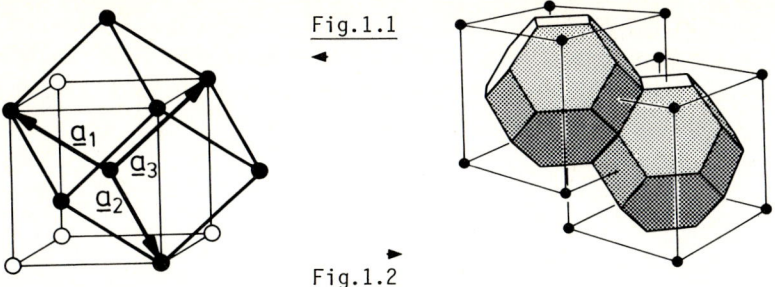

Fig.1.1

Fig.1.2

Fig.1.1. The three translational vectors a_1, a_2, a_3 which span the primitive cell of the body-centred cubic (bcc) structure. Also indicated is the conventional cubic unit cell

Fig.1.2. The Wigner-Seitz atomic polyhedron for the bcc structure

this space is constructed as the integer linear combination

$$G = m_1 b_1 + m_2 b_2 + m_3 b_3 \qquad (1.9)$$

of the set of translational vectors $\{b_j\}$ which satisfy

$$a_i \cdot b_j = 2\pi \delta_{ij} \quad . \qquad (1.10)$$

Regarded as functions of the wave vector k, the energy bands and wave functions have the translational symmetry of the reciprocal lattice, i.e.,

$$E(k) = E(k + G)$$

$$\psi(k,r) = \psi(k + G,r) \quad . \qquad (1.11)$$

Consequently all the non-equivalent k vectors may be confined to the primitive cell of volume $(2\pi)^3/\Omega$ in reciprocal space. Two examples of such cells, the Brillouin zones (BZ), which are the primitive Wigner-Seitz cells of the reciprocal lattice, are shown in Fig.1.3.

In addition to the Bloch vector, a complete description of the electronic states in a crystal requires a band index j, which may be defined such that

$$E_j(k) \le E_{j+1}(k) \quad . \qquad (1.12)$$

It is furthermore customary to label the individual bands according to the rotational symmetry of the k vector by group-theoretical means [1.7]. Two examples of band structures with symmetry labels are given in Figs.1.4,5. Others may be found in Chap.2.

At this stage one should mention some important quantities which may be derived immediately from the energy bands. The density of states in reciprocal space is uniform, and in the Brillouin zone each band has one state per

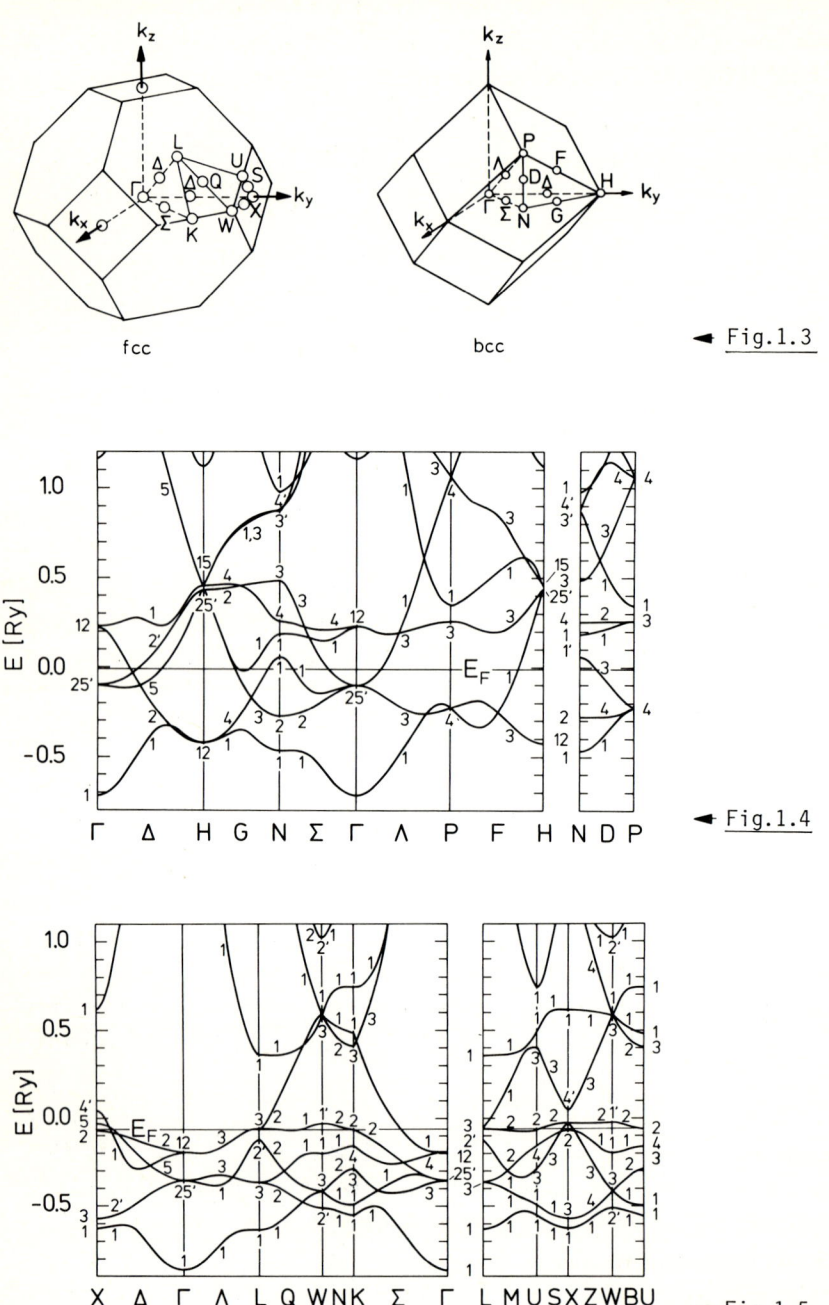

◄ Fig.1.3

◄ Fig.1.4

◄ Fig.1.5

(Captions to Figs.1.3-1.5 see opposite page)

primitive cell of the crystal lattice. The density of states in energy is therefore

$$N_\sigma(E) = (2\pi)^{-3}\Omega \sum_j \int_{BZ} d\mathbf{k} \; \delta[E - E_{j\sigma}(\mathbf{k})] \quad . \tag{1.13}$$

It is usual to include a factor 2 in (1.13) when the potential does not depend upon the spin σ of the electron. The number of states per spin is defined as

$$n_\sigma(E) = \int^E N_\sigma(E')dE' \tag{1.14}$$

and hence the Fermi level E_F which separates occupied and unoccupied states may be found from

$$n = \sum_\sigma \int^{E_F} N_\sigma(E)dE \quad , \tag{1.15}$$

where n is the number of electrons in the primitive cell of the crystal lattice.

In addition to the exact translational symmetry, the potential in a crystal possesses locally an approximate spherical symmetry. That is, in the vicinity of the atomic positions the potential is nearly spherically symmetric and resembles that of the corresponding free atom. Furthermore, in closely packed crystals the potential is extremely flat in the region between the atoms, and for such systems it is therefore nearly spherically symmetric even in the outer parts of the polyhedron.

A simplifying consequence of this approximate spherical symmetry of the potential inside each atomic polyhedron is that the wave function at energy E may be expressed approximately as

$$\Psi_j(\mathbf{k},\mathbf{r}) = \sum_{\ell m} b_{R\ell m}^{j\mathbf{k}} \psi_{R\ell}(E,|\mathbf{r} - \mathbf{R}|) i^\ell Y_\ell^m(\widehat{\mathbf{r} - \mathbf{R}}) \tag{1.16}$$

Fig.1.3. The Brillouin zones for the bcc and fcc structures. The conventional symmetry labels and the irreducible wedges, which by means of the rotational symmetry operations, i.e., the point group, may generate the full zone, are indicated

Fig.1.4. Self-consistent energy-band structure for bcc tungsten obtained by the LMTO method within the atomic-sphere approximation (ASA) using local-density theory for exchange and correlation. Relativistic effects are included except spin-orbit coupling which is neglected

Fig.1.5. Self-consistent energy-band structure for fcc platinum obtained similarly to that of tungsten on the previous figure

inside the polyhedron at **R**. Here Y_ℓ^m is a spherical harmonic and $\psi_{R\ell}$ a solution of the radial Schrödinger equation

$$\left[-\frac{d^2}{dr^2} + \frac{\ell(\ell + 1)}{r^2} + v(r) - E \right] r\psi_\ell(E,r) = 0 \tag{1.17}$$

for the spherically symmetric potential $v(r)$ at energy E. For reasons of notation included in (1.16) is a phase factor i^ℓ, and the phase convention of *Condon* and *Shortley* [1.8] is used for the spherical harmonics.

One important feature of (1.16) is that the radial and angular dependences have been separated inside the atomic polyhedron, and hence the well-known concepts of the central-field approximation are at our disposal to describe the system. We shall, in addition, adopt the "cellular" point of view that the difference between the one-electron spectrum in an atom and in a closely packed solid is primarily due to the different boundary conditions for the wave functions, and that the corresponding change in potential may be regarded as less important. We thus emphasise the relationship between the electronic states in atoms, and in molecules and solids.

1.3 Energy-Band Methods

Before presenting the linear method, let us briefly review how the energy-band problem has been tackled in the past. In this context we note that the traditional methods may be divided into those which express the wave functions as linear combinations of some fixed basis functions, say plane waves or atomic orbitals, and those like the cellular, APW, and KKR methods which employ matching of partial waves. As we shall see, both approaches have their strong and weak points.

Perhaps the most straightforward approach is to express the wave functions as Fourier series, but since an extremely large number of plane waves is required to expand the rapid oscillations near the nuclei, this method is useless in practice. It may, however, be made practicable for materials with broad conduction bands provided that the plane waves representing the conduction-band states are orthogonalised to the core states. In this form it is known as the orthogonalised plane-wave (OPW) method [1.9]. The orthogonalisation may be cast in the form of a repulsive contribution to the potential in the core region, and thus the OPW method leads to the widely used pseudopotential theory [1.10], which has the advantages of a weak potential and smooth pseudowave functions. If narrow bands like transition metal d bands are to be treated, one must add localised orbitals to the plane-wave basis

set as it is done, for instance, in the most recent development of the
pseudopotential method [1.11]. The resulting hybrid theory, however, is
much less elegant than traditional pseudopotential theory and may be cri-
ticised for being inconsistent with the smooth trends exhibited by the
energy-band structures as one proceeds through the periodic table. It is,
for instance, inconvenient to interpret the gradual lowering and narrow-
ing of the 3d band in the series K, Ca, Sc,..., Cu within a theory which
describes the d band of K by the d component of plane waves while that of
Cu is described by additional d orbitals.

In the "tight-binding" or LCAO method [1.12] one uses as basis functions the eigenfunctions $\chi_{n\ell m}$ of the bound states of the free atom, and the
wave function for an electron in the solid is then expressed in terms of the
Bloch sums

$$\psi_j(k,r) = \sum_R e^{ik \cdot R} \sum_{n\ell m} a^{jk}_{n\ell m} \chi_{n\ell m}(r - R) \tag{1.18}$$

of linear combinations of these atomic orbitals (LCAO's). By standard variational techniques one obtains a set of linear equations, i.e.,

$$(\underline{\underline{H}} - E\underline{\underline{O}}) \cdot a = 0 \quad , \tag{1.19}$$

in terms of the Hamiltonian $\underline{\underline{H}}$ and overlap $\underline{\underline{O}}$ matrices to determine the eigenvalues E and the expansion coefficients a.

The LCAO method is a typical fixed-basis method, and its advantages are
the local description based on the atomic viewpoint and the algebraic eigenvalue problem (1.19). Difficulties lie in the choice of a sufficiently small
and accurate basis set, and in the necessity to calculate a large number of
integrals involving potentials and orbitals centred at two and three different sites. It is therefore common to retain only the two-centre terms, and
treat these integrals as adjustable parameters in fitting procedures. Recently, the LCAO method has been revived and used in first-principles calculations [1.13].

We now turn to the partial-wave approach. _Wigner_ and _Seitz_ [1.14] suggested
that the spherical symmetry of the potential be extended all the way to the
boundaries of an atomic polyhedron. The wave functions in the solid can then
be described as the Bloch sum

$$\psi_j(k,r) = \sum_R e^{ik \cdot R} \sum_{\ell m} b^{jk}_{\ell m} \theta(r - R) \psi_\ell(E, |r - R|) i^\ell Y^m_\ell(\widehat{r - R}) \tag{1.20}$$

of functions of the form (1.16). In (1.20) we have assumed one atom per
primitive cell, and introduced θ which is unity inside the atomic polyhedron

and zero outside. Moreover, the energy-dependent radial part of the partial wave is obtained from numerical solutions of the radial Schrödinger equation (1.17). For a given Bloch vector $\mathbf{k}$, the one-electron energies $E_j(\mathbf{k})$ of the crystal are now those values of E for which a set of b coefficients can be found such that expansion (1.20) is continuous and differentiable across the boundary of the atomic polyhedron. It has appeared almost impossible to apply these boundary conditions rigorously, and the "cellular" method is little used in practice. It is, however, conceptually important and has given rise to the Wigner-Seitz rule concerning the extent of energy bands (Sect.2.3). Recently, the cellular method has been linearised and applied to a wide range of systems [1.15].

To avoid the troublesome boundary conditions encountered in the cellular method, *Slater* [1.16] suggested an alternative approach in 1937. In his augmented plane-wave (APW) method a sphere, the so-called *muffin-tin* (MT) *sphere*, is inscribed in each atomic polyhedron. Inside the sphere the potential is assumed to be spherically symmetric and the wave functions are expanded in the partial-wave solutions used by *Wigner* and *Seitz*. Outside the spheres the potential is assumed to be flat or slowly varying, and in this so-called *interstitial region* a plane-wave expansion is used. As a result, the boundary condition can be transformed into a matching of the wave functions and their first derivatives *at the sphere*. This proves to be a manageable problem, and the method has been one of the two most widely used techniques for calculating electronic properties of solids.

The *Korringa, Kohn, Rostoker* (KKR) method [1.17,18] employs an expansion inside the MT spheres similar to the cellular and APW methods. In the interstitial region between the spheres, however, the potential must be flat and the wave functions are expanded in phase-shifted spherical waves. The boundary condition can then be expressed as the condition for self-consistent multiple scattering between the muffin-tin spheres, or alternatively as the condition for destructive interference of the tails of these waves in the core region (Sect.2.1). This is the other most widely used computational technique in band theory.

The algebraic formulation of the matching condition differs for the various partial-wave methods, but in general the result is a set of linear, homogeneous equations of the form

$$\underline{\underline{M}}(E) \cdot \mathbf{b} = \mathbf{0} \ . \tag{1.21}$$

In contrast to the matrix $\underline{\underline{H}} - E\underline{\underline{O}}$ in (1.19), the secular matrix $\underline{\underline{M}}$ has a complicated, nonlinear energy dependence, and the one-electron energies E_j must be found individually by tracing the roots of the determinant of $\underline{\underline{M}}$ as a function of E. In self-consistent calculations where, in addition, the solutions $b_{\ell m}^{jk}$ are needed, these must be found by solving the linear equations at each individual root E_j.

Even for moderate size matrices $\underline{M}$, the partial-wave methods therefore require orders of magnitude more computer time than the solution of the eigenvalue problem (1.19). Furthermore, the formalisms of these methods are complicated, and perturbations are difficult to include because (1.21) are not derived from the variational principle for the one-electron Hamiltonian.

The partial-wave methods do, however, have two distinct advantages. Firstly, they provide solutions of arbitrary accuracy for a muffin-tin potential and, for close-packed systems, this makes them far more accurate than any traditional fixed-basis method. Secondly, the information about the potential enters (1.21) only via *a few* functions of energy, the logarithmic derivatives $\partial \ln|\psi_\ell(E,r)|/\partial \ln r$, at the muffin-tin sphere.

The linear methods devised by *Andersen* [1.19] are characterised by using fixed basis functions constructed from partial waves and their first energy derivatives obtained within the muffin-tin approximation to the potential. These methods therefore lead to secular equations (1.21) which are linear in energy, that is to *eigenvalue equations* of the form (1.19). When applied to a muffin-tin potential they use logarithmic-derivative parameters and provide solutions of arbitrary accuracy in a certain energy range. The linear methods thus combine the desirable features of the fixed-basis and partial-wave methods.

The linear muffin-tin orbital (LMTO) method described in detail in the following chapters employs a fixed basis set in the form of muffin-tin orbitals (MTO). A muffin-tin orbital is everywhere continuous and differentiable and, inside the MT spheres it is constructed from the partial waves $\psi_\ell(E_\nu,r)$, and their first energy derivatives

$$\dot{\psi}_\ell(E_\nu,r) \equiv [\partial \psi_\ell(E,r)/\partial E]_{E=E_\nu} \tag{1.22}$$

evaluated at a *fixed but arbitrary energy* E_ν. Outside the spheres the MTO is a spherical wave at fixed energy and of phase shift $\pi/2$.

If we use linear combinations of muffin-tin orbitals in a variational procedure we obtain the LMTO secular equations in the form (1.19) which will give all eigenvalues (and eigenvectors) at a given point in **k** space in one single diagonalisation. If, on the other hand, the so-called tail cancellation theorem is employed, we obtain the KKR equations which, within the so-called atomic-sphere approximation (ASA), offer a particularly simple and physically meaningful picture of the formation of energy bands. In practice, these KKR-ASA equations are more difficult to solve than the LMTO equations but mathematically the two methods are equivalent. For this reason we shall use the LMTO equations to calculate actual band structures and frequently use the KKR-ASA equations to interpret the results.

1.4 Brief History of Linear Methods

This monograph is based almost entirely on the work of *O.K. Andersen*. It is therefore appropriate to reveal the sources of the material presented, and at the same time give a brief history of the development of linear methods. At present several types of such methods are used, e.g. the linear muffin-tin orbitals (LMTO) method [1.19], the linear augmented plane-wave (LAPW) method [1.19], the augmented spherical-wave (ASW) method [1.20], and the linear rigorous cellular (LRC) method [1.15]. Of these the LMTO method, which was the earliest, will be our main concern.

The seeds for the development of linear methods may be found in the 1971 paper by *Andersen* [1.21] which contains a definition of muffin-tin orbitals, an addition theorem for tails of partial waves, and the tail cancellation theorem. Soon after, these ideas were developed into a practicable band-calculation method, the linear combination of muffin-tin orbitals (LCMTO) method [1.22, 23]. In the first attempt [1.22] the technique suffered from the appearance of false roots, but this shortcoming was subsequently removed by orthogonalisation of the MTO tails to the core states [1.23]. *Andersen's* unpublished *Trieste* notes [1.24] also date from this time, giving an elementary account of the linear theory including the atomic-sphere approximation and the concept of potential parameters.

In 1973 *Andersen* and *Woolley* [1.25] extended the LCMTO method to molecular calculations. At the end of their paper they introduced that choice of MTO tail, i.e. proportional to $\dot{\psi} = \partial\psi/\partial E$, which in a natural fashion ensured orthogonality to the core states and at the same time led to an accurate and elegant formulation of linear methods. The resulting $\psi,\dot{\psi}$ technique was immediately developed in a paper by *Andersen* [1.26] which, in a condensed form, contains most of what one need know about the simple concepts of linear band theory. Thus, we find here the KKR equation within the atomic-sphere approximation at this stage is called ASM; the LCMTO secular matrix, latter called the LMTO matrix; the energy-independent structure constants and the canonical bands; and the Laurent expansion of the logarithmic-derivative function and the corresponding potential parameters.

The theory underlying linear methods was presented rather extensively in the unpublished *Mont Tremblant* notes [1.27] and subsequently published in [1.19], which also contains the linear augmented plane-wave method. At the end of 1975 the LMTO method had been used in actual calculations by *Kasowski* [1.28-31], *Jepsen* [1.32], and *Jepsen* et al. [1.33]. The results showed that although the method is in principle approximate it has in practice an accuracy comparable to that normally obtained with the conventional APW and KKR methods. Computationally, the method was found to be orders of magnitude faster than the others in use at that time.

By 1976 the linear methods had come on age, and were used in a series of straightforward energy-band calculations which took advantage of the speed of the methods. *Koelling* and *Arbman* applied the LAPW method to copper [1.34], *Holtham* et al. calculated the Fermi surface of thallium [1.35], *Jan, Skriver* and co-workers used the LMTO method in a series of calculations on intermetallics including Pt_3Sn [1.36], MgHg and MgTl [1.37], Cu_2Sb [1.38], $PdTe_2$ [1.39], $AuCu_3$ [1.40], AgMg and AuMg [1.41], Al_2Cu [1.42] and NiSi [1.43], *Jarlborg* and *Arbman* applied the LMTO method to V_3Ga and other A15 compounds [1.44], and *Christensen* developed the first completely relativistic LMTO code which was subsequently used in calculations on Pd, Ag, Pt, and Au [1.45]. From this period date also the Chevrel-phase calculations by *Andersen* et al. [1.46], in which the physically simplifying assumptions of the LMTO-ASA and KKR-ASA methods are exploited to the full.

The concept of canonical bands [1.19] was used by *Pettifor* in a series of three papers [1.47-49] where he related the superconducting and cohesive properties of the 4d transition metals to the variation of ASA potential parameters as functions of volume and atomic number. By similar means *Duthie* and *Pettifor* [1.50] established a correlation with the d-band occupation numbers which explained the particular sequence of crystal structures found in the series of rare-earth metals.

Later developments of linear methods have been in the direction of self-consistent calculations of ground-state properties utilising local spin-density-functional formalism [1.51,52] for exchange and correlation. The basis of the self-consistency procedure was given in papers by *Madsen* et al. [1.53], *Poulsen* et al. [1.54] and *Andersen* and *Jepsen* [1.55], and was soon followed by results for the magnetic transition metals [1.56], the noble metals [1.57], some lanthanides [1.58], the actinides [1.59,60], and the 3d transition metal monoxides [1.61,62]. In this context one should also mention calculations of the electronic structure in transition metal compounds [1.63,64], A15 compounds [1.65,66], rare-earth borides [1.67], Chevrel

phases [1.68], actinide compounds [1.69], and semi-conductors [1.70,71].
Equations of state have been calculated for Cs [1.72], Be [1.73], Th [1.74]
and La [1.75].

The augmented spherical-wave method of *Williams* et al. [1.20] appeared in
1979 and is an efficient computational scheme to calculate self-consistent
electronic structures and ground-state properties of crystalline solids. According to its inventors it is a "direct descendant of the LMTO technique",
and a comparison will show that the two methods are indeed very similar.

Both techniques are based on the Rayleigh-Ritz variational principle using
energy-independent basis functions of the muffin-tin orbital type [1.21].

Both techniques employ a single fixed value of the kinetic energy variable κ^2, which in the LMTO-ASA method is taken to be zero, and in the ASW
is chosen to be small, e.g. -0.01 Ry, but not zero. For that reason the ASW
method does not include the important concept of canonical bands and the corresponding physically transparent description, and it does not, in contrast
to the LMTO method, have a scaling procedure which reduces the number of
energy-band calculations needed in a self-consistent calculation.

Both methods employ a linear approximation to the energy dependence of
the wave functions. In LMTO it is accomplished by a Taylor series in $\psi(E_\nu)$
and $\dot{\psi}(E_\nu)$ making the energy bands "exact" at the energy E_ν, which we may
choose freely. In ASW one expands in $\psi(C)$ and $\psi(V)$, i.e. one uses an energy
difference instead of a differential. Here C and V are the energies at which
the partial wave ψ joins on to a spherical Bessel and a spherical Hankel
function, respectively, and the energy bands are therefore "exact" at these
two energies. Now, C and V are determined by the potential and therefore can
not, in contrast to the arbitrary energy E in the LMTO method, be chosen so
as to minimise the errors in a particular energy range of interest. The ASW
may therefore be less accurate than the LMTO method in certain situations.

Both methods include a procedure for correcting some of the approximations made in the interstitial region. The ASW employs an elegant technique
in which the energy derivatives of the structure constants must be calculated. Similarly, the LMTO method requires an extra set of structure constants, but these will correct both for the approximations in the interstitial region *and* for the neglect of higher ℓ partial waves.

Recent applications of the ASW method include studies of heats of formation [1.76].

Although outside the scope of this book, one should also mention applications of the linear methods to thin films and surfaces [1.77-80], and to
molecules [1.81-85].

Recently, *Mackintosh* and *Andersen* [1.86] reviewed the LMTO-ASA method and *Götzel's* et al. applications of it to the simple, transition, and noble metals [1.87]. Some other results may be found in unpublished notes by *Andersen* [1.88]. An elementary review by *Andersen* of the linear methods and their applications has recently appeared in [1.89].

1.5 Organisation of the Book

There are two equally important aspects of the linear methods which I wish to emphasise here: physical transparency and computational efficiency. The two aspects will be represented by the KKR-ASA and the LMTO-ASA equations, respectively, which within the LMTO method may be regarded as Siamese twins. They are mathematically equivalent, they share the same so-called structure constants, and they supplement each other in the sense that the KKR-ASA equations have a mathematical form which leads to physically simple concepts while the LMTO-ASA equations are very efficient on a computer. With this duality in mind the content of the book is organised as follows.

In Chap.2 I deal with the simplest aspects of the LMTO method based upon the KKR-ASA equations. The intention is to familiarise the reader with the concepts and language used in linear theory. This is where I introduce structure constants, potential functions, canonical bands, and potential parameters, and where it is shown that the energy-band problem may be separated into a potential-dependent part and a crystal-structure-dependent part.

The real derivation of the LMTO method is in Chaps.5 and 6, but before we embark upon this I wish to emphasise the simple aspects of the method, e.g. the physical meaning of the potential parameters. In Chap.3 we therefore concentrate on the potential-dependent part of the energy-band problem by studying the energy dependence of the solutions to Schrödinger's equation inside a single sphere. This is where I introduce partial waves and their energy derivatives, i.e. $\phi, \dot{\phi}, \ldots$, which are subsequently used to construct energy-independent muffin-tin orbitals, Sect.5.4. We also expand the logarithmic derivative function $D(E)$ in a Laurent series, show that the coefficients in this expansion are related to the amplitudes of ϕ and $\dot{\phi}$ at the sphere, and present the trial function $\Phi(D)$ used in calculating the LMTO overlap integrals, Sect.5.7, and in a variational estimate of the energy function $E(D)$.

Chapter 4 is a slight digression in which I point out the physical significance of the four standard potential parameters used in the LMTO programme, and discuss the limitations of the parametrisations introduced. In addition, we derive free-electron potential parameters to be used in a correction to the atomic-sphere approximation, establish volume derivatives of some potential parameters for use in the calculation of ground-state properties, and end by giving a reasonably complete list of the many interrelations among the various potential parameters.

In Chap.5 we derive the LCMTO equations in a form not restricted to the atomic-sphere approximation, and use the ϕ, $\dot{\phi}$ technique introduced in Chap.3 to turn these equations into the linear muffin-tin orbital method. Here we also give a description of the partial waves and the muffin-tin orbitals for a single muffin-tin sphere, define the energy-independent muffin-tin orbitals and present the LMTO secular matrix in the form used in the actual programming, Sect.9.3.

In Chap.6 the atomic-sphere approximation is introduced and discussed, canonical structure constants are presented, and it is shown that the LMTO-ASA and KKR-ASA equations are mathematically equivalent in the sense that the KKR-ASA matrix is a factor of the LMTO-ASA secular matrix. In addition, we treat muffin-tin orbitals in the ASA, project out the ℓ character of the eigenvectors, derive expressions for the spherically averaged electron density, and develop a correction to the ASA.

At this stage we have completed the presentation of the LMTO method as an efficient procedure for solving the energy-band problem. *Inter alia* we have introduced a number of concepts which allow us to interpret the complete calculations.

In Chaps.7 and 8 it is shown how the LMTO method and the physically simple concepts contained in linear theory may be used in self-consistent calculations to estimate ground-state properties of metals and compounds. Here we treat the local-density approximation to the functional formalism of *Hohenberg*, *Kohn*, and *Sham*, and the force relation derived by *Andersen* together with an accurate and a first-order pressure relation. In addition, the LMTO-ASA and KKR-ASA methods are generalised to the case of many atoms per cell.

Chapter 9 contains a manual for a series of computer codes based upon the theory presented in the first 8 chapters of the book. With the programmes and the examples given there the user should be able to perform full-scale self-consistent calculations of his own. Finally, the book contains a table of self-consistent potential parameters which together with the LMTO programme will allow the user to reproduce the self-consistent energy bands of 61 metals at normal volume.

2. Canonical Band Theory

It is the concept of canonical bands which raises the LMTO method above the level of being just a new procedure for calculating energy-band structures. We shall therefore immediately give an account of the principles behind canonical band theory based upon the so-called KKR-ASA equations. The important ingredients are volume- and energy-independent *structure constants*, and *parameters* which contain information relating to the one-electron *potential*. When brought together, these quantities completely specify the energy-band structure of a given material in a given crystal structure. This canonical description, valid for closely packed crystalline solids, gives a physically simple picture of energy-band formation. In addition, it gives rise to a convenient scaling principle for state densities and it allows for analytic model calculations which may be used to understand, for instance, cohesive and magnetic properties.

The presentation in this chapter is based upon [2.1], and the reviews by *Andersen* and *Jepsen* [2.2], and *Mackintosh* and *Andersen* [2.3].

2.1 Muffin-Tin Orbitals and Tail Cancellation

Formally the energy-band structure for an infinite crystal is defined to be the eigenvalues $E_j(\mathbf{k})$ of the one-electron Schrödinger equation (1.4) obtained as functions of the Bloch vector $\mathbf{k}$. Physically, this definition is of course not very illuminating and I shall therefore now give the simplest possible derivation of a condition for the formation of energy bands, which has a very appealing physical interpretation.

The starting point is the energy-dependent orbital

$$\chi_{\ell m}(E,\mathbf{r}) = i^{\ell}Y_{\ell}^{m}(\hat{r}) \begin{cases} \psi_{\ell}(E,r) + p_{\ell}(E)(r/S)^{\ell} & r < S \\ \\ (S/r)^{\ell+1} & r > S \end{cases} \qquad (2.1)$$

where $\psi_{\ell}(E,r)$ is a solution of the radial Schrödinger equation (1.17) inside the atomic sphere of radius S given by (1.8). This orbital is regular, conti-

nuous and differentiable in all space and is essentially what is called a
muffin-tin orbital [2.3]. The so-called potential function $p_\ell(E,r)$ plus the
normalisation of the radial wave function $\psi_\ell(E,r)$ are determined by the requirement of continuity and differentiability at the sphere, i.e.

$$p_\ell(E) = \frac{D_\ell(E) + \ell + 1}{D_\ell(E) - \ell} \quad . \tag{2.2}$$

Here,

$$D_\ell(E) = \frac{S}{\psi_\ell(E,S)} \left. \frac{\partial \psi_\ell(E,r)}{\partial r} \right|_{r=S} \tag{2.3}$$

is the logarithmic-derivative function to which we shall return in Sect.2.3.

The tail of the orbital, $(S/r)^{\ell+1}$, is chosen to have zero kinetic energy,
i.e. it is a solution of Laplace's equation $\nabla^2 \chi = 0$. Consequently, the tail
centred at $\mathbf{R}$ may be expanded around the origin in terms of functions of the
form $i^\ell Y_\ell^m(\hat{r})(r/S)^\ell$, giving

$$\sum_{\mathbf{R} \neq 0} e^{i\mathbf{k}\cdot\mathbf{R}} \left(\frac{S}{|\mathbf{r} - \mathbf{R}|} \right)^{\ell+1} i^\ell Y_\ell^m(\widehat{\mathbf{r} - \mathbf{R}})$$

$$= \sum_{\ell' m'} \frac{-1}{2(2\ell' + 1)} \left(\frac{r}{S} \right)^{\ell'} i^{\ell'} Y_{\ell'}^{m'}(\hat{r}) s^{\mathbf{k}}_{\ell' m', \ell m} \tag{2.4}$$

which converges inside a sphere passing through the nearest-neighbour posi-
tions. The expansion coefficients $s^{\mathbf{k}}_{\ell' m', \ell m}$ in this Bloch sum of tails are the
canonical structure constants, which will play an important role in the LMTO
method. We shall return to them in Sect.2.2.

We are now in a position to obtain the condition that the linear combination of Bloch sums

$$\sum_{\ell m} a^{j\mathbf{k}}_{\ell m} \sum_{\mathbf{R}} e^{i\mathbf{k}\cdot\mathbf{R}} \chi_{\ell m}(E, \mathbf{r} - \mathbf{R}) \tag{2.5}$$

is a solution of the Schrödinger equation for the crystal. Inside the atomic
sphere the first term in the muffin-tin orbital $i^\ell Y_\ell^m(\hat{r}) \psi_\ell(E,r)$ is already
a solution of the Schrödinger equation and therefore the correct one-centre
expansion in the sphere at the origin and hence inside any other sphere will
be

$$\sum_{\ell m} a^{j\mathbf{k}}_{\ell m} i^\ell Y_\ell^m(\hat{r}) \psi_\ell(E,r) \quad , \tag{2.6}$$

provided that the tails from all the other spheres cancel the term

18

$$\sum_{\ell m} a_{\ell m}^{jk} i^{\ell} Y_{\ell}^{m}(\hat{r}) p_{\ell}(E)(r/S)^{\ell} \quad . \tag{2.7}$$

By means of the one-centre expansion (2.4) of the Bloch sum of MTO tails, the required *tail cancellation* is seen to occur if

$$\sum_{\ell m} \left[P_{\ell}(E) \delta_{\ell'\ell} \delta_{m'm} - S_{\ell'm',\ell m}^{k} \right] a_{\ell m}^{jk} = 0 \quad , \tag{2.8}$$

where the potential function $P_{\ell}(E)$ is defined by

$$P_{\ell}(E) = 2(2\ell + 1) \frac{D_{\ell}(E) + \ell + 1}{D_{\ell}(E) - \ell} \quad . \tag{2.9}$$

The condition that the Bloch sum of muffin-tin orbitals be a wave function for the crystal thus leads to the so-called KKR-ASA equations (2.8). They form a set of linear, homogeneous equations which have non-trivial solutions for the eigenvectors $a_{\ell m}^{jk}$ at those values of $E = E_{j}^{k}$ for which the determinant of the coefficient matrix vanishes, i.e.

$$\det \left[P_{\ell}(E) \delta_{\ell'\ell} \delta_{m'm} - S_{\ell'm',\ell m}^{k} \right] = 0 \quad . \tag{2.10}$$

The equations clearly exhibit two kind of terms: potential functions $P_{\ell}(E)$, and structure constants $S_{\ell'm',\ell m}^{k}$. While $P_{\ell}(E)$ is a function of energy which depends only on the potential inside the atomic sphere, the structure matrix $S_{\ell'm,\ell m}^{k}$ is a function of k which depends only on the crystal structure and not on the lattice constant. The KKR-ASA equations therefore establish the link between the potential and the structure-dependent parts of the energy-band problem which are otherwise completely decoupled within the ASA, and provide the connection between E and k which is the energy-band structure.

The physical interpretation of the KKR-ASA equations is that the energy-band problem may be approximated by a boundary-value problem in which the surrounding lattice through the structure constants imposes a k-dependent and non-spherically symmetric boundary condition on the solutions $P_{\ell}(E)$ inside the atomic Wigner-Seitz sphere. This interpretation is illustrated in Fig.2.1.

In the following sections we shall discuss the structure- and potential-dependent parts of the energy-band problem separately and introduce the concepts of canonical band theory.

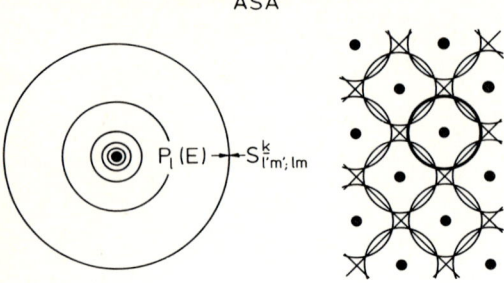

ASA

Atomic W-S Sphere   Crystal Structure

Fig.2.1. The connection between the crystal structure and the atomic sphere viewed as a crystal-structure-dependent boundary condition $S^{k}_{\ell'm',\ell m}$ on the solutions $P_{\ell}(E)$ to the Schrödinger equation (1.4) inside the sphere

2.2 Structure Constants and Canonical Bands

The structure matrix $S^{k}_{\ell'm',\ell m}$ of the KKR-ASA equations is given by the lattice sum

$$S^{k}_{\ell'm',\ell m} = g_{\ell'm',\ell m} \sum_{R \neq 0} e^{ik \cdot R} \left(\frac{S}{R}\right)^{\ell''+1} [\sqrt{4\pi} i^{\ell''} Y^{m''}_{\ell''}(\hat{R})]^{*} \quad , \qquad (2.11)$$

where $\ell'' = \ell + \ell$ and $m'' = m' - m$. We shall derive this expression in Sect. 6.4. The matrix (2.11) is Hermitian and has the form shown in Fig.2.2a. Because of the assumed spherical symmetry the potential function $P_{\ell}(E)$ which enters the KKR-ASA equations does not depend on the m quantum number, and one may transform the structure matrix from the m representation to an ℓi representation where the subblocks $S^{k}_{\ell m',\ell m}$ are diagonalised as indicated in Fig.2.2b. The $2\ell + 1$ diagonal elements $S^{k}_{\ell i}$ of each subblock are the *unhybridised* or *pure, canonical ℓ bands*.

The above unitary transformation is independent of potential and atomic volume, and leaves the form of the KKR-ASA equations invariant. If we therefore neglect hybridisation, i.e. set the elements of $S^{k}_{\ell'i',\ell i}$ with $\ell \neq \ell'$ to zero, the pure nℓ energy band $E_{n\ell i}(k)$ is simply found as the n'th solution of

$$P_{\ell}(E) = S^{k}_{\ell i} \quad , \qquad (2.12)$$

which is merely a monotonic mapping of the canonical bands on to an energy scale specified by the n'th branch of the potential function. To solve (2.12) with respect to energy we need a parametrisation of the potential function. When this is obtained, (2.12) will lead to an *unhybridised* or *pure ℓ-energy-band structure*.

Examples of canonical s-, p-, d-, and f-band structures are given in Figs.2.3-5. Other examples may be found in [2.2-5]. In Tables 2.1-3 the

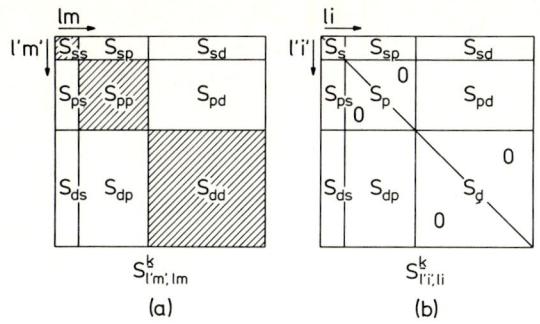

Fig.2.2. Lay-out for the canonical structure constant
matrix in the ℓm representation (a), and in the ℓi representation (b). In the latter
the diagonal exhibits the canonical s, p, and d bands

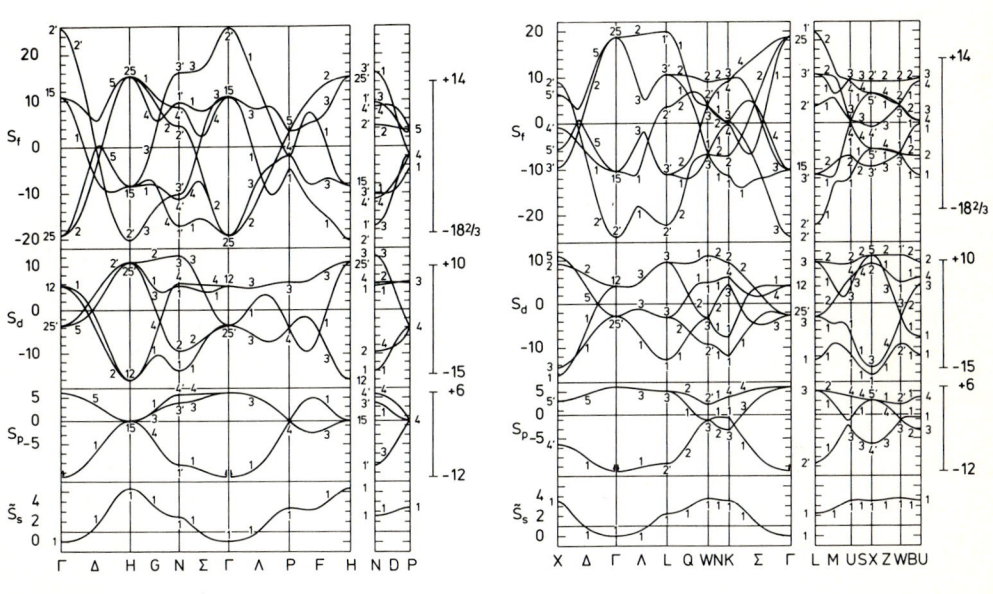

Fig.2.3

Fig.2.4

Fig.2.3. Canonical s, p, d, and f bands for the bcc structure with conventional symmetry labels. The symmetry lines and points are those of the irreducible wedge, Fig.1.3 and S is defined in (2.14)

Fig.2.4. Canonical s, p, d, and f bands for the fcc structure with conventional symmetry labels. The symmetry lines and points are those of the irreducible wedge, Fig.1.3 and Š is defined in (2.14)

canonical bands at symmetry points and on some symmetry lines in the fcc,
bcc, and hcp crystal structures are listed. These figures and tables may be
used to construct a first estimate of the energy-band structure in an fcc,
bcc or hcp crystal from the parameters of the potential function. Several
such estimates for other crystal structures exist in the literature [2.6-11],
and we shall return to this aspect in Sect.2.4 when we have presented the
necessary potential parameters.

21

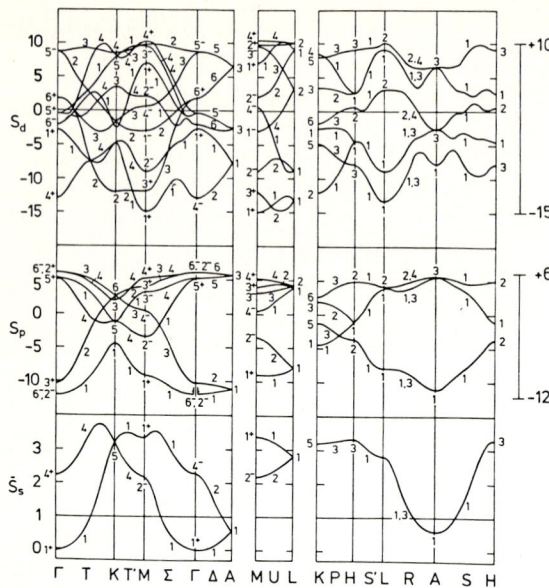

Fig.2.5. Canonical s, p, and d bands for the hcp structure with conventional symmetry labels. $\check{S}$ is defined in (2.14)

Table 2.1. Canonical bands $S^{k}_{\ell i}$ at points of high symmetry in the bcc structure

point[a]

ℓ	Γ (0,0,0)		H (2,0,0)		N(1,1,0)		P(1,1,1)		Δ(1,0,0)		Λ(1/2,1/2,1/2)	
	2'	25.666	25		3'	16.072	5		5		1	12.719
	15		25	15.192	1'	9.480	5	3.491	5	5.747	3	
f	15	10.692	25		4'	8.639	5		2'	1.350	3	8.269
	15		15		2'	4.515	4	-1.940	5		1	- 4.741
	25		15	- 8.440	3'	-10.218	4		5	-0.516	3	
	25	-19.250	15		4'	-11.288	1	-4.655	2	-5.273	3	- 6.811
	25		2'	-20.257	1'	-17.199			1	-6.539	2	-10.892
	12		25'		3	12.293	3		2'	3.841	3	
	12	5.395	25'	10.793	4	6.062	3	6.213	5		3	5.190
d	25'		25'		1	5.396	4	-4.142	5	0.825	1	2.833
	25'	- 3.597	12		2	- 9.605	4		1	-1.238	3	
	25'		12	-16.189	1	-14.146			2	-4.254	3	- 6.606
	15		15		4'	5.688	4		5		3	
p	15	6.000	15	0.000	3'	3.834	4	0.000	5	3.469	3	4.838
	15	(-12.000)	15		1'	- 9.522	4		1	-6.939	1	- 9.676
s	1	- ∞	1	2.004	1	1.458	1	1.721	1		1	

[a]Units of π/a.

The canonical bands have a number of properties which follow from (2.11), which we shall now discuss. The proofs may be found in [2.1.13].

A pure canonical s band diverges at the centre of the Brillouin zone, i.e.

$$S_s(\mathbf{k}) \to - 6(kS)^{-2} + \text{const.} \quad , \quad \text{for} \quad k \to 0 \quad . \tag{2.13}$$

22

Table 2.2. Canonical bands $S_{\ell i}^{k}$ at points of high symmetry in the fcc structure

point[a]

ℓ	Γ(0,0,0)		X(0,2,0)		W(1,2,0)		L(1,1,1)		K($1\frac{1}{2},1\frac{1}{2}$,0)		Δ(0,1,0)	
	25		2'	9.019	2	8.976	1'	19.817	3	9.612	5	7.390
	25	18.684	5'		2'	3.982	3'		4	8.966	5	
	25		5'	6.333	3		3'	10.622	3	0.332	2	4.516
f	15		4'	− 0.849	3	3.620	2'	3.525	4	0.121	5	
	15	−10.380	5'		1	− 6.587	3'		1	− 0.331	5	−2.150
	15		5'	− 5.654	3		3'	−11.198	2	− 7.206	1	−6.549
	2'	−24.912	3'	− 9.528	3	− 6.805	2'	−22.190	1	−11.494	2'	−8.445
	12		5		1'	10.936	3		2	9.241	2	6.636
	12	4.125	5	10.927	1	4.891	3	9.515	4	5.967	5	
d	25		2	9.126	3		3		3	4.393	5	3.768
	25'	− 2.750	3	−14.589	3	− 3.261	3	− 3.057	1	− 7.658	1	−5.653
	25'		1	−16.391	2	− 9.306	1	−12.917	1	−11.942	2'	−8.520
	15		5'		2'	2.259	3		4	3.840	5	
p	15	6.000	5'	3.106	3		3	5.177	1	0.593	5	4.600
	15	(−12.000)	4'	− 6.212	3	− 1.129	2'	−10.355	3	− 3.247	1	−9.200
s	1	− ∞	1	1.763	1	1.809	1	1.366	1	1.769	1	−0.598

[a]Units of π/a.

Table 2.3. Canonical bands $S_{\ell i}^{k}$ at points of high symmetry in the hcp structure

point[a]

ℓ	Γ(0,0,0)		K($\frac{4}{3}$,0,0)		M($\frac{4}{3},\frac{2}{3}$,0)		L($\frac{4}{3},\frac{2}{3},\frac{a}{c}$)		A(0,0,$\frac{a}{c}$)		H($\frac{4}{3}$,0,$\frac{a}{c}$)	
	5⁻		4	8.41	4⁺	10.36	2		3		3	
	5⁻	8.61	5		2⁺	9.65	2	10.06	3		3	9.11
	6⁺		5	7.93	3⁻	9.53	1		3	6.51	1	
	6⁺	1.88	3	3.50	1⁺	7.09	1	9.09	3		1	2.68
	5⁺		6		2⁻	2.06	2		3		2	
d	5⁺	− 0.47	6	− 1.80	4⁻	0.52	2	3.17	3		2	0.60
	6⁻		1	− 2.60	1⁻	− 2.99	1		3	− 2.63	1	
	6⁻	− 2.14	5		2⁻	− 9.12	1	− 8.88	3		1	−4.48
	1⁺	− 2.86	5	− 4.78	3⁺	−12.68	1		1	− 7.76	3	−7.91
	4⁻	−12.91	2	−12.01	1⁺	−15.01	1	−13.44	1		3	
	6⁻		6		4⁺	5.180	2		3		2	4.807
	6⁻	6.000	6	2.546	3⁺	3.928	2	4.149	3	5.888	2	
	2⁻	6.000	3	1.871	3⁻	3.107	1		3		1	−1.082
p	5⁺		5		4⁻	0.437	1	3.815	3		1	
	5⁺	5.180	5	− 1.273	2⁻	− 3.544	1		1		2	−3.725
	3⁺	−10.362	1	− 4.417	1⁺	− 9.109	1	− 7.964	1	−11.175	2	
s	4⁻	1.356	5		1⁺	1.731	1		1		3	
	1⁺	− ∞	5	1.694	2⁻	1.334	1	− 1.580	1	− 1.907	3	1.718

[a]Units of π/a in hexagonal coordinates, $\mathbf{k}_a$, $\mathbf{k}_b$ are at an angle of 120°.

For that reason we have used a free-electron-like scale[1]

$$\tilde{S}_s = [1 - (2/\pi)^2 S_s]^{-1} \tag{2.14}$$

to plot the canonical s bands in Figs.2.3-5. The divergence of the s structure constant is matched by a similar divergence in the s potential function.
Specifically, we find from (3.42) that

$$P_s(E) \rightarrow \frac{-6}{(E - V_s)\tau_s S^2} \quad , \quad \text{for} \quad E \rightarrow V_s \tag{2.15}$$

and hence the energy band, of course, behaves like the parabola

$$E_s(\mathbf{k}) = V_s + k^2/\tau_s \quad , \quad \text{for} \quad k \rightarrow 0 \tag{2.16}$$

of mass τ_s and minimum energy V_s.

The longitudinal branch of the canonical p band is discontinuous at the
centre of the zone, i.e. it tends towards the value -12 rather than +6. This
behaviour, which is intimately connected with the requirement that the canonical bands be independent of the scale of the lattice, might seem patholo-
gical. However, the longitudinal p branch hybridises strongly with any s
band, and thereby eventually becomes perfectly continuous at the centre of
the zone. Furthermore, real energy-band structures do in fact show a "soft"
longitudinal p band, whose dispersion depends sensitively on whether the s
band with which it hybridises lies above or below it, e.g. the Cu p Δ_1 band
in fcc Cu looks quite different from the Cl p Δ_1 band in NaCl.

The centre of gravity of a canonical band with $\ell > 0$ is zero at each
value of the Bloch vector $\mathbf{k}$, i.e.

$$\sum_{i=1}^{2\ell+1} S_{\ell i}^{\mathbf{k}} = 0 \quad . \tag{2.17}$$

The average over the Brillouin zone therefore vanishes, i.e.

$$\sum_{i=1}^{2\ell+1} \int_{BZ} S_{\ell i}^{\mathbf{k}} \, d\mathbf{k} = 0 \tag{2.18}$$

and this also holds for $\ell = 0$.

The width of a canonical band may be estimated from the second moment

$$S_\ell^2 = \frac{1}{2\ell + 1} \sum_{i=1}^{2\ell+1} \frac{\Omega}{(2\pi)^3} \int_{BZ} (S_{\ell i}^{\mathbf{k}})^2 \, d\mathbf{k}$$

---

[1] The form (2.14) is taken from (2.29), the γ parameter is obtained from
(2.26), with $\mu = 1$ and $(C-V)S^2 = \pi^2/4$, Table 4.3.

$$= 2^{\ell+2}(2\ell + 1)\frac{(2\ell + 1)(2\ell + 3)\ldots(4\ell - 1)}{1 \times 2 \times \ldots \times \ell}\sum_{R \neq 0}\left(\frac{S}{R}\right)^{2(2\ell+1)} \tag{2.19}$$

which depends only upon the radial distribution of the atoms in the crystal. Within a few per cent the second moments are independent of the bcc, fcc, and hcp crystal structure and on the average one finds S_ℓ^2 = 29, 46, and 81 for ℓ = 1,2, and 3. If one assumes an idealised, rectangular state density of width w and height $(2\ell + 1)/w$ positioned around S = 0, one finds $w = \sqrt{12S_\ell^2}$. With this estimate the widths of the canonical bands are 19, 23, and 31 for p, d, and f states, respectively.

The second moment of the canonical s band diverges, but the maximum values of S_s may be estimated under the assumption that (2.13) holds throughout the Brillouin zone. If we assume this zone to be a sphere of radius $k_F = (9\pi/2)^{1/3}/S$ and determine the constant in (2.13) such that the first moment vanishes, as it must according to (2.18), we find

$$\max S_s \approx 12(2/9\pi)^{2/3} = 2.05 \quad . \tag{2.20}$$

Hence, the canonical s band ranges from $-\infty$ to approximately 2 and $\tilde{S}_s$ defined in (2.14) from 0 to 5.9.

A comparison with Figs.2.3-5 and Tables 2.1-3 shows that the extent of the canonical bands estimated above is quite accurate. We shall return to this question in the next section when we discuss the Wigner-Seitz rule concerning the width of the ℓ band.

2.3 Potential Function and the Wigner-Seitz Rule

The potential function $P_\ell(E)$ which appears in the KKR-ASA equations is uniquely related to the logarithmic derivative functions $D_\ell(E)$ through the definition (2.9).

It is well known that the logarithmic derivative (2.3) is a monotonically decreasing function of energy except at its singularities. Hence, the potential function (2.9) is an increasing function of energy, and the two functions have the forms shown schematically in Fig.2.6. For each ℓ they consist of periods in energy labelled by the principal quantum number n, and separated by the energies $V_{n\ell}$ defined by (2.21) below. The advantage of working with $P_\ell(E)$ rather than with $D_\ell(E)$ is that the poles of the former function are outside the range of the ℓ band.

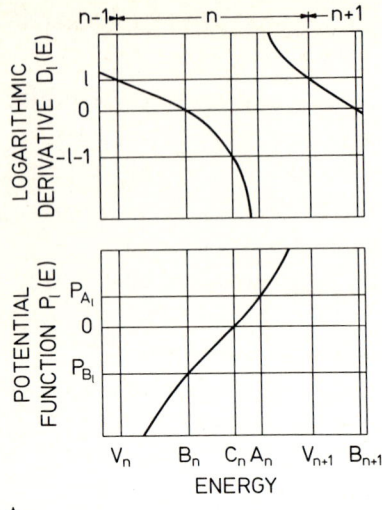

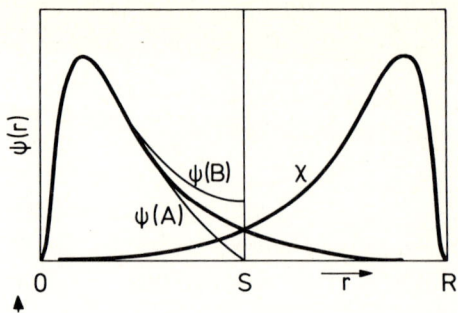

Fig.2.7. Atomic orbitals χ and partial waves ψ in a diatomic molecule. The bonding state inside the sphere centred at the origin 0 is $\chi(r) + \chi(r - R)$ in terms of the atomic orbitals and $\psi(B,r)$ in terms of the partial waves. Similarly, the antibonding state is $\chi(r) - \chi(r - R) \simeq \psi(A,r)$

Fig.2.6. The n'th period of the logarithmic-derivative function $D_\ell(E)$, and the corresponding potential function $P_\ell(E)$. The values P_{A_ℓ} and P_{B_ℓ} equal ℓ and $-2(2\ell + 1)(\ell + 1)/\ell$, respectively

The energies $V_{n\ell}$ separating the periods are defined by

$$D_\ell(V_{n\ell}) = \ell \qquad\qquad (2.21)$$

and, within each period, we further define the three parameters B_ℓ, C_ℓ, and A_ℓ through

$$D_\ell(B_\ell) = 0$$

$$D_\ell(C_\ell) = -\ell-1$$

$$D_\ell(A_\ell) = -\infty \quad . \qquad\qquad (2.22)$$

As we shall now explain, these four energies represent the square-well pseudopotential, the bottom, the centre, and the top of the $n\ell$ band, respectively, and thus give the first crude approximation to the properties of this band.

The intuitive Wigner-Seitz rule used, for instance, in the renormalised atom approximation [2.12] states that an ℓ band is formed in the energy range where the logarithmic derivative is negative. The physical origin of this rule may be appreciated by appeal to the diatomic H_2 molecule. As shown in Fig.2.7 there are in this case two states: a bonding state $\psi(B,r)$ corresponding to the boundary condition $\psi'(B,S) = 0$, i.e. $D(B) = 0$, and an antibonding state $\psi(A,r)$ corresponding to the boundary condition $\psi(A,S) = 0$, i.e. $D(A) - -\infty$. The bonding state concentrates the electrons in the region between the atoms and has the lower energy.

26

Table 2.4. Canonical bandwidths (P scale) obtained from the second moment w_M, the canonical bands (Tables 2.1,2) w_S, and the Wigner-Seitz rule (2.23) w_{WS}, for fcc and bcc crystal structures

	w_M^{bcc}	w_S^{bcc}	w_{WS}	w_S^{fcc}	w_M^{fcc}
s	$-\infty$,2.05	$-\infty$,2.00	$-\infty$,2	$-\infty$,1.81	$-\infty$,2.05
p	18.8	15.7	18	16.4	18.7
d	23.8	28.5	25	27.3	23.5
f	31.8	45.9	32.7	44.7	30.5

In a solid the states of lowest or highest energy, B or A respectively, and with predominant ℓ character, correspond to bonding or antibonding between most of the nearest neighbours. In addition, a band of states between B and A exists arising from all possible combinations of bonding and antibonding between nearest neighbours throughout the crystal.

In terms of the potential function the Wigner-Seitz rule corresponds to the range

$$- 2 \frac{2\ell + 1}{\ell} (\ell + 1) < P_\ell < 2(2\ell + 1) \tag{2.23}$$

which yields an ℓ band width on the "P scale" of $2(2\ell + 1)^2/\ell$. According to the KKR-ASA equations (2.12) for the unhybridised case, the limits (2.23) should correspond to the extremes of the canonical bands. This is exactly true for the bottom of the s band and approximately true in all other cases, as may be seen from Figs.2.3-5 and in Table 2.4. The canonical bands therefore confirm the intuitive Wigner-Seitz rule.

It follows from the above discussion and the relation (2.12) that one may call the energies B_ℓ and A_ℓ the bottom and the top, respectively, of the ℓ band. Similarly, the facts that the first moment of the canonical bands vanishes and that the potential function evaluated at $D_\ell(C_\ell) = -\ell-1$ is zero lead us to call C_ℓ the centre of the ℓ band. Finally, we note that the energy $V_s = B_s$ is the bottom of the s band. In the free-electron case all V_ℓ with n = 0 take the value of the uniform potential independent of ℓ, and therefore the parameter V_ℓ is named the square-well pseudopotential.

2.4 Potential Parameters, Unhybridised, and Hybridised Bands

The ASA potential parameters may be introduced to solve the unhybridised KKR-ASA equations (2.12) explicitly. This is a simple procedure both in principle and practice, although it has given rise to some confusion. The reasons are that the parametrisation may be done in a number of ways, that the sets of parameters which are convenient differ from context to context, and that the many interrelations among the potential parameters appear difficult to grasp. As usual, the only real way out of the confusion is to obtain practical experience with the material. This we shall do here and postpone the derivations and exact definitions until Chap.3.

We shall first introduce the parametrisation in the simplest possible manner, using some of the parameters already defined. It turns out that three parameters are needed to describe the energy dependence of the potential function $P_\ell(E)$ with reasonable accuracy over an energy range of the order of $A_\ell - B_\ell$. Hence, on the basis of hindsight (Sect.3.4) we construct the following expression

$$P_\ell(E) \approx \frac{1}{\gamma_\ell} \frac{E - C_\ell}{E - V_\ell} \qquad (2.24)$$

which has the correct behaviour at V_ℓ and C_ℓ (Fig.2.6). We note that by enforcing this behaviour we have exhausted two of the three parameters and that the remaining one, i.e. γ_ℓ, cannot be defined such that the expression is correct at both B_ℓ and A_ℓ. Instead one may, for instance, introduce the definition (4.16) of γ_ℓ (explained later), and the parametrised function (2.24) can then be used to obtain estimates of the bottom B_ℓ and the top A_ℓ of the ℓ band.

The parametrisation may be inverted to give the function E(P) inverse to P(E), i.e.

$$E(P) \approx C + (C - V) \frac{\gamma P}{1 - \gamma P}$$

and equivalently $\qquad\qquad\qquad\qquad\qquad\qquad\qquad\qquad (2.25)$

$$E(P) \approx V + (C - V) \frac{1}{1 - \gamma P} \quad .$$

Here and in the following we shall often drop the subscript ℓ. Instead of the parameter $(C - V)$, which is obviously connected to the bandwidth, one may introduce the band mass parameters μ and τ

$$\mu \approx 1/[\gamma(C - V)S^2] \qquad (2.26)$$

$$\tau \approx 2(2\ell + 1)^2(2\ell + 3)\gamma/[(C - V)S^2] \qquad (2.27)$$

obtained from (4.9,10,16,17). Here we should mention that μ and τ are unity for free electrons. From (2.12,25,26,27) the pure or unhybridised bands may be obtained directly in terms of the canonical bands and potential parameters as the simple forms

$$E_{\ell i}(\mathbf{k}) \approx C_\ell + \frac{1}{\mu_\ell S^2} \frac{S_{\ell i}^{\mathbf{k}}}{1 - \gamma_\ell S_{\ell i}^{\mathbf{k}}} \tag{2.28}$$

and equivalently

$$E_{\ell i}(\mathbf{k}) \approx V_\ell + \frac{2(2\ell + 1)^2(2\ell + 3)}{\tau_\ell S^2} \frac{1}{\gamma_\ell^{-1} - S_{\ell i}^{\mathbf{k}}} . \tag{2.29}$$

These expressions represent monotonic mappings of the canonical bands from an S or P scale to an energy scale. The interpretation of (2.28) is that the pure ℓ band is obtained from the canonical ℓ-band structure by fixing the position through C_ℓ, scaling it by $\mu_\ell S^2$, and distorting it by γ_ℓ. In the case of d bands the distortion is small, and if it is neglected we find from (2.28) and the Wigner-Seitz rule (2.23) that the bandwidth W_d on an energy scale is related to the band mass μ_d, as $W_d = 25/\mu_d S^2$. Similarly, we verify by means of (2.29,13) that the bottom of the pure s band falls at V_s and has the mass τ_s (2.16).

Conceptually, (2.28,29) are very satisfactory in that they give rise to the simple picture of energy-band formation shown in Fig.2.8. From a practical point of view they may also be used in analytic model calculations of band structures and related properties in a variety of materials. Hence, although the parametrisation presented so far is of limited accuracy, the expressions derived are extremely useful as tools for interpretation.

We shall illustrate the content of the present section with the examples of the 3d band transition metal chromium [2.13] in its bcc paramagnetic phase and the 6s6p band metal thallium [2.6] in its hcp phase.

A comparison between the unhybrised bands of Cr, Fig.2.9, and the canonical bcc bands of Fig.2.3 shows clearly how energy bands are formed by the simple scaling contained in (2.28,29). Furthermore, the distortion parameter

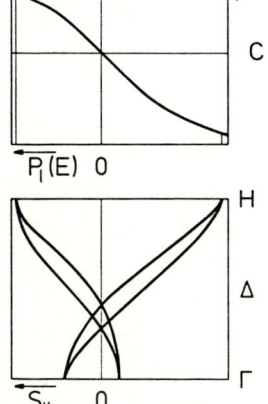

Fig.2.8. Illustration of how the canonical bands $S_{\ell i}$ along one symmetry direction are transformed into unhybridised energy bands by the non-linear scaling prescribed by (2.12). The process is most easily understood if instead of the potential function $P_\ell(E)$ one considers its inverse parametrised by (2.25)

29

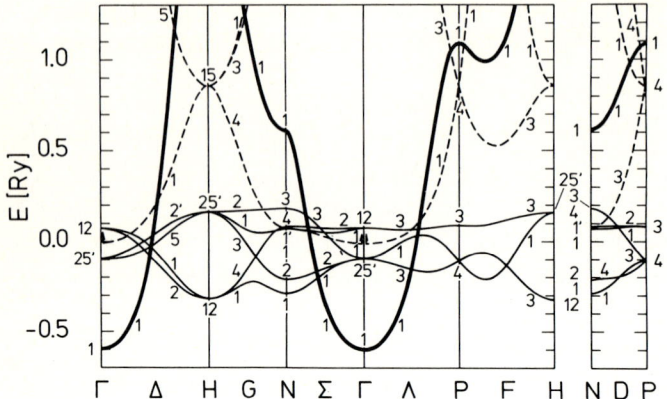

Fig.2.9. Unhybridised band structure for non-magnetic chromium (s band: heavy line; p band: broken line; d band: thin line). The energy bands are synthesised from the canonical s, p, and d bands shown in Fig.2.3 by means of the potential parameters given in Table 4.1. The scaling function used in the actual calculation was the variational estimate (3.50), and (4.18), both explained below. Within the accuracy of the figure we might equally well have used either (2.28,29)

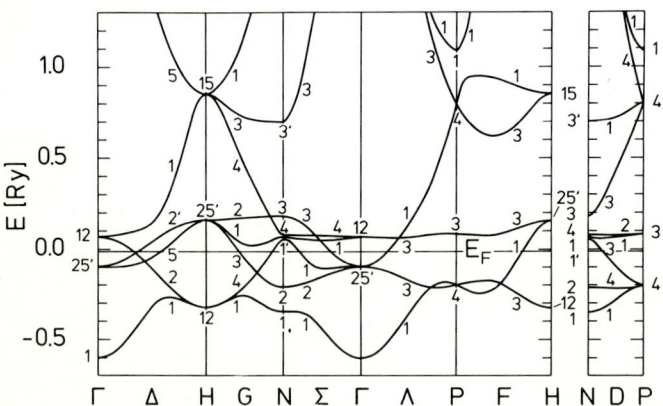

Fig.2.10. Self-consistent, fully hybridised energy-band structure for non-magnetic chromium obtained as for tungsten and platinum, Figs.1.4,5. Comparison with the unhybridised bands in the previous figure gives a feeling for the accuracy one may obtain within the very simple unhybridised canonical band theory. It also gives an idea of the effect and importance of hybridisation, defined essentially as the difference between the two figures

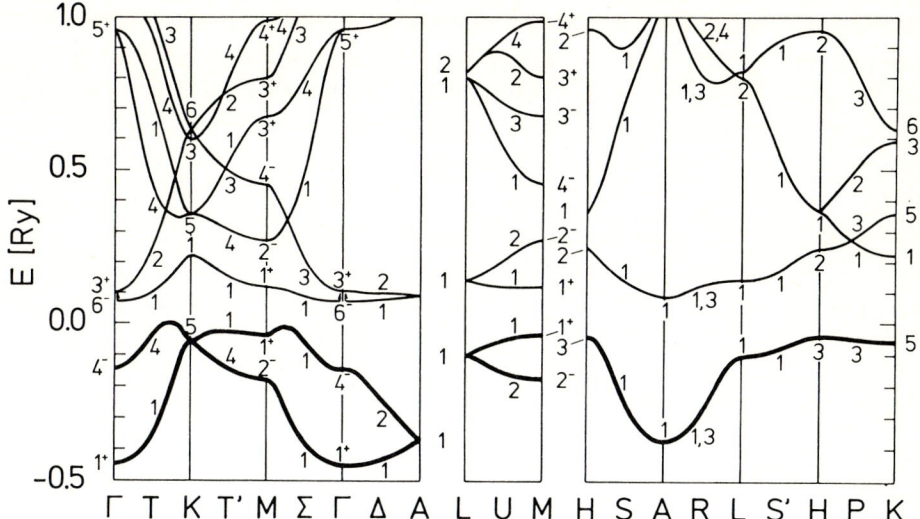

Fig.2.11. Unhybridised band structure for hcp thallium as obtained in [2.6]

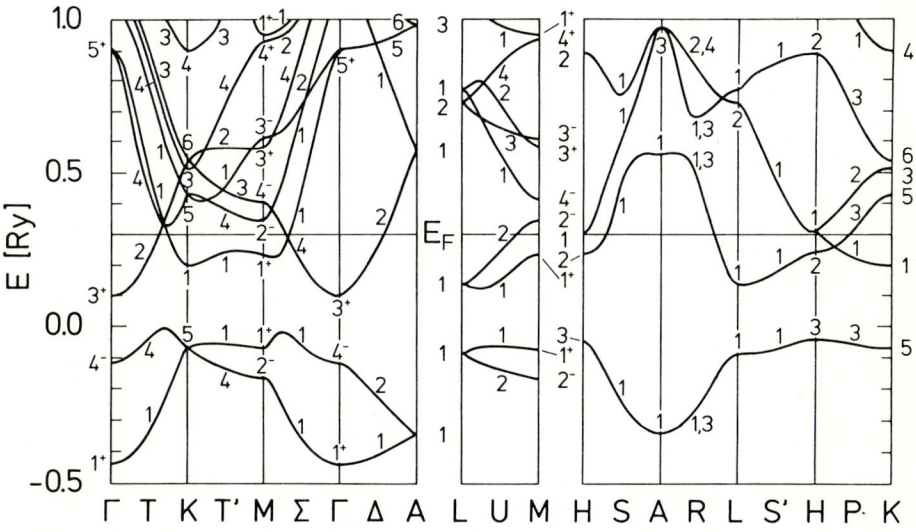

Fig.2.12. Fully hybridised energy-band structure for thallium as given in [2.6]. As in the case of chromium, comparison with the previous figure indicates the usefulness of the computationally simple unhybridised bands and the effect of hybridisation. Spin-orbit coupling has been neglected

γ_d is small [(4.16) and Table 4.1 give γ_d = 0.006] and the Cr 3d band may be recognised as an almost undistorted version of the corresponding canonical d band.

When hybridisation is taken into account by including structure constants with $\ell \neq \ell'$ in the KKR-ASA equations (2.8), bands with similar symmetry labels are not allowed to cross and, instead, hybridisation gaps are created. This is the case of strong hybridisation. In addition, bands with similar symmetry labels which do not cross in the unhybridised case repel each other when hybridisation is included. This is called weak hybridisation. Several examples of these effects may be found in the comparison between unhybridised and fully hybridised band structures, i.e. between Figs.2.9,10 for Cr and between Figs. 2.11,12 for Tl.

The usefullness and accuracy of the *unhybridised* bands, whose calculation requires no matrix diagonalisation but merely scaling, may be appreciated by a comparison of Figs.2.9,10 and Figs.2.11,12. Thus, it is possible in the fully hybridised band structures to recognise large portions of relatively undisturbed, pure bands, a fact that may be used to establish the origin and the symmetry labels of the hybridised band structure. In this connection one should note that a description like the one offered by the unhybridised bands in Figs.2.9,11 may be obtained from Figs.2.3-5 for all bcc, fcc, and hcp solids with little computational effort, i.e. almost by hand.

2.5 Hybridised Canonical Theory

It is obvious that the effect of hybridisation, as seen for instance in Fig. 2.10, depends upon the relative position of the unhybridised energy bands. Therefore, it does not seem possible to define *hybridised canonical* bands which are independent of the potential and the lattice constant, and which can be transformed into hybridised energy bands simply by scaling. However, inspection of the KKR-ASA equations (2.8) reveals that the energy, lattice-parameter, and potential dependences enter only through the 4 potential functions $P_s(E)$, $P_p(E)$, $P_d(E)$, and $P_f(E)$. Hence, if we regard the 4-dimensional vector $\mathbf{P} \equiv \{P_s, P_p, P_d, P_f\}$ as the independent variable we may, for a given $\mathbf{k}$, plot the surface $f_j^{\mathbf{k}}(\mathbf{P}) = 0$ in 4-dimensional $\mathbf{P}$ space, for which the KKR-ASA determinant (2.10) vanishes. This "hyper" band structure now constitutes the canonical bands which in principle may be computed once and for all for a given crystal structure and which are independent of the potential and the lattice constant.

The hyper band structure is, in fact, just another representation of the canonical structure matrix (on- *and* off-diagonal blocks) and it may be used to generate the fully hybridised energy-band structure for any specified vector $P(E) = \{P_s(E),\ldots,P_f(E)\}$ of potential functions through

$$f_j^k(P(E)) = 0 \quad . \tag{2.30}$$

The formation of an energy-band structure may be viewed as follows. The crystal structure defines a set of surfaces in P space — the canonical surfaces. The potential in turn defines a path $P(E)$ in P space along which E is the parameter. The values of E for which the potential path intersects the canonical surfaces then constitute the energy-band structure. When hybridisation is neglected

$$f_j^k(P) = \prod_{j=\ell i} (P_\ell - S_{\ell i}^k) \quad , \tag{2.31}$$

whereby (2.30) reduces to (2.12) and we obtain the simple unhybridised picture outlined in the previous section.

The above-mentioned canonical hyper band structure $f_j^k(P)$ is a function of too many variables to be useful in practice, except in its contracted form where it is integrated over the Brillouin zone. We shall return to this formulation in the following section when we discuss state densities.

Since the hyper band structure is so complicated we must resort to performing band-structure calculations for each given potential, and it is the purpose of this book to show how such calculations are done in an efficient manner. But, before we do this we should like to outline a very recent development [2.14] which shows how the important scaling relations (2.28,29) for the unhybridised energy-band structure may be generalised to include hybridisation.

According to *Andersen* [2.14], the hybridised energy-band structure may be obtained simply by diagonalising one of the following Hamiltonian matrices

$$H_{\ell'm',\ell m}^k = C_\ell \delta_{\ell'\ell} \delta_{m'm}$$

$$+ (\mu_\ell, S^2)^{-1/2} [S^k (1 - \gamma S^k)^{-1}]_{\ell'm',\ell m} (\mu_\ell S^2)^{-1/2} \tag{2.32}$$

or

$$H_{\ell'm',\ell m}^k = V_\ell \delta_{\ell'\ell} \delta_{m'm}$$

$$+ \Gamma_{\ell'}^{1/2} [\gamma^{-1} - S^k]_{\ell'm',\ell m}^{-1} \Gamma_\ell^{1/2} \quad , \tag{2.33}$$

where

$$\Gamma_\ell \equiv 2(2\ell + 1)^2(2\ell + 3)/(\tau_\ell S^2) \quad . \tag{2.34}$$

In (2.32) we must invert the matrix

$$1 - \gamma S^k \equiv \delta_{\ell'\ell}\delta_{m'm} - \gamma_{\ell'}S^k_{\ell'm',\ell m} \tag{2.35}$$

and multiply by the structure matrix S^k.

If instead we wish to use (2.33) we must invert the matrix

$$\gamma^{-1} - S^k \equiv \gamma_\ell^{-1}\delta_{\ell'\ell}\delta_{m'm} - S^k_{\ell'm',\ell m} \quad . \tag{2.36}$$

Expressions (2.32,33) are clearly the matrix generalisation of the unhybridised scaling relations (2.28,29), and they are much simpler than the conventional LMTO equations. They are, however, also slightly less accurate, and therefore we consider here the more versatile, conventional LMTO formalism.

2.6 State Densities and Energy Scaling

The description of the formation of energy bands contained in (2.28) and illustrated in Fig.2.7 constitutes a *scaling principle* according to which the unhybridised band structure of any close-packed solid of a given crystal structure may be synthesised from the same canonical bands. Hence, the unhybridised energy bands of all elemental metals with, for instance, fcc structure may be obtained from the fcc canonical bands shown in Fig.2.4 once their one-electron potentials (or potential parameters) are known.

Often one is not interested in the dispersion of the energy bands but merely in their state density (1.13) or in their ℓ-projected state density (per spin):

$$N_\ell(E) = (2\pi)^{-3}\Omega \sum_j \int_{BZ} dk \sum_m |b^{jk}_{\ell m}|^2 \delta(E - E_j(k)) \quad . \tag{2.37}$$

Here $b^{jk}_{\ell m}$ is the coefficient in the cellular expansion (1.20) when the partial waves ϕ_ℓ are normalised to unity in the polyhedron (sphere). For instance, the density of conduction electrons, spherically averaged in the atomic sphere, is given by

$$n(r) = (4\pi)^{-1} \sum_\ell \int^{E_F} \phi_\ell^2(E,r) 2N_\ell(E)dE \quad , \tag{2.38}$$

where ϕ_ℓ is a normalised partial wave. Since the electron density is the only quantity needed to construct a one-electron potential, the ℓ-projected state densities carry all the structural information necessary to perform self-consistent energy-band calculations.

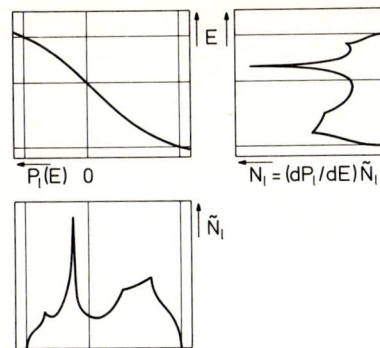

Fig.2.13. Illustration of how a projected state density N_ℓ, which includes hybridisation, may be scaled into a hybridised canonical ℓ-state density $\tilde{N}_\ell$ by the potential function $P_\ell(E)$. The relevant equations are (2.12), which relates energy and structure constants, and (2.40), which relates ℓ-projected energy and canonical state densities

In the approximation that hybridisation may be neglected, the ℓ-projected state density may, of course, be obtained from the state density

$$\tilde{N}_\ell(P_\ell) \equiv (2\pi)^{-3}\Omega \sum_j \int_{BZ} d\mathbf{k}\ \delta(P_\ell - S^{\mathbf{k}}_{\ell j}) \tag{2.39}$$

for the unhybridised, canonical ℓ band by the scaling

$$N_\ell(E) = \tilde{N}_\ell[P_\ell(E)]\dot{P}_\ell(E) \quad, \tag{2.40}$$

where $\dot{P}_\ell = dP_\ell/dE$. This is so because the integrated state density, i.e. the number of states $n(E)$ defined in (1.14), equals the $\mathbf{k}$-space volume contained within the constant-energy surface $E_{\ell j}(\mathbf{k}) = E$ divided by the volume of the Brillouin zone. This is the same as the volume inside the identical, canonical surface $S^{\mathbf{k}}_{\ell j} = P_\ell$, and hence

$$n_\ell(E) = \tilde{n}_\ell(P_\ell(E)) \quad. \tag{2.41}$$

The scaling (2.40) of the state density is illustrated in Fig.2.13.

When hybridisation is included, it follows from the discussion of the canonical hyper band structure in the previous section that the *number-of-states functions* $n(\mathbf{P})$ is *canonical*. That is, a value $\mathbf{P}$ of the potential vector uniquely specifies the surface in $\mathbf{k}$ space for which the KKR-ASA determinant (2.10) vanishes, and hence the volume included within that surface. One may then prove [2.15] that the ℓ-projected state density (2.37) for a given potential, as specified by a path $\mathbf{P}(E)$ may be obtained from the canonical number-of-states function through

$$N_\ell(E) = [\partial n(\mathbf{P})/\partial P_\ell]_{\mathbf{P}(E)}\dot{P}_\ell(E) \quad. \tag{2.42}$$

In principle, the canonical number-of-states function $n(\mathbf{P})$ contains all the structural information needed to perform self-consistent band calculations for a given crystal structure. No diagonalisations but only scalings according to (2.42) need be performed. However, n(**P**) is a function of 4 variables and, presumably, difficult to tabulate. In practice, therefore, one cannot avoid performing band-structure calculations for various potentials but even so the scaling principle may still be useful in the following sense.

Let us assume that we have calculated the energy bands and corresponding eigenvectors, and hence obtained the ℓ-projected state densities $N_\ell^o(E)$ for fcc Ni. Now we want the ℓ-projected state densities for a slightly different atomic-sphere potential, say for the next iteration towards a self-consistent Ni potential, or for Ni at a different lattice constant, or for fcc Pd or Rh. If the new potential $v(r)$ is so similar to the original potential $v^o(r)$ that

$$[\partial n(\mathbf{P})/\partial P_\ell]_{\mathbf{P}(E)} \approx [\partial n(\mathbf{P})/\partial P_\ell]_{\mathbf{P}^o(E_\ell^o)} \quad , \tag{2.43}$$

where E_ℓ^o is determined by

$$P_\ell(E) = P_\ell^o(E_\ell^o) \quad , \tag{2.44}$$

i.e. that the canonical state densities $\tilde{N}_\ell(P_\ell) = \partial n(\mathbf{P})/\partial P_\ell$ along the two paths specified are approximately similar, then the desired state density is approximately given by

$$N_\ell(E) \approx N_\ell^o(E_\ell^o) \frac{\dot{P}_\ell(E)}{\dot{P}_\ell^o(E_\ell^o)} \quad . \tag{2.45}$$

In other words we may transform the ℓ-projected state densities of Ni from an energy scale to a canonical scale using (2.24) or a similar expression with Ni potential parameters. The canonical state densities constructed in this way will include hybridisation appropriate to Ni. By forward scaling, using the energy functions (2.25) with some trial potential parameters, one may then obtain new projected state densities. These may subsequently be used to construct an electron density, a one-electron potential and corresponding new potential parameters, which in turn leads to new projected state densities, etc. The whole cycle may be iterated to self-consistency using only the state densities of Ni and the scaling principle. This approximation which consists of neclecting differential hybridisation is sufficient for a number of applications and forms the basis of an extremely efficient self-consistency procedure.

To avoid misunderstanding I mention that the above scaling cycle is used in the self-consistency procedure mainly to reduce the number of band-structure calculations needed. If one wants very accurate self-consistent bands one must include an energy-band calculation at the end of each self-consistent scaling cycle. However, the scaling procedure is so efficient that fully converged bands of most metals may be obtained with only one or two band calculations included in the complete self-consistency procedure.

36

3. One-Electron States in a Single Sphere

The essence of the introductory chapters is that the energy-band problem
may be separated into two parts: one which depends on the one-electron potential and the atomic volume, and the other which depends on the crystal
symmetry. It is therefore natural first to study the one-electron states
in a single sphere, then to place such spheres on a regular lattice and
establish the boundary conditions which follow from the crystal symmetry,
and finally to introduce the approximations leading to the LMTO method. This
will be the programme for the following four chapters. Hence, we start the
derivation of the linear methods by considering the potential-dependent part
of the band problem. In this connection, we shall study the energy dependence of the solutions of Schrödinger's equation in a spherically symmetric
system of radius S and show that in a limited range, i.e. locally around a
fixed energy, a few physically significant parameters suffice to describe
the energy dependence of not only the logarithmic derivatives at S but also
of the wave functions and the LMTO overlap integrals. As it turns out, this
limited range is larger than the relevant bandwidths, and the potential
parameters, four for each angular-momentum quantum number ℓ, together with
the structure constants, are all that is needed to perform approximate but
still accurate band calculations over the energy range in which one is normally interested.

The original material for the present chapter may be found in the unpublished notes by *Andersen* [3.1] and in a condensed form in the papers by
Andersen and *Woolley* [3.2] and by *Andersen* [3.3].

3.1 Radial Basis Functions

The differential equation which governs the radial part of the one-electron
wave functions inside a single atomic or muffin-tin sphere is the radial
Schrödinger equation (1.17). This we write symbolically in the form

$$(H - E)|\psi_\ell(E,r)\rangle = 0 \quad . \tag{3.1}$$

The normalisation of $\psi_\ell(E,r)$ will not be specified at present but we shall define a radial function normalised to unity in the sphere, i.e.

$$\phi_\ell(E,r) \equiv <\psi_\ell^2(E,r)>^{-\frac{1}{2}}\psi_\ell(E,r) \quad , \tag{3.2}$$

where the normalisation integral is

$$<\psi_\ell^2(E,r)> \equiv \int \psi_{\ell m}^*(E,\mathbf{r})\psi_{\ell m}(E,\mathbf{r})d\mathbf{r}$$

$$= \int_0^S \psi_\ell^2(E,r)r^2 dr \iint |i^\ell Y_\ell^m(\hat{r})|^2 d\Omega$$

$$= \int_0^S \psi_\ell^2(E,r)r^2 dr \quad . \tag{3.3}$$

In (3.3) we have used the well-known normalisation of spherical harmonics.

In the LMTO method, attention is focussed upon an energy range centred around some energy E_ν which we are free to choose to suit the problem at hand. Therefore, for each value of ℓ we choose an $E_{\nu\ell}$, and use the energy-independent basis set formed by the normalised radial function

$$\phi_{\nu\ell}(r) \equiv \phi_\ell(E_{\nu\ell},r) \tag{3.4}$$

and its energy derivative

$$\dot{\phi}_{\nu\ell}(r) \equiv [\partial\phi_\ell(E,r)/\partial E]_{E_{\nu\ell}} \quad . \tag{3.5}$$

The corresponding radial logarithmic derivatives at the sphere boundary are

$$D_{\nu\ell} = S\phi_{\nu\ell}'(S)/\phi_{\nu\ell}(S) = S\psi_{\nu\ell}'(S)/\psi_{\nu\ell}(S)$$

$$D_{\dot{\nu}\ell} = S\dot{\phi}_{\nu\ell}'(S)/\dot{\phi}_{\nu\ell}(S) \quad , \tag{3.6}$$

where $' \equiv \partial/\partial r$. In the following we shall discuss the energy and radial dependences of each radial function, and we therefore drop the subscript ℓ.

According to (3.2,3), the normalisation integral $<\phi^2(E)>$ in the sphere is unity, i.e.

$$<\phi(E)|\phi(E)> = 1 \quad . \tag{3.7}$$

Consequently $\phi_\nu(r)$ and $\dot{\phi}_\nu(r)$ are orthogonal, as may be seen from the first of the following relations obtained by successive differentiations of (3.7) with respect to energy:

$$<\dot{\phi}_\nu|\phi_\nu> = 0 \tag{3.8}$$

$$<\dot{\phi}_\nu^2> = - <\phi_\nu|\ddot{\phi}_\nu> \quad . \tag{3.9}$$

We note that the (unspecified) normalisation of $\psi_\ell(E,r)$ may be energy dependent, and therefore $\psi_\nu(r)$ and $\dot{\psi}_\nu(r)$ are in general not orthogonal. Regard-
less of the normalisation of $\psi_\nu(r)$, however, $\dot{\psi}_\nu(r)$ is always *some linear
combination of* $\phi_\nu(r)$ *and* $\dot{\phi}_\nu(r)$ as differentiation of (3.2) will show (see,
e.g. (5.22) below).

The local expansion of the energy dependence of the normalised partial
wave (3.2) is simply the Taylor series

$$\phi(E,r) = \phi_\nu(r) + \varepsilon\dot{\phi}_\nu(r) + (1/2)\varepsilon^2\ddot{\phi}_\nu(r) + o(\varepsilon^2) \quad , \tag{3.10}$$

where $\varepsilon = E - E_{\nu\ell}$ and o means terms of higher order.

If we briefly return to the discussion of the diatomic molecule (Sect.
2.3 and Fig.2.7) we realise that the difference $\psi(B,r) - \psi(A,r)$ between the
bonding and the antibonding state in the LCAO description must be supplied
by the tail of the orbital $\chi(r-R)$ centred at the other site. Since it turns
out that the *linear* part of the Taylor expansion for $\psi(E,r)$ is well converged
in the range from B to A, provided that E_ν is chosen appropriately, it seems
natural to *augment* (i.e. substitute) the tail inside the overlapped atom by
the energy derivative function $\dot{\psi}_\nu(r)$. We shall return to this augmentation
in more technical terms in Sect.5.4 when we discuss the energy-independent
muffin-tin orbitals.

3.2 Partial Waves and Their Energy Derivatives

With later use in mind we now proceed to establish relations between the
value of the u'th energy derivative function $\overset{u}{\phi}_\nu(S)$ at the sphere, the corres-
ponding logarithmic derivative $D\{\overset{u}{\phi}_\nu\}$, and the integral $\langle\phi_\nu|\overset{u}{\phi}_\nu\rangle$ in the sphere.
Differentiations of the Schrödinger equation

$$(H - E)|\psi\rangle = 0 \tag{3.11}$$

with respect to energy give

$$(H - E_\nu)|\overset{u}{\psi}_\nu\rangle = u|\overset{u-1}{\psi}_\nu\rangle \quad . \tag{3.12}$$

From Green's second identity we find

$$\underset{\text{sphere}}{\iiint} [\psi_1^*(\mathbf{r})\nabla^2\psi_2(\mathbf{r}) - \psi_2^*(\mathbf{r})\nabla^2\psi_1(\mathbf{r})]d\mathbf{r}$$

$$= \langle\psi_2|H - E|\psi_1\rangle - \langle\psi_1|H - E|\psi_2\rangle$$

$$= [D_2 - D_1]S\psi_1(S)\psi_2(S)$$

$$= \underset{\text{surface}}{\iint} \left[\psi_1^*(\mathbf{r})\frac{\partial\psi_2(\mathbf{r})}{\partial n} - \psi_2^*(\mathbf{r})\frac{\partial\psi_1(\mathbf{r})}{\partial n}\right]S^2 d\hat{\Omega} \quad . \tag{3.13}$$

Here $\psi_1(\mathbf{r})$ and $\psi_2(\mathbf{r})$ are arbitrary functions of the form $\psi_\ell(r)i^\ell Y_\ell^m(\hat{r})$ with the same ℓ and m and with the radial logarithmic derivatives D_1 and D_2, respectively. Furthermore, $\partial/\partial n$ means differentiation with respect to the normal to the surface of the sphere and in the first step we have inserted the Hamiltonian $H = -\nabla^2 + v(r)$, and used $\langle\psi_1|E - v(r)|\psi_2\rangle = \langle\psi_2|E - v(r)|\psi_1\rangle$. Setting now $\psi_2(r) = \psi_\nu(r)$ and $\psi_1(r) = \overset{u}{\psi}_\nu(r)$, we obtain

$$\langle\psi_\nu|H - E_\nu|\overset{u}{\psi}_\nu\rangle - \langle\overset{u}{\psi}_\nu|H - E_\nu|\psi_\nu\rangle$$

$$= [D_\nu - D\{\overset{u}{\psi}_\nu\}]S\psi_\nu(S)\overset{u}{\psi}_\nu(S) \tag{3.14}$$

and with the help of (3.11,12) we find our first result:

$$u\langle\psi_\nu|\overset{u-1}{\psi}_\nu\rangle = \langle\psi_\nu|H - E_\nu|\overset{u}{\psi}_\nu\rangle$$

$$= [D_\nu - D\{\overset{u}{\psi}_\nu\}]S\psi_\nu(S)\overset{u}{\psi}_\nu(S) \quad . \tag{3.15}$$

A similar procedure with $\psi_2(r) = \dot{\psi}_\nu(r)$ and $\psi_1(r) = \ddot{\psi}_\nu(r)$ gives

$$2\langle\dot{\psi}_\nu^2\rangle = \langle\dot{\psi}_\nu|H - E_\nu|\ddot{\psi}_\nu\rangle$$

$$= \langle\ddot{\psi}_\nu|\psi_\nu\rangle + [D\{\dot{\psi}\} - D\{\ddot{\psi}\}]S\dot{\psi}_\nu(S)\ddot{\psi}_\nu(S) \quad . \tag{3.16}$$

If we adopt the normalisation (3.7) in the sphere, we find from (3.8,15) with u = 1

$$D_{\dot{\nu}} = D_\nu - [S\phi_\nu(S)\dot{\phi}_\nu(S)]^{-1} \quad , \tag{3.17}$$

and with u = 2

$$D\{\ddot{\phi}_\nu\} = D_\nu \quad . \tag{3.18}$$

Similarly, with the help of (3.9,17,18), (3.16) gives the result

$$3\langle\dot{\phi}_\nu^2\rangle = -\ddot{\phi}_\nu(S)/\phi_\nu(S) \quad . \tag{3.19}$$

Finally, we prove that $\dot{\phi}_\nu(r)$ is orthogonal to the core states $\phi_n(r)$ defined by

$$H|\phi_n\rangle = E_n|\phi_n\rangle$$

$$D\{\phi_n\} = D_\nu \qquad n \neq \nu$$

$$\phi_n(S) \ll \phi_\nu(S) \quad . \tag{3.20}$$

Clearly, ϕ_n and ϕ_ν are orthogonal since they satisfy the same boundary condition, and consequently

$$0 = \langle\phi_n|\phi_\nu\rangle = \langle\phi_n|H - E_\nu|\dot{\phi}_\nu\rangle \quad , \tag{3.21}$$

where we used (3.12). The functions appearing in the last integral may now be interchanged using (3.14), with ϕ_n substituted for ϕ_ν to give

$$0 = <\dot{\phi}_\nu|H - E_\nu|\phi_n> + [D_\nu - D_n^\bullet]S\phi_n(S)\dot{\phi}_\nu(S)$$

$$= (E_n - E_\nu)<\dot{\phi}_\nu|\phi_n> + \phi_n(S)/\phi_\nu(S) \quad , \tag{3.22}$$

where we used (3.17). With (3.20) we obtain the desired result:

$$<\dot{\phi}_\nu|\phi_n> = 0 \qquad n \neq \nu \quad . \tag{3.23}$$

Hence, any linear combination of the orthogonal functions $\phi_\nu(r)$ and $\dot{\phi}_\nu(r)$, e.g. $\dot{\psi}_\nu(r)$, is orthogonal to the core states. Therefore, the augmentation of the muffin-tin orbital by a $\dot{\psi}_\nu(r)$ function (mentioned at the end of the previous section) is one particular way of orthogonalising the orbital to the core states of the neighbouring atoms.

3.3 Logarithmic Derivative and Laurent Expansion

The energy dependence of the logarithmic derivative (2.3) may be established by numerical integration of the Schrödinger equation (1.17) and evaluation of (2.3) over a range of energies. An example of this energy dependence is shown schematically in Fig.2.6, and below we shall show how the ν'th period of $D_\ell(E)$ may be approximated by a Laurent series.

The radial Schrödinger equation and the boundary condition

$$D_\ell(E_{n\ell}) = D_{\nu\ell} \tag{3.24}$$

may be used to define a complete set of orthogonal eigenfunctions $\phi_{n\ell}(r)$ and corresponding eigenvalues $E_{n\ell}$, $n = 0\cdots\nu\cdots\infty$. Here $D_{\nu\ell}$ is some number, at present chosen arbitrarily, and $\phi_{n\ell}(r)$ is normalised according to (3.7). The formal expansion of the partial wave

$$\psi(E,r) = \sum_{n=0}^{\infty} <\psi(E)|\phi_n>\phi_n(r) \tag{3.25}$$

converges at the sphere to $\psi(E,S)$, although its logarithmic derivative is D_ν rather than $D_\ell(E)$.

The expansion coefficients $<\psi(E)|\phi_n>$ may be related to the logarithmic derivative $D(E)$ by means of (3.13), i.e.

$$<\psi(E)|H - E|\phi_n> - <\phi_n|H - E|\psi(E)>$$

$$= [D(E) - D_\nu]S\psi(E,S)\phi_n(S) \tag{3.26}$$

41

into which we insert the Schrödinger equation (3.1) for $\psi(E,r)$ and $\phi_n(r)$ and find

$$(E - E_n)<\psi(E)|\phi_n> = -[D(E) - D_\nu]S\psi(E,S)\phi_n(S) \quad . \tag{3.27}$$

Expansion (3.25) becomes

$$\psi(E,r) = -[D(E) - D_\nu]\psi(E,S) \sum_{n=0}^{\infty} \frac{S\phi_n(S)}{E - E_n} \phi_n(r) \quad , \tag{3.28}$$

which for $r = S$ gives the resonance series

$$\frac{1}{D(E) - D_\nu} = - \sum_{n=0}^{\infty} \frac{S\phi_n^2(S)}{E - E_n} \tag{3.29}$$

for the logarithmic-derivative function.

Relation (3.29) shows that the properties of $[D(E) - D_\nu]^{-1}$ are entirely described by a set of eigenvalues E_n and the corresponding residues $S\phi_n^2(S)$. Since the partial probability density integrated over the sphere surface $4\pi S^2\phi_n^2(S)$ is exceedingly small for core states, the corresponding resonances are extremely sharp.

The form (3.29) suggests that we try to expand the logarithmic derivative as a Laurent series in $\varepsilon = (E - E_\nu)$

$$[D(E) - D_\nu]^{-1} = -(mS^2\varepsilon)^{-1} + a + bS^2\varepsilon + o(\varepsilon) \quad , \tag{3.30}$$

where $(mS^2\varepsilon)^{-1}$ is the $n = \nu$ term in (3.29), and $a + bS^2\varepsilon + o(\varepsilon)$ is the contribution from the remainder of the sum. Since this expansion turns out to be accurate to within one percent of the relevant bandwidth over the entire range of the $\nu\ell$ band, the important potential parameters are $D_{\nu\ell}$, m_ℓ, a_ℓ, and b_ℓ.

The Taylor expansion (3.10) of the normalised partial wave may be inserted into (3.27) with $n = \nu$, and we find

$$-[D(E) - D_\nu][\phi_\nu(S) + \varepsilon\dot\phi_\nu(S) + (1/2)\varepsilon^2\ddot\phi_\nu(S)]S\phi_\nu(S)$$

$$= \varepsilon<[\phi_\nu + \varepsilon\dot\phi_\nu + (1/2)\varepsilon^2\ddot\phi_\nu]|\phi_\nu>$$

$$= \varepsilon[1 - (1/2)\varepsilon^2<\dot\phi_\nu^2>] \quad , \tag{3.31}$$

where we have used (3.7-9). If we now keep terms to second order in ε and use (3.19), we obtain an alternative version of expansion (3.30)

$$[D(E) - D_\nu]^{-1} = -\left(\frac{S\phi_\nu^2}{\varepsilon} + S\psi_\nu\dot\phi_\nu + S\psi_\nu^2<\dot\phi_\nu^2>\varepsilon \right) + o(\varepsilon) \tag{3.32}$$

written in terms of the amplitudes ϕ_ν and $\dot\phi_\nu$ introduced as a short-hand for $\phi_\nu(S)$ and $\dot\phi_\nu(S)$. This allows the identification

$$m = (S^3\phi_\nu^2)^{-1} \tag{3.33}$$

familiar from (3.29,30) and

$$a = -S\phi_\nu\dot\phi_\nu \tag{3.34}$$

$$b = S^{-1}\phi_\nu^2<\dot\phi_\nu^2> \tag{3.35}$$

of the parameters appearing in (3.30).

Historically the three parameters m, a, and b were the first to be used in band calculations using the LMTO method, and they also appear in various parts of the computer programmes presented in Chap.9. However, for the purpose of comparing different materials, they depend too strongly on $E_{\nu\ell}$ to be useful, and in Chap.4 we therefore introduce physically more significant parameters for comparisons between elements.

3.4 Potential Function and Bandwidth

It follows from the energy scaling outlined in Sect.2.6 that the slope of the potential function $P_\ell(E)$ provides a measure of the inverse of the width of the ℓ band. To parametrise $P_\ell(E)$ and establish a local approximation to the bandwidth we therefore wish to calculate

$$\dot P_\ell(E) = \frac{dP_\ell(D)}{dD}\, \dot D_\ell(E) \quad . \tag{3.36}$$

Now, dP_ℓ/dD is evaluated directly from (2.17)

$$\frac{dP_\ell(D)}{dD} = -2\frac{(2\ell + 1)^2}{(D - \ell)^2} \quad , \tag{3.37}$$

while $\dot D_\ell(E)$ is obtained from (3.27) by letting $E_{n\ell}$ approach E and letting ψ be the normalised partial wave, i.e.

$$-\dot D_\ell(E) = \frac{1}{S\phi_\ell^2(E)} \equiv m_\ell(E)S^2 \quad . \tag{3.38}$$

This definition of $m_\ell(E)$ is equivalent to (3.33). At the centre of the band C_ℓ, defined in (2.22), we find

$$\mu_\ell S^2 \equiv 2m_\ell(C_\ell)S^2 = \frac{2}{S\phi_\ell^2(C_\ell)} = \dot P_\ell(C_\ell) \quad , \tag{3.39}$$

43

where we have defined the intrinsic mass μ_ℓ of the ℓ band and used (3.36-38). Similarly, we define the mass at V_ℓ

$$\tau_\ell S^2 \equiv (2\ell + 3)m_\ell(V_\ell)S^2 = \frac{2\ell + 3}{S\phi_\ell^2(V_\ell)} \quad . \tag{3.40}$$

The factors 2 and $2\ell + 3$ in the definitions ensure that μ and τ are unity for free electrons, Sect.4.4.

Relation (3.39) suggests the parametrisation

$$P_\ell(E) = \mu_\ell S^2(E - C_\ell) \tag{3.41}$$

which is obviously the linear approximation to the scaling expression (2.28). More accurate parametrisations may be obtained by means of the first two terms of the Laurent series (3.32) with $D_\nu = -\ell - 1$ or $D_\nu = \ell$ and the definition (2.9) of the potential function. We find

$$P(E) = \begin{cases} \dfrac{\mu S^2(E - C)}{1 + \mu S^2(E - C)\gamma(C)} & D_\nu = -\ell-1 \\[4mm] -\dfrac{2(2\ell + 1)^2(2\ell + 3)}{\tau S^2(E - V)} + \dfrac{1}{\gamma(V)} & D_\nu = \ell \end{cases}$$

$$\gamma^{-1}(C) = 2(2\ell + 1)/[1 + (2\ell + 1)S\dot\phi]$$

$$\gamma^{-1}(V) = 2(2\ell + 1) \times [1 - (2\ell + 1)S\dot\phi] \quad , \tag{3.42}$$

which are the inverse of the scaling laws (2.28,29). Here we should mention that the expressions for γ given above are consistent with definition (4.16) below.

According to the Wigner-Seitz rule (Sect.2.3) the ℓ bandwidth on the "P scale" is $2(2\ell + 1)^2/\ell$. In addition, the maximum value of P_s is found to be 2. Hence, if we expand $P(E)$ around C when $\ell \neq 0$ and around V when $\ell = 0$, we find the following simple, approximate expressions for the bandwidths:

$$W_\ell = \frac{2}{\mu_\ell S^2}\frac{(2\ell + 1)^2}{\ell} \qquad \ell \neq 0$$

$$W_s = \frac{1}{\tau_s S^2}\frac{6}{(1/\gamma_s) - 2} \qquad \ell = 0 \tag{3.43}$$

which are therefore proportional either to $1/\mu S^2$ or $1/\tau S^2$.

3.5 Matrix Elements and Variational Estimate of Energies

So far we have obtained parametrisations of the logarithmic derivative and potential functions which are appropriate when the Schrödinger equation is regarded as a differential equation, and which allow us to find D_ℓ and P_ℓ whenever E is given. In the ASA, however, Schrödinger's equation is treated as an eigenvalue problem subject to boundary conditions in the form of specified logarithmic derivatives at the sphere. Therefore, we need to find a parametrisation of the function $E_\ell(D)$ inverse to $D_\ell(E)$, valid around $E_{\nu\ell}$.

To do this, we use the Rayleigh-Ritz variational principle in connection with the radial trial function of arbitrary logarithmic derivative D at the sphere boundary defined by the linear combination

$$\Phi(D,r) = \phi_\nu(r) + \omega(D)\dot\phi_\nu(r) \quad , \tag{3.44}$$

where

$$\omega(D) = -\frac{\phi_\nu}{\dot\phi_\nu}\frac{D - D_\nu}{D - D_{\dot\nu}} \quad . \tag{3.45}$$

Here we used the notation $\phi_\nu \equiv \phi_\nu(S)$ and $\dot\phi_\nu \equiv \dot\phi_\nu(S)$, and the definitions (3.6) of D_ν and $D_{\dot\nu}$. Using similar notation for $\Phi(D,S)$ we find

$$\Phi(D) = \phi_\nu \frac{D_\nu - D_{\dot\nu}}{D - D_{\dot\nu}} \quad . \tag{3.46}$$

Within the basis

$$\Phi_{\ell m}(D,\mathbf{r}) = i^\ell Y_\ell^m(\hat{r})\Phi_\ell(D,r) \tag{3.47}$$

the matrix elements of the Hamiltonian in the sphere become

$$<\Phi_{\ell'm'}(D')|H - E_{\nu\ell}|\Phi_{\ell m}(D)> = \omega_\ell(D)\delta_{\ell'\ell}\delta_{m'm} \quad . \tag{3.48}$$

Here we have used (3.11,12). Similarly, the overlap matrix elements become

$$<\Phi_{\ell'm'}(D')|\Phi_{\ell m}(D)> = \delta_{\ell'\ell}\delta_{m'm}[1 + \omega_\ell(D')\omega_\ell(D)<\dot\phi_{\nu\ell}^2>] \quad , \tag{3.49}$$

where we have made use of (3.8,9).

With the above matrix elements we obtain the following variational estimate

$$E_\ell(D) = E_{\nu\ell} + <\Phi_{\ell m}(D)|H - E_{\nu\ell}|\Phi_{\ell m}(D)>/<\Phi_{\ell m}^2(D)>$$

$$= E_{\nu\ell} + \frac{\omega_\ell(D)}{1 + <\dot\phi_{\nu\ell}^2>\omega_\ell^2(D)} + o(\varepsilon^3) \tag{3.50}$$

of the $\nu\ell$'th branch of the function inverse to $D_\ell(E)$. Comparison with (3.10) shows that the trial function (3.44) is correct to first order in ε, whereby the variational energy (3.50) is correct to order ε^3.

The variational estimate is consistent with our previous development because from the two first terms of the Laurent series (3.32) we obtain

$$E(D) = E_\nu - \frac{S\phi_\nu^2}{(D - D_\nu)^{-1} + S\phi_\nu\dot{\phi}_\nu} + o(\varepsilon^2)$$

$$= E_\nu + \omega(D) + o(\varepsilon^2) \quad , \tag{3.51}$$

where $\omega(D)$ has been introduced by means of (3.17,45). If we multiply the three-term Laurent series by ε and insert (3.51) we find

$$E(D) = E_\nu + \omega(D)[1 - <\dot{\phi}_\nu^2>(E(D) - E_\nu)^2] + o(\varepsilon^3)$$

$$= E_\nu + \omega(D)[1 - <\dot{\phi}_\nu^2>\omega^2(D)] + o(\varepsilon^3) \tag{3.52}$$

which is equal to the variational expression (3.50) to third order in ε, and thus confirms the accuracy of that estimate.

The parametrisations represented by (3.30,50,51) are much more general than the simple forms (2.24,28,29). Firstly, they are expansions around the energy E_ν, which may be chosen to suit the particular problem at hand. This is in contrast to the energies V and C, which are given by the potential and cannot be suitably adjusted. The freedom to choose E_ν is particularly important in the case of weak hybridisation, where the interesting energy region is either above or below the $n\ell$ band. Secondly, the energy function $E_\ell(D)$ may be made accurate to third order simply by "switching on" the small parameter $<\dot{\phi}_\nu^2>$.

The variational estimate $E(D)$ given by (3.50) is a single-valued function of D confined to the energy window

$$E_\nu - (1/2)<\dot{\phi}_\nu^2>^{-\frac{1}{2}} \leq E(D) \leq E_\nu + (1/2)<\dot{\phi}_\nu^2>^{-\frac{1}{2}} \tag{3.53}$$

of width $<\dot{\phi}_\nu^2>^{-\frac{1}{2}}$, and it has in addition an unphysical behaviour in a range around $D_{\dot{\nu}}$. This difficulty is fundamental and stems from the fact that $D(E)$ has many branches labelled by the principal quantum number n. The inverse of the logarithmic derivative function is consequently a multivalued function of D. Hence, the single-valued variational estimate must at some D value switch from one branch of the logarithmic derivative function to another. The switching occurs at $D = D_{\dot{\nu}}$, and in a limited range around $D_{\dot{\nu}}$ the variational estimate has an erroneous positive slope. The second-order estimate $\omega(D)$ also changes branch at $D = D_{\dot{\nu}}$, but in this case the energy window is infinitely

46

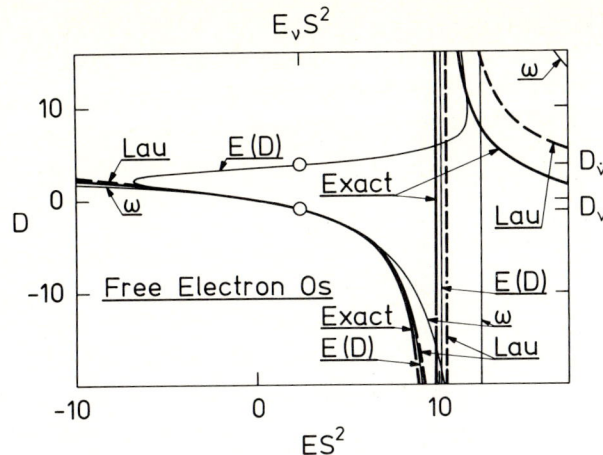

Fig.3.1. Estimates of the logarithmic-derivative function for free s electrons compared with the exact result $D(x) = x \cot x - 1$, $x = S\sqrt{E}$ explained in Sect. 4.4. The curve labelled ω is the second-order estimate (3.51), $E(D)$ is the third-order estimate (3.50), while Lau is the Laurent expansion (3.30) valid to third order in $(E - E_\nu)S^2$. The potential parameters used in the three estimates are derived in Sect.4.4 and listed in Table 4.4. The two open circles in the figure refer to the points $(E_\nu S^2, D_\nu)$ and $(E_\nu S^2, D_{\dot\nu})$, where $E_\nu S^2$ is $K_\nu^2 S^2$ of Table 4.4

wide and the unphysical positive slope has vanished. Expression (3.51) may therefore be used where the unphysical behaviour of the variational estimate cannot be tolerated. We shall return to this question in Sects.4.2,3 when we consider the 3d band of chromium in detail.

The results of this and Sect.3.3 are collated in Fig.3.1 and compared to an exact calculation of the logarithmic derivative function for free s electrons. The most striking features of this comparison are the very accurate description provided by both the Laurent expansion (3.30) and the variational estimate (3.50) in the region of negative D, and the quite erroneous estimate given by the variational expression in the region around $D_{\dot\nu}$.

4. Physically Significant Parameters

We have now completed the parametrisation of the non-structurally dependent part of the energy-band problem. In the process, we found that for each ℓ quantum number only four parameters D_ν, ϕ_ν, $\dot\phi_\nu$, and $\langle\dot\phi_\nu^2\rangle$ were needed to provide an accurate description of the logarithmic derivative function $D(E)$ and its inverse $E(D)$ in the region around (D_ν, E_ν). From the point of view of physical understanding, however, these parameters are not very useful because their values depend critically upon the choice of $E_{\nu\ell}$, and in practice we prefer to work with the following equivalent set of four per ℓ value

$$\omega(-) \equiv \omega(-\ell-1)$$
$$S\Phi^2(-) \equiv S\Phi^2(-\ell-1)$$
$$\Phi(-)/\Phi(+) \equiv \Phi(-\ell-1)/\Phi(\ell)$$
$$\langle\dot\phi_\nu^2\rangle^{-\frac{1}{2}} \tag{4.1}$$

which we shall refer to as the standard parameters. In (4.1), + and - refer to the boundary conditions $D = \ell$ and $D = -\ell-1$, respectively, which are the logarithmic derivatives of the spherical Bessel and Neumann functions in the $\kappa^2 = 0$ limit used in the atomic-sphere approximation (Chap.6).

The three first parameters in the standard set may be found in terms of the original D_ν, ϕ_ν, and $\dot\phi_\nu$ by means of (3.45,46,17), while the fourth parameter gives the width of the energy window (3.53). We prefer to use these standard potential parameters because they depend little upon the choice of E_ν, have physically simple interpretations, and vary in a systematic way from element to element across the periodic table.

As an example of a set of standard parameters, Table 4.1 lists all the potential-dependent information needed to perform an energy-band calculation for (non-magnetic) chromium metal. In the following, chromium is used as an example when we discuss the physical significance of each of the four potential parameters (4.1). At the end of the chapter we derive free-electron potential parameters, give expressions for the volume derivatives of some se-

Table 4.1. Standard potential parameters for non-magnetic chromium at
S = 2.684 a.u., from [4.1]. Here E_ν is relative to the Coulomb potential
at S. This is a natural choice since for a monoatomic material in the ASA
the spheres are neutral and their charge density is spherically symmetric.
Relative to this energy zero, $v(S)$ = -0.8181 Ry

	E_ν [Ry]	$\omega(-)$ [Ry]	$S\Phi^2(-)$ [Ry]	$\Phi(-)/\Phi(+)$	$\langle\dot{\phi}_\nu^2\rangle^{-\frac{1}{2}}$ [Ry]
S	-0.3565	0.2082	0.3963	0.8681	5.5564
p	-0.1973	1.0802	0.3620	0.7105	6.5181
d	-0.1792	0.1520	0.0406	0.0632	1.0041

lected potential parameters, and finally present a list of useful expressions
relating the many different potential parameters defined in various contexts.

4.1 The Four Potential Parameters

The first parameter $\omega_\ell(-)$ in Table 4.1 is defined by (3.45), and according to
(3.51) it represents the second-order estimate of the position of the band
centre relative to $E_{\nu\ell}$, i.e.

$$\hat{C}_\ell = E_{\nu\ell} + \omega_\ell(-) \quad . \tag{4.2}$$

Similarly, for the square-well pseudopotential

$$\hat{V}_\ell = E_{\nu\ell} + \omega_\ell(+) \tag{4.3}$$

and an expression for $\omega_\ell(+)$ in terms of the parameters in the table is given
by (4.17) below.

The parameter $S\Phi^2(-)$ is defined through (3.46), and is proportional to the
bandwidth of the ℓ band. To see this connection, we calculate $\dot{D}_\ell^{-1}$ in the
following fashion from the variational expression (3.50)

$$\frac{dE(D)}{dD} = \frac{d}{d\omega} \frac{\omega(D)}{1 + \langle\dot{\phi}_\nu^2\rangle\omega^2(D)} \frac{d\omega(D)}{dD}$$

$$= \frac{1 - \langle\dot{\phi}_\nu^2\rangle\omega^2(D)}{[1 + \langle\dot{\phi}_\nu^2\rangle\omega^2(D)]^2} \frac{d\omega(D)}{dD} \quad . \tag{4.4}$$

From (3.45,46,17) it follows that $\omega(D)$ and $\Phi(D)$ are related by

$$\frac{d\omega(D)}{dD} = -S\Phi^2(D) \tag{4.5}$$

and a second-order estimate of mS^2 defined in (3.38) is therefore

$$m(D)s^2 = \frac{1}{S\phi^2(D)} \frac{[1 + <\dot{\phi}_\nu^2>\omega^2(D)]^2}{1 - <\dot{\phi}_\nu^2>\omega^2(D)} \quad . \tag{4.6}$$

The choice $D = - \ell - 1$ leads to a similar estimate of the intrinsic mass defined in (3.39):

$$\mu s^2 = \frac{2}{S\phi^2(-)} \frac{[1 + <\dot{\phi}_\nu^2>\omega^2(-)]^2}{1 - <\dot{\phi}_\nu^2>\omega^2(-)} \quad , \tag{4.7}$$

and for $D = \ell$ we have an estimate of the intrinsic mass at V_ℓ, i.e.

$$\tau s^2 = \frac{2\ell + 3}{S\phi^2(+)} \frac{[1 + <\dot{\phi}_\nu^2>\omega^2(+)]^2}{1 - <\dot{\phi}_\nu^2>\omega^2(+)} \quad . \tag{4.8}$$

If $<\dot{\phi}_\nu^2>$ can be neglected we have the first order estimates

$$\hat{\mu}s^2 = 2[S\phi^2(-)]^{-1} \tag{4.9}$$

$$\hat{\tau}s^2 = (2\ell + 3)[S\phi^2(+)]^{-1} \tag{4.10}$$

analogous to (4.2,3). By means of (3.43,4.16) we now find the simple expressions

$$\hat{W}_\ell = S\phi^2(-) \frac{(2\ell + 1)^2}{\ell} \qquad \ell \neq 0$$

$$\hat{W}_s = S\phi^2(-) \left(\frac{\phi(+)}{\phi(-)}\right)^2 \left(\frac{\phi(+)}{\phi(-)} - 1\right)^{-1} \qquad \ell = 0 \tag{4.11}$$

for the width of the ℓ band, indicating the above mentioned proportionality to $S\phi^2(-)$. If we evaluate the estimates (4.11), $\phi(-)/\phi(+) \simeq 0.8$ for free s electrons, we find that $\hat{W}_\ell = (6, 9, 13, 16) \times S\phi^2(-)$ for s, p, d, and f electrons, respectively. Hence, as a rule of thumb the width of the ℓ band is of the order of $10\, S\phi^2(-)$.

To see the significance of $\phi(+)/\phi(-)$ we must first derive several relations connecting the potential parameters. With the boundary conditions $D = \ell$ and $D = - \ell - 1$, we find from (3.44) the equations

$$\phi(+) = \phi_\nu + \omega(+)\dot{\phi}_\nu$$

$$\phi(-) = \phi_\nu + \omega(-)\dot{\phi}_\nu \tag{4.12}$$

which may be solved to give

$$\phi_\nu = \frac{\omega(+)\phi(-) - \omega(-)\phi(+)}{\omega(1) \qquad \omega(-)}$$

50

$$\dot{\phi}_\nu = \frac{\Phi(+) - \Phi(-)}{\omega(+) - \omega(-)} \quad .$$

(4.13)

Similarly, from (3.46)

$$D_\nu = \ell + (2\ell + 1) \left[\frac{\Phi(+)\omega(-)}{\Phi(-)\omega(+)} - 1\right]^{-1}$$

$$D_{\dot{\nu}} = \ell + (2\ell + 1) \left[\frac{\Phi(+)}{\Phi(-)} - 1\right]^{-1} \quad ,$$

(4.14)

which together with (4.13) may be inserted into (3.45) to give an alternative expression

$$\frac{\omega(D) - \omega(-)}{\omega(D) - \omega(+)} = \frac{\Phi(-)}{\Phi(+)} \frac{D + \ell + 1}{D - \ell}$$

$$= \gamma P \quad .$$

(4.15)

In the last step we used (2.9) and defined the parameter

$$\gamma \equiv \frac{\Phi(-)}{\Phi(+)} [2(2\ell + 1)]^{-1} \quad .$$

(4.16)

The Wronskian relation (3.17) combined with (4.13,14) gives the final relation needed:

$$\omega(-) - \omega(+) = (2\ell + 1)S\Phi(-)\Phi(+)$$

$$= \hat{C} - \hat{V} \quad .$$

(4.17)

By means of (4.16,17) we may now solve (4.15) to give

$$\omega(P) = \omega(-) + (1/2)S\Phi^2(-) \frac{P}{1 - \gamma P} \quad ,$$

(4.18)

which shows that $\Phi(-)/\Phi(+)$ governs the distortion contained in the scaling from canonical bands to energy bands. If we take $D_\nu = -\ell - 1$, i.e. $\omega(-) = 0$, we regain the simple scaling law (2.28).

The final parameter $\langle\dot{\phi}_\nu^2\rangle^{-\frac{1}{2}}$ determines the width of the energy window, i.e. the energy range over which the variational estimate (3.50) is valid, cf. the discussion following equation (3.53). We shall return to $\langle\dot{\phi}^2\rangle^{-\frac{1}{2}}$ in Sect.4.3 where we consider the case of the Cr 3d band.

To give a feeling for orders of magnitude, Table 4.2 lists some selected potential parameters for chromium calculated from the entries in Table 4.1 by means of (3.50,4.7,8,14). We shall return to them when we consider volume derivatives in Sect.4.5.

Table 4.2. Band energies, band masses and the logarithmic derivative of $\dot{\phi}$ for chromium at S = 2.684 a.u.

	V_ℓ [Ry]	C_ℓ [Ry]	τ_ℓ	μ_ℓ	$D_{\dot{\nu}}$
s	-0.60	-0.15	0.80	0.70	6.58
p	-0.64	0.85	0.98	0.83	8.36
d	-	-0.03	-	7.33	2.34

4.2 How to Choose $E_{\nu\ell}$

The energy $E_{\nu\ell}$ around which the expansion is performed is in our particular example taken to be the centre of gravity of the *occupied* part of the ℓ band. This choice leads to accurate charge densities to be used in a self-consistency procedure. Of other possible choices of $E_{\nu\ell}$ the choice $E_{\nu\ell} = C_\ell$ results in the best overall energy bands, while $E_{\nu\ell} = E_F$ gives the correct Fermi surface and the correct Fermi velocities. In general $E_{\nu\ell}$ can be fixed at any energy to suit the problem at hand. There is, however, one situation which we shall now discuss where one may obtain incorrect bands. Fortunately, this case is easy to identify and the error may be corrected without losing the freedom to choose $E_{\nu\ell}$.

If, for some particular values of ℓ, $E_{\nu\ell}$ is positioned in the range $A_{n-1,\ell} - B_{n,\ell}$ between the top of the $(n-1)\ell$ band and the bottom of the $n\ell$ band, an LMTO calculation may yield a steep band which connects parts of the low-lying $(n-1)\ell$ band with parts of the high-lying $n\ell$ band. The reason is that the LMTO, which uses the smallest possible basis set with only one principal quantum number for each value of ℓ, can give only *one* set of ℓ bands. Hence, when E_ν is chosen between two bands of the same ℓ, the LMTO method develops schizophrenia and tries to present parts of both bands.

It is generally possible to circumvent the problem by fixing $E_{\nu\ell}$ such that the unphysical band disappears. As explained below, this requires that $D_{\dot{\nu}}$ is positive, and usually means that $E_{\nu\ell}$ must be fixed far above the pertinent energy range. Therefore the weak hybridisation which is important in such cases will be inaccurately described and for that reason the solution given below is generally preferable.

The problem may be analysed in terms of the theory of band formation outlined in Chap.2, using the more accurate formalism of Chap.3. The origin of the disease can be traced to the pole at $P = 1/\gamma$ in the $\omega(P)$ function (4.18). Usually, this pole is not seen in an energy-band calculation because $1/\gamma$ is outside the range of P or S values given by (2.23), where bands are formed according to the Wigner-Seitz rule. However, if $1/\gamma$ is inside this range the pole in the $\omega(P)$ function splits the canonical ℓ band into two well-separated parts simulating the $(n - 1)\ell$ and $n\ell$ bands, respectively, and if subsequently the variational expression (3.50) is used these two parts will again be connected by a steep, unphysical band. Hence, the first step in the cure is to set the potential parameter $\langle\dot{\phi}_\nu^2\rangle$ for the ℓ value in question equal to zero, whereby the unphysical band will be swept out of the energy range of interest. Unfortunately, this precaution is not always sufficient because the hybridisation included in an LMTO calculation may also cause the unphysical effect we are trying to avoid by setting $\langle\dot{\phi}_\nu^2\rangle = 0$. Hence, we must take the extra step of choosing $1/\gamma$ outside the range given by (2.23).

To see how this may be done in practice we write γ in the form

$$\gamma = \frac{1}{2(2\ell + 1)} \frac{D_{\dot{\nu}} - \ell}{D_{\dot{\nu}} + \ell + 1}$$

$$= 1/P(D_{\dot{\nu}}) \quad , \tag{4.19}$$

which may be obtained by means of (4.14,16) and where we have used the definition (2.9) of the potential function in the last step. From (4.19) we realise that $1/\gamma$ will be outside the range (2.23) if $D_{\dot{\nu}}$ is positive. This also follows from the fact that the range we wish to avoid is the range where energy bands are formed, i.e. where the argument, *in casu* $D_{\dot{\nu}}$, of the potential function is negative. Hence, the symptom of the disease is a negative $D_{\dot{\nu}}$, and the cure is to fix $D_{\dot{\nu}}$ equal to some $D_{\dot{\nu}}^* > 0$, such that $-\ell/(\ell + 1)$ $< \phi^*(-)/\phi^*(+) < 1$, but keeping ϕ_ν and D_ν. The corresponding value of $\dot{\phi}_\nu$, $\dot{\phi}_\nu^*$ may be obtained from (3.17), and then the new standard parameters $\omega^*(-)$, $S\phi^*(-)$, and $\phi^*(-)/\phi^*(+)$ follow from (3.45,46) while $\langle\dot{\phi}_\nu^{2*}\rangle$ is set to zero for simplicity. With this choice of parameters the effect of the weakly contributing ℓ is correct to first order in ε, i.e.

$$\frac{1}{D^*(E) - D_\nu} = \frac{S\phi_\nu^*}{E - E_\nu} + S\phi_\nu\dot{\phi}_\nu^* + o(\varepsilon) \quad . \tag{4.20}$$

The fact that the LMTO method can give only one set of ℓ bands in any one calculation does not exclude the use of the method in cases where more than one principal quantum number gives rise to a band of ℓ character. In such a situation one simply divides the energy range into panels, performs LMTO calculations with potential parameters appropriate to each panel, and pieces the individual bands together to form the complete band structure. That this in reality is the most efficient way of obtaining such bands may be seen from the fact that is easier to diagonalise N 9 by 9 matrices than one 9N by 9N matrix.

4.3 Chromium 3d Bands: An Example

From the entries in Table 4.1 and (3.50,4.16,18) it is simple to establish the overall picture of the energy bands of chromium presented in Fig.4.1. In this connection one should note the following points.

Firstly, as discussed below, the LMTO method with $\kappa^2 = 0$ has errors of order $(E - v(S))^2$ and should be used only up to 1-2 Ry above the potential $v(S)$ at the sphere boundary. In the present case $v(S) \simeq -0.8$ Ry and the LMTO method can be used well above the top A_d of the d band.

Secondly, from the point of view of potential construction and ground-state properties, the interesting energy range includes only the occupied part of the band structure. This extends from B_s to approximately C_d, and is thus included in the range of satisfactory accuracy of the LMTO method.

Thirdly, the variational estimate (3.50) of the 3d logarithmic derivative function is valid in a range of $<\dot\phi_{\nu d}^2>^{-\frac{1}{2}} = 1.0$ Ry around $E_{\nu d}$ which spans the bottom of the s band, and therefore includes the occupied bands.

It follows that the interesting part of the Cr band structure can be generated accuarately by the LMTO method, with the parametrisation of only one period of each ℓ-logarithmic derivative function. This situation is typical for most metals and intermetallic compounds.

We saw in connection with Fig.3.1 that the price for the accurate variational description of E(D) for negative D's was unphysical behaviour around $D_{\overset{\bullet}{\nu}}$. To elucidate the situation, we plot in Fig.4.2 the second-order estimate (3.51) for the 3d period of the Cr logarithmic derivative function by means of (4.13) and Table 4.1. The $\omega(D)$ function has asymptotes at $D = D_{\overset{\bullet}{\nu}} = D_\nu - 1/S\phi_\nu\dot\phi_\nu$ and at $E = \hat{A} = E_\nu - \phi_\nu/\dot\phi_\nu$, and according to (4.5) is a decreasing function of D except at $D = D_{\overset{\bullet}{\nu}}$. The third-order estimate (3.50) plotted in Fig.4.3 has extrema at

$$E - E_\nu = \pm(1/2)<\dot\phi_\nu^2>^{-\frac{1}{2}}$$

$$\frac{1}{D - D_\nu} - \frac{1}{D_{\overset{\bullet}{\nu}} - D_\nu} = \mp s\phi_\nu^2<\dot\phi_\nu^2>^{\frac{1}{2}} \quad , \tag{4.21}$$

and around $D_{\overset{\bullet}{\nu}}$ it has a positive slope. This does no harm in the band calculation because $D_{\overset{\bullet}{\nu}}$ is positive (Table 4.2).

The potential function P(E) implied by (2.9,3.50) is plotted in Fig.4.4. According to the Wigner-Seitz rule the 3d band is formed in the range from

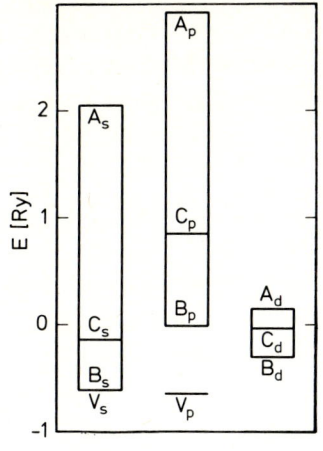

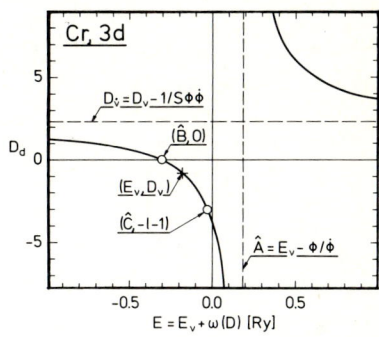

Fig.4.1. Estimates of the extent (bottom: B, top: A) and position (centre: C) of the s, p, and d bands of non-magnetic chromium obtained by the third-order expression (3.50), definitions (2.22), and the potential parameters in Table 4.1. Expression (3.50) spans only the square-well pseudopotentials V for the s and p bands, i.e. V_d is outside the energy window

Fig.4.2. Second-order estimate (3.51) of the function E(D) inverse to the logarithmic-derivative function D(E) for the 3d band in chromium. The second-order estimates of energies are denoted by a ^

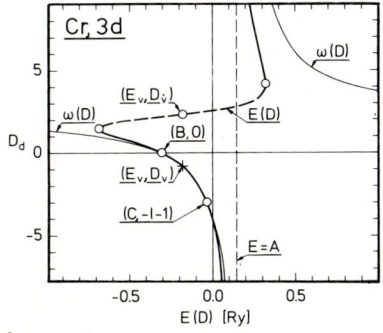

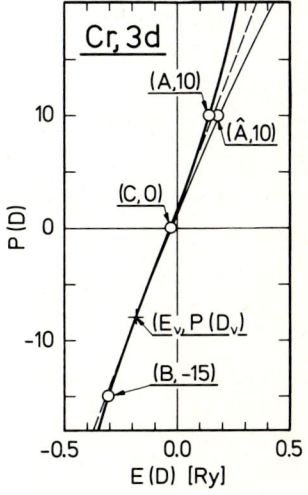

Fig.4.3. Third-order (variational) estimate (3.50) of the energy function E(D) for the 3d band in chromium. The unphysical range of positive slope is indicated by a broken line. For comparison, the second-order estimate ω(D) given in the previous figure is also included

Fig.4.4. Estimates of the 3d potential function P(E) for chromium. The heavy line is the third-order estimate obtained from (2.9,3.50), the thin line is the second-order estimate obtained from (2.9,3.51), and the broken line is the linear approximation (3.41)

P = -15 to P = 10. Outside that range the d states influence the band structure only through (weak) hybridisation. If we move along P(D) from below,
this corresponds to going along E(D) from D approximately unity, out to -∞
and back from +∞, without ever passing the forbidden range around $D_\nu^\bullet$. Thus,
the variational estimate spans the range of logarithmic derivatives encountered in practice.

The potential function in Fig.4.4 has a smooth shape in its entire range
of interest. In fact, the simple linear approximation (3.41) provides a good
description in the energy range where bands are formed. The reason is that
γ_d [(4.16,18) and Table 4.1] is small, and consequently the d bands of the
3d, 4d, and 5d transition metals are to a first approximation linear scalings of the canonical bands described in Sect.2.2. This fact may be used to
construct simple analytic models of the properties of these metals. The distortion parameters for s and p states are an order of magnitude larger than
γ_d, and therefore the first-order approximation is less satisfactory in those
cases.

4.4 Free-Electron Potential Parameters

The potential parameters for the free-electron case are interesting for
several reasons. First of all they are easy to calculate analytically from
standard expressions for the spherical Bessel functions [4.2] and therefore
useful in order-of-magnitude estimates. Secondly, they may be used in empty-
lattice tests of the LMTO method in order to indicate the accuracy of that
method in various applications. Such tests are also useful for programme de-
bugging purposes. Thirdly, muffin-tin orbitals with free-electron parameters
are used in Sect.6.9 to derive a correction to the atomic-sphere approximation.

For a potential which is constant with the value V_{FE} in the entire space,
the radial Schrödinger equation (1.17) reduces to the Bessel equation

$$\frac{d^2}{dx^2} x\phi_\ell = \left[\frac{\ell(\ell + 1)}{x^2} - 1\right] x\phi_\ell \tag{4.22}$$

of argument

$$x = Kr = r\sqrt{E - V_{FE}} \quad . \tag{4.23}$$

The radial wave functions of kinetic energy K^2 are the normalised spherical
Bessel functions

$$\phi_\ell(K^2,r) = j_\ell(x)<j_\ell^2>^{-\frac{1}{2}} \tag{4.24}$$

which are regular at $r = 0$. The logarithmic derivative functions are

$$D_\ell(K^2,r) \equiv D_\ell\{j_\ell(Kr)\} = xj'_\ell(x)/j_\ell(x) \quad . \tag{4.25}$$

From standard relations for the derivatives of the spherical Bessel functions [4.2] we find

$$D_\ell(E) + \ell + 1 = xj_{\ell-1}(x)/j_\ell(x)$$

$$D_\ell(E) - \ell = - xj_{\ell+1}(x)/j_\ell(x) \quad . \tag{4.26}$$

Hence, we obtain

$$s^2(V_{n\ell} - V_{FE}) = X^2_{n,\ell+1}$$

$$s^2(C_{n\ell} - V_{VE}) = X^2_{n+1,\ell-1}$$

$$s^2(A_{n\ell} - V_{FE}) = X^2_{n+1,\ell} \tag{4.27}$$

for the band energies defined in Sect.2.3. The zeros $X_{n,\ell}$ of the Bessel functions satisfy the relations

$$j_\ell(X_{n,\ell}) = 0$$

$$X_{n,-1} = (n - 1/2)\pi$$

$$X_{n,0} = n\pi \quad , \qquad n \neq 0 \quad , \tag{4.28}$$

where we used the convention that $n = 1$ corresponds to the first positive zero of the Bessel function $j_\ell(x)$. Some values for the band energies (4.27) are given in Table 4.3.

For $\ell = 0$

$$j_0(x) = \sin(x)/x$$

$$D_0(x) = x \cot(x) - 1 \quad . \tag{4.29}$$

Furthermore, (4.26) gives

$$(D_{\ell+1} + \ell + 1 + 1)(D_\ell - \ell) = -x^2 \tag{4.30}$$

which may be used in connection with (4.29) to obtain by recursion the logarithmic derivative functions for not too small values of x^2 and not too large values of ℓ.

The local free-electron potential parameters may be found by differentiation with respect to energy of the logarithmic derivatives. To see this we multiply the Taylor expansion of $D(\varepsilon = E - E_\nu)$

$$D - D_\nu = \dot{D}\varepsilon + (1/2)\ddot{D}\varepsilon^2 + (1/6)\dddot{D}\varepsilon^3 + o(\varepsilon^3) \tag{4.31}$$

Table 4.3. Free-electron band energies obtained from (4.27). The constant potential is assumed to be zero, i.e. $V_{FE} = 0$

n,ℓ	$s^2 V_{nℓ}$	$s^2 C_{nℓ}$	$s^2 A_{nℓ}$
0,s	0	$\pi^2/4$	π^2
0,p	0	π^2	20.19
0,d	0	20.19	33.22
1,s	20.19	22.21	39.48
1,p	33.22	39.48	59.68
1,d	48.83	59.68	82.72

with the Laurent expansion (3.30), retaining terms to second order in ε, and obtain

$$1 = - \frac{\dot{D}}{mS^2} + \left[\dot{D}a - (1/2) \frac{\ddot{D}}{mS^2} \right] \varepsilon$$

$$+ \left[\dot{D}bS^2 + (1/2)\ddot{D}a - (1/6) \frac{\dddot{D}}{mS^2} \right] \varepsilon^2 + o(\varepsilon^2) \quad . \tag{4.32}$$

Since this must hold for any ε, then

$$mS^2 = -\dot{D}$$

$$a = (1/2) \frac{\ddot{D}}{mS^2 \dot{D}}$$

$$bS^2 = (1/6) \frac{\dddot{D}}{mS^2 \dot{D}} - (1/2) \frac{\ddot{D}}{\dot{D}} a \quad . \tag{4.33}$$

The radial Schrödinger equation may be written in the form

$$- S \frac{\partial D}{\partial S} = (D + ℓ + 1)(D - ℓ) + [E - v(S)]S^2 \quad . \tag{4.34}$$

Furthermore, with E and S as the independent variables we find for free electrons

$$\dot{D} = \frac{\partial D}{\partial E} = \frac{S}{2E} \frac{\partial D}{\partial S} \quad , \tag{4.35}$$

here specialised to the case $V_{FE} = 0$, cf. (4.23), whereby

$$m = (1/2) \left[\frac{(D + ℓ + 1)(D - ℓ)}{x^2} + 1 \right] \quad . \tag{4.36}$$

It follows from (4.33) that $2amS^2 = \dot{m}/m$ and hence

$$ma = \frac{1}{4x^2} \left(\frac{1}{m} - 2D - 3 \right) \quad . \tag{4.37}$$

Finally, $b = \ddot{m}/(6m^2 S^4) - ma^2$ so that differentiating (4.36) twice gives

$$b = \frac{1}{6x^2} [1 - a(2D + 5)] - ma^2 \quad . \tag{4.38}$$

The ϕ, $\dot{\phi}$ parameters may now be found from (3.33-35) and in turn the standard parameters (Table 4.1) may be obtained from (3.17,45,46) or Sect.4.6.

For $D_{\nu\ell} = -\ell - 1$, we realise from (4.36) that $m_\ell = 1/2$, and hence the mass at the centre of the band, $\mu_{n\ell}$, as defined by (3.39) is unity for free electrons for all values of n and ℓ, i.e.

$$\mu_{n\ell}^{FE} = 1 \quad . \tag{4.39}$$

For $D_{\nu\ell} = \ell$ and $x^2 \neq 0$, i.e. $n \neq 0$, we find that $\tau_{n\ell}$ defined by (3.40) is equal to $(2\ell + 3)/2$. Finally, in the limit $x^2 \to 0$ the Bessel functions behave as r^ℓ [cf. the asymptotic forms (5.8)] and therefore $D_\ell = \ell$. Since $D_{\ell+1} = \ell + 1$ the limit of $(D_\ell - \ell)/x^2$ to be used in (4.36) is found from (4.30) to be $-1/(2\ell + 3)$, whereby $m = 1/(2\ell + 3)$ and $\tau_{0\ell} = 1$. Hence

$$\tau_{0\ell}^{FE} = 1$$

$$\tau_{n\ell}^{FE} = (2\ell + 3)/2 \quad ; \quad n \neq 0 \quad . \tag{4.40}$$

The free-electron potential parameters are particularly simple when $D_{\nu\ell} = -\ell - 1$ or $D_{\nu\ell} = \ell$. In the former case they may be found in a straightforward manner from (4.36-38), since $S^2(C_{n\ell} - V_{FE}) = x_{n+1,\ell-1}^2 \neq 0$ (Table 4.3). The result is shown in Table 4.4 where the a, b, m parameters have been converted to standard parameters.

For the case $\ell = 0$, $n = \nu = 0$ we must once more resort to the comparison of Laurent and Taylor expansions. Hence, by means of the recursion relation (4.30), we compare the Laurent expansion for $(D_\ell - \ell)^{-1}$ with the Taylor expansion for $D_{\ell+1} + (\ell + 1) + 1$. We find $m_\ell(V_{0\ell}) = (2\ell + 3)^{-1}$, $a_\ell(V_{0\ell}) = (2\ell + 5)^{-1}$, and $b_\ell(V_{0\ell}) = (2\ell + 5)^{-2}(2\ell + 7)^{-1}$. The corresponding standard parameters are shown in Table 4.5.

It is rather apparent in the present section as well as elsewhere in this chapter that the values $D = -(\ell + 1)$ and $D = \ell$ are of particular significance in the theory of potential parameters. This feature stems from the radial Schrödinger equation, as is easily seen in (4.34). From there it can be traced to the centrifugal term $\ell(\ell + 1)/r^2$ in (1.17). Hence, the fact that

Table 4.4. Standard potential parameters for free electrons. $D_{\nu\ell} = -\ell - 1$, $\nu = 0$, $V_{FE} = 0$

	$s^2 \kappa_\nu^2$	$s^2 \omega(-)$	$s^3 \phi^2(-)$	$\phi(-)/\phi(+)$	$s^2 \langle \dot{\phi}_\nu \rangle^{-\frac{1}{2}}$
	$x_{n+1,\ell-1}^2$	0	2	$1 - \dfrac{(2\ell+1)^2}{2x_{n+1,\ell-1}^2}$	$\left[\dfrac{1}{12x_{n+1,\ell-1}^2} \left\{ 1 - \dfrac{(2\ell+1)(2\ell+9)}{4x_{n+1,\ell-1}^2} \right\} \right]^{-\frac{1}{2}}$
s	2.47	0	2	0.80	18
p	9.87	0	2	0.54	27
d	20.19	0	2	0.38	35

Table 4.5. Standard potential parameters for free electrons. $D_{\nu\ell} = \ell$, $\nu = 0$, $V_{FE} = 0$

	$s^2 \kappa_\nu^2$	$s^2 \omega(-)$	$s^3 \phi^2(-)$	$\phi(-)/\phi(+)$	$s^2 \langle \dot{\phi}^2 \rangle^{-\frac{1}{2}}$
	0	$\dfrac{(2\ell+1)(2\ell+5)}{2}$	$\dfrac{(2\ell+5)^2}{4(2\ell+3)}$	$\dfrac{2\ell+5}{2(2\ell+3)}$	$(2\ell+5)\sqrt{(2\ell+3)(2\ell+7)}$
s	0	2.5	2.08	0.83	23
p	0	10.5	2.45	0.70	47
d	0	22.5	2.89	0.64	79

spherical symmetry leads to an extra contribution to the effective potential attaches a special meaning to the two numbers ℓ and $\ell + 1$.

4.5 Volume Derivatives of Potential Parameters

We shall briefly discuss the volume derivatives of a few important potential parameters for a fixed potential. Our starting point is the radial Schrödinger equation in the form

$$- S \frac{\partial D}{\partial S} = (D + \ell + 1)(D - \ell) + (E - v)S^2 \quad , \tag{4.41}$$

where $v = v(S)$ is the potential at a sphere of radius S. With this potential frozen there are three variables D, E, and S in the problem, whereby

$$\left. \frac{\partial E}{\partial S} \right|_D = - \left. \frac{\partial E}{\partial D} \right|_S \left. \frac{\partial D}{\partial S} \right|_E = - \frac{1}{\dot{D}} \left. \frac{\partial D}{\partial S} \right|_E \quad . \tag{4.42}$$

Hence, by means of (4.41,3.38) we find

$$- S \frac{\partial E}{\partial S} = S \phi^2 \left[(D + \ell + 1)(D - \ell) + (E - v)S^2 \right] \quad , \tag{4.43}$$

which for the choices $D = -\ell - 1$ and $D = \ell$ and with (3.39,40) gives

$$\frac{\partial C}{\partial \ln(S)} = -2\frac{C - v}{\mu}$$

$$\frac{\partial V}{\partial \ln(S)} = -(2\ell + 1)\frac{V - v}{\tau} \quad . \tag{4.44}$$

From the second-order estimates (4.2,9) and the entries in Table 4.1 we find that at normal volume $\partial C_s/\partial \ln S \sim -0.5\, S^2$ and $\partial C_d/\partial \ln S \sim -0.06\, S^2$. Hence we have substantiated the notion that for a transition metal the position of the d band changes much slower with volume than that of the s band. In the present language the reason is the much higher mass μ or smaller amplitude $S\phi^2(-)$ of the d orbitals.

Since according to (3.38) $\dot{D}(E) = -[S\phi^2]^{-1} = -m(E)S^2$, the variation of the band mass may be found from

$$\frac{S}{\dot{D}}\frac{d\dot{D}}{dS}\bigg|_D = \frac{S}{\dot{D}}\frac{\partial^2 D}{\partial S \partial E} + \frac{S}{\dot{D}}\frac{\partial^2 D}{\partial E^2}\frac{dE}{dS}\bigg|_D \quad . \tag{4.45}$$

The first term on the right-hand side may be found from (4.41), the second term from the Laurent expansion (3.32), i.e.

$$\frac{\partial^2 D}{\partial E^2} = \frac{2S\phi\dot{\phi}}{(S\phi^2)^2} \tag{4.46}$$

and the third from (4.43). We obtain

$$\frac{S}{\dot{D}}\frac{d\dot{D}}{dS}\bigg|_D = -(2D + 1) + S^3\phi^2 + 2S\phi\dot{\phi}[(D + \ell + 1)(D - \ell) + (E - v)S^2] \quad . \tag{4.47}$$

By means of (3.17) for $S\phi\dot{\phi}$, definitions (3.39,40) for μ and τ, and the boundary conditions $D = -\ell - 1$ and $D = \ell$, we have

$$\frac{\partial \ln(\mu S^2)}{\partial \ln(S)} = (2\ell + 1) + \frac{2}{\mu} - \frac{2(C - v)S^2}{\dot{D}_\nu + \ell + 1}$$

$$\frac{\partial \ln(\tau S^2)}{\partial \ln(S)} = -(2\ell + 1) + \frac{2\ell + 3}{\tau} - \frac{2(V - v)S^2}{\dot{D}_\nu - \ell} \quad . \tag{4.48}$$

Since the change in the bandwidth W is

$$\frac{\partial \ln(W)}{\partial \ln(S)} = -\frac{\partial \ln(\mu S^2)}{\partial \ln(S)} \quad , \tag{4.49}$$

61

then (4.48) provides the dependence of the bandwidth on the atomic volume. For the d bands in chromium we find from Table 4.2, $v \sim -0.8$ Ry, that $\partial \ln(\mu S^2)/\partial \ln(S) = 3.2$. Hence the "ideal" behaviour " $\sim S^{-5}$ is somewhat modified when we use frozen potentials. Self-consistent calculations give $d\ln(\mu S^2)/d\ln(S) = 4.5$, which is much closer to ideal behaviour. It is, however, the modified behaviour which enters our estimates of cohesive properties, so we shall return to (4.44,48) when we consider ground-state properties in Chap.7.

4.6 Potential Parameter Relations

The many interrelationships which exist among the potential parameters derived in this chapter may be difficult to absorb at first sight. Therefore is given below a number of such relations grouped together in a relatively systematic way. No proofs will be supplied, but all the expressions given are based upon the theory outlined in the preceeding sections.

D_ν

$\phi_\nu = \phi(E_\nu)$

$\dot{\phi}_\nu = \dot{\phi}(E_\nu)$

$<\dot{\phi}_\nu^2> = \int_0^2 \dot{\phi}^2(E_\nu, r)r^2 dr$

Basic quantities from the Schrödinger equation

$$\omega(-) = -\frac{\phi_\nu - \ell - 1 - D_\nu}{\dot{\phi}_\nu - \ell - 1 - D_\nu^{\cdot}}$$

$$S\phi^2(-) = S\phi_\nu^2\left(\frac{D_\nu - D_\nu^{\cdot}}{-\ell - 1 - D_\nu^{\cdot}}\right)^2$$

$$\frac{\Phi(-)}{\Phi(+)} = \frac{D_\nu^{\cdot} - \ell}{D_\nu^{\cdot} + \ell + 1}$$

$<\dot{\phi}_\nu^2>^{-\frac{1}{2}}$

Standard parameters

$$\phi = \Phi(-)\left\{1 + \frac{\omega(-)}{(2\ell + 1)(S\phi^2(-))}\left[1 - \frac{\Phi(-)}{\Phi(+)}\right]\right\}$$

$$\dot{\phi} = \frac{1}{(2\ell + 1)S\Phi(-)}\left[\frac{\Phi(-)}{\Phi(+)} - 1\right]$$

Conversion from standard parameters to ϕ, $\dot{\phi}$ paramters

$$D_\nu = \ell + (2\ell + 1) \left[\frac{\Phi(+)}{\Phi(-)} \frac{\omega(-)}{\omega(+)} - 1 \right]^{-1}$$

$$<\dot\phi_\nu^2>$$

$$D_{\dot\nu} = \ell - \frac{2\ell + 1}{1 - \Phi(+)/\Phi(-)}$$

$D_{\dot\nu}$ from Φ ration

$$D_\nu$$
$$mS^2 = (S\phi_\nu^2)^{-1}$$
$$a = -S\phi_\nu\dot\phi_\nu = (D_{\dot\nu} - D_\nu)^{-1}$$
$$bS^2 = S\phi_\nu^2 <\dot\phi_\nu^2>$$

a, b, m parameters

$$D_\nu$$
$$\phi_\nu = 1/\sqrt{mS^3}$$
$$\dot\phi_\nu = -a\sqrt{mS}$$
$$<\dot\phi_\nu^2> = mbS^4$$

Conversion from
a, b, m parameters
to ϕ, $\dot\phi$ parameters

$$(D_\nu - D_{\dot\nu})S\phi_\nu\dot\phi_\nu = 1$$
$$\omega(D_2) - \omega(D_1) = -S\Phi(D_1)\Phi(D_2)(D_2 - D_1)$$
$$\omega(-) - \omega(+) = (2\ell + 1)S\Phi(-)\Phi(+)$$

Wronskian
relations

$$\frac{d\omega(D)}{dD} = -S\Phi^2(D)$$

$$\frac{\omega(D) - \omega(-)}{\omega(D) - \omega(+)} = \frac{\Phi(-)}{\Phi(+)} \frac{D + \ell + 1}{D - \ell}$$

$$\Phi(D) = \frac{(2\ell + 1)\Phi(-)\Phi(+)}{(D + \ell + 1)\Phi(-) - (D - \ell)\Phi(+)}$$

$$S\Phi^2(D) = \frac{1}{mS^2} \left[\frac{1}{1 + (D_\nu - D)a} \right]^2$$

Expressions for
general D

$$\omega(D) = - \frac{\phi_\nu}{\dot\phi_\nu} \frac{D - D_\nu}{D - D_{\dot\nu}}$$

Alternative ex-
pressions for the

63

$$\omega(D) = \frac{S\phi_\nu^2}{(D - D_\nu)^{-1} + S\phi_\nu\dot\phi_\nu}$$

$$\omega(D) = S\phi_\nu^2 \frac{(\dot D_\nu - D_\nu)(D - D_\nu)}{D - \dot D_\nu}$$

$$\omega(D) = \frac{1}{mS^2[a - (D - D_\nu)^{-1}]}$$

$$\gamma = \frac{1}{2(2\ell + 1)} \frac{\Phi(-)}{\Phi(+)}$$

$$\gamma = \frac{1}{2(2\ell + 1)} \frac{\dot D_\nu - \ell}{\dot D_\nu + \ell + 1}$$

$$\gamma = 1/P(\dot D_\nu)$$

$$\gamma = \frac{\omega(-) - \omega(+)}{2(2\ell + 1)^2 S\phi^2(+)}$$

$$\gamma = \frac{1}{2} \frac{S\phi^2(-)}{\omega(-) - \omega(+)}$$

$$= \frac{1}{2(2\ell + 1)} \frac{1 + (D_\nu - \ell)a}{1 + (D_\nu + \ell + 1)a}$$

---

5. The Linear Method

In this chapter we shall use the Rayleigh-Ritz variational principle in conjunction with energy-independent muffin-tin orbitals (MTO's) to derive the LMTO equations, which will have the form (1.19) of a generalised eigenvalue problem and hence provide solutions to the energy-band problem in a computationally efficient manner. First we define energy-dependent muffin-tin orbitals similar to those used in Sect.2.1. Then we present that choice of *augmented* spherical waves which will make the muffin-tin orbitals simultaneously orthogonal to the core states and energy independent. The augmentation is based upon the ϕ, $\dot{\phi}$ formalism developed in Chap.3. Both kinds of muffin-tin orbitals obey a convenient expansion theorem, whose expansion coefficients are the structure constants which contain all the necessary information about the crystal symmetry. The final derivation of the LMTO Hamiltonian and overlap matrices is based upon muffin-tin orbitals with tails which, before augmentation, are solutions of the Helmholtz wave equation rather than the Laplace equation, as in the atomic-sphere approximation (Chap.6). Thereby the results of the present chapter may be applied not only to the case of energy-band calculations, where the atomic-sphere approximation has the necessary accuracy, but also to cases of molecules and clusters where one may need basis functions more versatile than those given within the atomic-sphere approximation.

The material here is taken mainly from the unpublished notes by *Andersen* [5.1] and the paper by *Andersen* and *Woolley* [5.2].

5.1 Partial Waves for a Single Muffin-Tin

We begin by approximating the crystal potential $v(\mathbf{r})$ appearing in the Schrödinger equation (1.4) by a so-called muffin-tin potential which is defined to be spherically symmetric within spheres of radius S_{MT} and to have a constant value V_{MTZ}, the muffin-tin zero, in the interstitial region between the spheres, Fig.5.1. This kind of potential is designed to facilitate

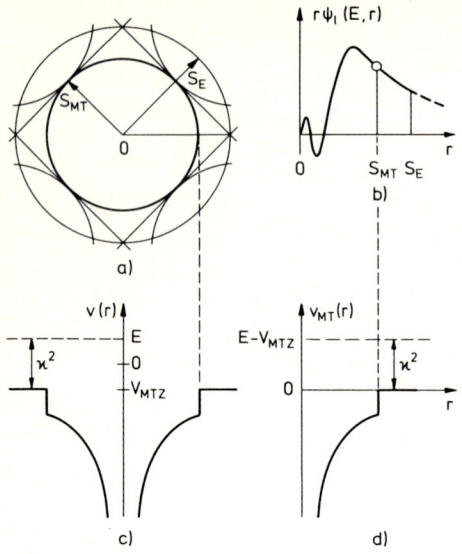

Fig.5.1. The muffin-tin approximation. a): the unit cell, the muffin-
tin sphere of radius S_{MT}, and the
escribed sphere of radius S_E.
b): the radial wave function.
c): the muffin-tin part of the
crystal potential $v(r)$. d): the
muffin-tin potential $v_{MT}(r)$ (5.1)

matching of wave functions from cell to cell through the assumption that the electrons propagate freely between spheres with a constant wave number $\kappa = \sqrt{E - V_{MTZ}}$. This assumption is valid only if the wavelength $2\pi/\kappa$ is large compared to the "thickness" of the interstitial region, i.e. the distance $S_{MT} - S_E$ in Fig.5.1. Since in most applications we are interested only in those electrons which barely move from cell to cell, that is with kinetic energy κ^2, between approximately -1 Ry to 1 Ry, the wave-number criterion is satisfied in all closely packed solids, and the muffin-tin construction is therefore employed in both the KKR and the APW methods.

In the following we consider for simplicity a crystal with one atom per primitive cell, and within a single muffin-tin well (Fig.5.1) we define the potential

$$
v_{MT}(r) = \begin{cases} v(r) - V_{MTZ} & r \leq S_{MT} \\ 0 & r \geq S_{MT} \end{cases} . \tag{5.1}
$$

Here $v(r)$ is the spherically symmetric part of the crystal potential. The Hamiltonian minus the energy for a system of superimposed muffin-tin wells is

$$
H - E = - \nabla^2 + \sum_R v_{MT}(|\mathbf{r} - \mathbf{R}|) - \kappa^2 , \tag{5.2}
$$

where the sum extends over the crystal, and the kinetic energy κ^2 in the interstitial region is defined by

$$
\kappa^2 \equiv E - V_{MTZ} . \tag{5.3}
$$

We now seek the solutions of Schrödinger's differential equation

$$[- \nabla^2 + v_{MT}(r) - \kappa^2]\psi_L(E,\mathbf{r}) = 0 \qquad (5.4)$$

for all values of κ^2, i.e. both in the continuum and in the bound-state regime, for an electron moving in the potential from an isolated muffin-tin well embedded in the flat potential V_{MTZ}. In this case the spherical symmetry extends throughout all space and the wave functions are

$$\psi_L(E,\mathbf{r}) = i^\ell Y_\ell^m(\hat{r})\psi_\ell(E,r) \quad , \qquad (5.5)$$

where the ℓ,m quantum numbers have been combined into one subscript L. For reasons of notation we have included a phase factor i^ℓ, and the spherical harmonics $Y_\ell^m(\hat{r})$ have the phase defined by *Condon* and *Shortley* [5.3].

Inside the muffin-tin well the radial part $\psi_\ell(E,r)$ must be regular at the origin (specifically $\psi \sim r^\ell$) in order to be normalisable. It is obtained by numerical integration of the radial Schrödinger equation

$$\left[- \frac{d^2}{dr^2} + \frac{\ell(\ell + 1)}{r^2} + v_{MT}(r) - \kappa^2 \right] r\psi_\ell(E,r) = 0 \quad . \qquad (5.6)$$

In the region of constant potential the solutions of (5.4) are spherical waves with wave number κ, and their radial parts satisfy (5.6) with $V_{MT}(r) = 0$, i.e.

$$\left[- \frac{d^2}{dr^2} + \frac{\ell(\ell + 1)}{r^2} - \kappa^2 \right] r y_\ell(\kappa r) = 0 \quad . \qquad (5.7)$$

This well-known Helmholtz wave equation has two linearly independent solutions which we may take to be the spherical Bessel $j_\ell(\kappa r)$ and Neumann $n_\ell(\kappa r)$ functions, respectively. From the small κr limits

$$\left. \begin{aligned} j_\ell(\kappa r) &\to (\kappa r)^\ell/(2\ell + 1)!! \\[2mm] n_\ell(\kappa r) &\to -(2\ell - 1)!!/(\kappa r)^{\ell+1} \end{aligned} \right\} \quad \kappa r \to 0 \qquad (5.8)$$

where the double factorial is defined by $!! = 1 \cdot 3 \cdot 5 \cdot \ldots$ and $-1!! = 1$, and from the asymptotic forms

$$\left. \begin{aligned} j_\ell(\kappa r) &\to \frac{\sin(\kappa r - \ell\pi/2)}{\kappa r} \\[2mm] n_\ell(\kappa r) &\to -\frac{\cos(\kappa r - \ell\pi/2)}{\kappa r} \end{aligned} \right\} \quad r \to \infty \quad , \qquad (5.9)$$

67

it follows that only $j_\ell(\kappa r)$ is regular at the origin while both $j_\ell(\kappa r)$ and $n_\ell(\kappa r)$ are regular at infinity.

Here and in the following we write expressions explicitly for the case of positive kinetic energy κ^2. This is the range of unbounded continuum states. When the kinetic energy in the interstitial region is negative, i.e. $\kappa = i|\kappa| = i\sqrt{V_{MTZ} - E}$, and bound states may be formed, the spherical Neumann function n_ℓ should be substituted everywhere, including (5.11) below, by $-ih_\ell^{(1)} = n_\ell - ij_\ell$. Here $h_\ell^{(1)}$ is the Hankel function of the first kind [5.4] which has the asymptotic form $i^{-\ell} e^{-|\kappa|r}/|\kappa|r$ and hence is regular at infinity.

The partial wave

$$
\psi_L(E,\kappa,r) = i^\ell Y_\ell^m(\hat{r}) \begin{cases} \psi_\ell(E,r) & r \leq S_{MT} \\ \\ \kappa[n_\ell(\kappa r) - \cot(n_\ell)j_\ell(\kappa r)] & r \geq S_{MT} \end{cases} \tag{5.10}
$$

is formed by attaching the solution of the interstitial region, i.e. the tail, and the solution inside the MT sphere to each other at the sphere boundary. Since the Schrödinger equation is of second order, $\psi_L(E,\kappa,r)$ is a solution of (5.4) at energy E in the entire space, provided $\psi_\ell(E,S_{MT})$ and the constant of integration $\cot(n_\ell)$ are chosen in such a way that the partial wave is everywhere continuous and differentiable. This requires that

$$
\cot(n_\ell(E,\kappa)) = \frac{n_\ell(\kappa r)}{j_\ell(\kappa r)} \cdot \frac{D_\ell(E) - \kappa n_\ell'(\kappa r)/n_\ell(\kappa r)}{D_\ell(E) - \kappa j_\ell'(\kappa r)/j_\ell(\kappa r)}\Bigg|_{r=S_{MT}} \tag{5.11}
$$

where the logarithmic derivative $D_\ell(E)$ is defined by (2.3). For positive κ^2 values (5.11) defines the conventional phase shifts n_ℓ and the asymptotic form of the partial wave (5.10), i.e.

$$
\psi_\ell(E,\kappa,r) \sim -\frac{\sin(\kappa r + n_\ell - \ell\pi/2)}{r \sin(n_\ell)} \quad , \tag{5.12}
$$

then follows from (5.9). This shows that the solution in the large r limit may be regarded as a free-space spherical wave shifted by the phase shift n_ℓ due to the muffin-tin potential.

Solution (5.10) is unbounded but delta-function normalisable when κ^2 is positive, but if κ^2 is negative $\psi_L(E,\kappa,r)$ can be normalised only at the eigenvalues of the single well, where the constant of integration (5.11), obtained by substituting $-ih_\ell^{(1)}$ for n_ℓ, is zero. For this reason these partial waves are not well suited as basis functions.

68

5.2 Muffin-Tin Orbitals

Muffin-tin orbitals were introduced by *Andersen* [5.5] to obtain basis functions which are approximately independent of energy, reasonably localised, and normalisable for all values of κ^2. This is partially accomplished by adding a spherical Bessel function that cancels the divergent part of $\psi_\ell(E,r)$ and simultaneously reduces the energy- and potential dependence of the tails to that given by (5.3).

The muffin-tin orbital is therefore

$$\chi_L(E,\kappa,\mathbf{r}) = i^\ell Y_\ell^m(\hat{r}) \begin{cases} \psi_\ell(E,r) + \kappa \cot(n_\ell)j(\kappa r) & r \le S_{MT} \\ \\ \kappa n_\ell(\kappa r) & r \ge S_{MT} \end{cases} \tag{5.13}$$

and an important feature of this definition is that the functions inside the well are regular at the origin, while the tail $\kappa n_\ell(\kappa r)$ is regular at infinity. If the kinetic energy κ^2 is negative, the spherical Neumann function in (5.13) as well as in (5.11) should be substituted by $-ih_\ell^{(1)}$ which decays as $e^{-|\kappa| r}/|\kappa| r$. Owing to the appearance of the spherical Bessel function inside the muffin tin, the MTO's defined in this fashion are not eigenfunctions for the muffin-tin potential $v_{MT}(r)$, except at bound states and resonances, where the constant of integration (5.11) is zero. However, the Bloch sums of $\psi_L(E,\kappa,\mathbf{r})$ and $\chi_L(E,\kappa,\mathbf{r})$ are identical because their difference is a Bloch sum of spherical Bessel functions, which is zero except at the free-electron parabola $|\mathbf{k} + \mathbf{G}|^2 = \kappa^2$.

We already encountered muffin-tin orbitals in Sect.2.1 when we derived the KKR-ASA equations. There they were defined in a form valid for a fixed κ^2 equal to zero. The connection between the present definition (5.13), valid for a general κ^2, and the earlier definition (2.1) follows immediately from the small κr limits (5.8) of the spherical Bessel and Neumann functions and a suitable renormalisation of the orbitals. Hence, $(r/S)^{-\ell-1}$ stems from the Neumann function, $(r/S)^\ell$ from the Bessel function and $p_\ell(E)$ is the equivalent of $\cot(n_\ell)$.

If we approximate the crystal potential by an array of non-overlapping muffin-tin wells as in (5.2), the energy-dependent muffin-tin orbitals (5.13) may be used in conjunction with the tail-cancellation theorem to obtain the so-called KKR equations. These have the form (1.21) and provide exact solutions for muffin-tin geometry. Computationally, however, they are rather inefficient and it is therefore desirable to develop a method based upon the variational principle and a fixed basis set, which leads to the computatinally efficient eigenvalue problem (1.19).

The LMTO method is such a fixed-basis method based upon the variational principle. However, to use the muffin-tin orbitals in a variational procedure they should be made energy independent, so that the eigenvalue equation (1.19) is linear in energy, and orthogonal to the core states, so that the eigenvalues will converge to non-core values. In Sect.5.4 both objectives are accomplished by a suitable augmentation of the spherical Bessel and Neumann functions.

5.3 Expansion Theorem for MTO Tails

One reason for choosing the tails of muffin-tin orbitals as solutions of the translationally invariant Helmholtz wave equation (5.7) is the extremely simple expansion theorem

$$n_L(\kappa, \mathbf{r} - \mathbf{R}) = 4\pi \sum_{L'} \sum_{L''} C_{LL'L''} j_{L'}(\kappa, \mathbf{r} - \mathbf{R'}) n^*_{L''}(\kappa, \mathbf{R} - \mathbf{R'}) \qquad (5.14)$$

which these functions obey [5.6]. This expansion is valid inside the sphere centred at $\mathbf{R'}$ and passing through $\mathbf{R}$, i.e. for $|\mathbf{r} - \mathbf{R}| < |\mathbf{R} - \mathbf{R'}|$, Fig.5.2. The Gaunt coefficients are defined by

$$C_{LL'L''} = \int Y^m_\ell(\hat{\kappa}) Y^{m'}_{\ell'}(\hat{\kappa})^* Y^{m''}_{\ell''}(\hat{\kappa}) d\hat{\kappa}$$

$$= \left(\frac{2\ell'' + 1}{4\pi}\right)^{\frac{1}{2}} c^{\ell''}(\ell'm'; \ell m) \qquad (5.15)$$

and $c^{\ell''}(\ell'm; \ell m)$ are tabulated by *Condon* and *Shortley* [5.3]. $C_{LL'L''}$ vanish unless $m'' = m' - m$, and the ℓ'' summation in (5.14) includes only the few terms with $\ell'' = |\ell - \ell'|$, $|\ell - \ell'| + 2, \ldots, (\ell + \ell')$.

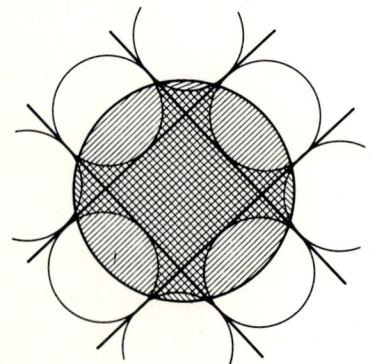

Fig.5.2. The convergence area for the expansion theorem (5.14) (hatched area), and the one-centre expansion (5.29) (cross-hatched area)

The expansion theorem means that the tail of the muffin-tin orbital
(5.13), i.e. a spherical Neumann function including the angular part $i^\ell Y_\ell^m(\hat{r})$,
as in (5.5), positioned at $\mathbf{R}$ may be expanded in terms of spherical Bessel
functions centred at $\mathbf{R'}$. The reason for the functional form (for simplicity
$\mathbf{R'} = \mathbf{0}$)

$$j_{L'}(\kappa,\mathbf{r}) = j_{\ell'}(\kappa r) i^{\ell'} Y_{\ell'}^{m'}(\hat{r}) \tag{5.16}$$

is that the Neumann function centred at $\mathbf{R}$ is regular at the origin and
therefore is expanded in the regular solutions of the wave equation only.
Consequently, inside any muffin-tin sphere the tails from the other spheres
will have the same functional form as the term proportional to $\cot(n_\ell)$. An expression analogous to (5.14) may be obtained for any linear combination of
Bessel and Neumann functions provided n is substituted by the linear combination in question.

5.4 Energy-Independent Muffin-Tin Orbitals

So far we have not discussed in detail the energy (in)-dependence of the
muffin-tin orbital. As (5.13) stands, its amplitude $\kappa n_\ell(\kappa S)$ and logarithmic
derivative $\kappa S n'_\ell(\kappa S)/n_\ell(\kappa S)$ at the sphere, as well as its tail, depend on
energy only through κ. If we could therefore disregard (5.3) and fix κ at
some suitable value, the energy dependence of the tail would be suppressed.
In addition, the amplitude and slope of the muffin-tin orbital at the sphere
would become energy independent, and that part of the orbital inside the
sphere would be so clamped at the boundary as to be virtually energy independent.

By construction, the partial wave (5.10) and the muffin-tin orbital (5.13)
are everywhere continuous and differentiable, and the phase shifts obtained
from (5.11) maintain continuity and differentiability for any value of E and
κ^2 independently of a possible relation between these two quantities. Furthermore, the partial wave is constructed to be a solution of the Schrödinger
equation (5.4) in the entire space, whereby E and κ^2 must satisfy (5.3). In
that case, κ^2 constitutes the exact kinetic energy in the region of constant
potential, and the phase shifts are functions of E only.

If follows that continuous and differentiable wave functions may be defined even where (5.3) is not obeyed and the tail is no longer an exact solution in the region of constant potential. If, furthermore, the variational
principle is used, the value of κ^2 becomes of secondary importance, and one
may therefore consider disregarding (5.3) altogether. In the following we

shall keep the energy E of the orbital and the wave number κ of the tail·as completely separate entities, fix κ at the characteristic value and use linear combinations of such muffin-tin orbitals in a variational procedure. Usually we shall work with a single κ value, but κ may be treated as an additional variational parameter if necessary.

We shall now augment the muffin-tin orbital (5.13) and show that the augmented muffin-tin orbital

$$\chi_L(E,\kappa,\mathbf{r}) = i^\ell Y_\ell^m(\hat{r}) \begin{cases} \psi_\ell(E,r) + \kappa\cot(n_\ell)J_\ell(\kappa r) & r \leq S_{MT} \\ \kappa N_\ell(\kappa r) & r \geq S_{MT} \end{cases} \tag{5.17}$$

for a particular choice of the augmented spherical Bessel and Neumann functions $J_\ell(\kappa r)$ and $N_\ell(\kappa r)$ may be made energy independent around a fixed energy E_ν to first order in $(E - E_\nu)$. At the same time, the muffin-tin orbital becomes orthogonal to the core states, ensuring that the LMTO method does not converge to core eigenvalues. In connection with the augmentation one should realise that once κ has been fixed, the spherical Bessel and Neumann functions lose their special significance as exact solutions of the Schrödinger equation (5.6) in the region of constant potential. Hence, if desired, they may be replaced, i.e. augmented, by more appropriate functions which are attached to them at the sphere in a continuous and differentiable fashion. As shown below, the energy derivative of the partial wave $\dot{\psi}_\ell$ is a particularly convenient choice for such an augmentation.

To arrive at suitable definitions of J_ℓ and N_ℓ we simply disregard (5.3), fix κ, and demand that the energy derivative of the muffin-tin orbital (5.17)

$$\dot{\chi}_\ell(E,\kappa,r) = \dot{\psi}_\ell(E,r) + \kappa\dot{\cot}(n_\ell(E))J_\ell(\kappa r) \quad r \leq S_{MT} \tag{5.18}$$

be zero at $E = E_\nu$. This leads to the following possible definition of the *augmented* spherical Bessel function

$$J_\ell(\kappa r) = \begin{cases} -\dot{\psi}_\ell(E_\nu,r)/\kappa\dot{\cot}(n_\ell(E_\nu)) & r \leq S_{MT} \\ j_\ell(\kappa r) & r \geq S_{MT} \end{cases} \tag{5.19}$$

which will make the muffin-tin orbital (5.17) energy independent, i.e. $\dot{\chi}(E_\nu,\kappa,r) = 0$, to first order in $(E - E_\nu)$.

From the continuity and differentiability of the MTO (5.13), it follows that

$$n_\ell(\kappa r) = \psi_\ell(E,r) + \kappa\cot(n_\ell(E))j_\ell(\kappa r) \tag{5.20}$$

72

near the sphere boundary. Therefore

$$0 = \dot{\psi}_\ell(E_\nu,r) + \kappa \dot{\cot}(n_\ell(E_\nu))j_\ell(\kappa r) \tag{5.21}$$

holds to first order in $(r - S_{MT})$, showing that (5.19) is continous and differentiable at $r = S_{MT}$. This result is a direct consequence of the normalisation which is implied in definition (5.13) and which is characterised by
an energy-independent amplitude $n(\kappa S_{MT})$ and an energy-independent logarithmic derivative $D\{n_\ell\}$ at the sphere.

The energy derivative $\dot{\psi}_\ell(E_\nu,r)$ appearing in (5.19,21) may be found from the
definitions of the *normalised* partial wave ϕ (3.2) and the trial function Φ
(3.44). We obtain

$$\dot{\psi}_\ell(E_\nu,r) = {<\psi_\ell^2}\dot{(E_\nu)>}^{\frac{1}{2}}\phi_{\nu\ell}(r) + {<\psi_\ell^2(E_\nu)>}^{\frac{1}{2}}\dot{\phi}_{\nu\ell}(r)$$

$$= {<\psi_\ell^2}\dot{(E_\nu)>}^{\frac{1}{2}}\Phi(D\{\dot{\psi}_\ell\},r) \quad . \tag{5.22}$$

It follows from (5.21) that $\dot{\psi}_\ell(E_\nu)$ and $j_\ell(\kappa r)$ will have the same logarithmic
derivative at S_{MT} and consequently

$$\dot{\psi}_\ell(E_\nu,r) = {<\psi_\ell^2}\dot{(E_\nu)>}^{\frac{1}{2}}\Phi(D\{j_\ell\},r) \quad . \tag{5.23}$$

The augmented Bessel function defined through (5.19,21,23) has several
desirable properties. It is energy independent, it is everywhere continuous
and differentiable, it is orthogonal to the core states of its own muffin-
tin well as shown by (3.23), and it is finally proportional to the function
$\Phi(D\{j\},r)$.

It now remains to define the tail $N_\ell(\kappa r)$ of the augmented muffin-tin orbital in a suitable form. Here we recall that the original spherical Bessel
and Neumann functions obey the expansion theorem (5.14), and it is therefore
natural to require that the augmented functions J_ℓ and N_ℓ also satisfy this
theorem. Hence, we are led to the definition

$$N_L(\kappa,\mathbf{r-R})$$

$$= \begin{cases} 4\pi \sum_{L'} \sum_{L''} C_{LL'L''} J_{L'}(\kappa,\mathbf{r-R'})n_{L''}^*(\kappa,\mathbf{R-R'}) & \begin{cases} |\mathbf{r-R'}| \leq S_{MT} \\ \forall \mathbf{R'} \neq \mathbf{R} \end{cases} \\ n_L(\kappa,\mathbf{r-R}) & \text{otherwise} \quad . \end{cases} \tag{5.24}$$

With the properties of $J_\ell(\kappa r)$ in mind (i.e. continuity, differentiability,
orthogonality to core states) and the addition theorem at $|\mathbf{r-R'}| = S_{MT}$, we
are satisfied that $N_L(\kappa,\mathbf{r-R})$ is everywhere continuous and differentiable,
and furthermore orthogonal to the core states of all muffin-tin wells except
that centred at $\mathbf{R}$.

The results of this section are collated by rewriting (5.17) in the form

$$
\chi_L(\kappa,r) = i^\ell Y_\ell^m(\hat{r})
\begin{cases}
\dfrac{\kappa n_\ell(\kappa S)}{\Phi_\ell(D\{n_\ell\},S)}\, \Phi_\ell(D\{n_\ell\},r) & r \leq S_{MT} \\[3ex]
\kappa N_\ell(\kappa r) & r \geq S_{MT}
\end{cases}
\tag{5.25}
$$

where the augmented spherical Neumann function is given by (5.24) and the augmented spherical Bessel function by

$$
J_\ell(\kappa r) =
\begin{cases}
\dfrac{j_\ell(\kappa S)}{\Phi(D\{j_\ell\},S)}\, \Phi(D\{j_\ell\},r) & r \leq S_{MT} \\[3ex]
j_\ell(\kappa r) & r \geq S_{MT}
\end{cases}
\tag{5.26}
$$

This may be verified as follows: χ has to be that linear combination of ϕ and $\dot{\phi}$ which has the logarithmic derivative $D\{n\}$ (5.17,19,22). It is therefore proportional to $\Phi(D\{n\},r)$. In addition it has the amplitude $\kappa n(\kappa S_{MT})$ at the sphere. Similarly, J is propotional to $\Phi(D\{j\},r)$ and has the amplitude $j(\kappa S_{MT})$.

It follows from the above that the *augmented* muffin-tin orbital (5.25) is everywhere continuous and differentiable, and orthogonal to the core states of all muffin tins. Hence, these orbitals are well suited for use in connection with the variational principle.

5.5 One-Centre Expansion and Structure Constants

A wave function for a MT potential of a non-overlapping array of MT wells, $v_{MT}(|r-R|)$, centred at sites R of a three-dimensional periodic lattice and embedded in the same constant potential as in (5.2) may be written as the linear combination of MTO's

$$
\Psi(E,r) = \sum_L \alpha_L^k \chi_L^k(E,\kappa,r) \quad,
\tag{5.27}
$$

where we introduced the Bloch sum of MTO's

$$
\chi_L^k(E,\kappa,r) = \sum_R e^{ik\cdot R} \chi_L(E,\kappa,r-R) \quad.
\tag{5.28}
$$

Below we show that as an alternative to the multi-centre expansion (5.28), the wave function may be written in terms of a one-centre expansion of the form

74

$$\chi_L^{\mathbf{k}}(E,\kappa,\mathbf{r}) = \chi_L(E,\kappa,\mathbf{r}) + \sum_{L'} J_{L'}(\kappa,\mathbf{r})B_{L'L}^{\mathbf{k}}(\kappa) \tag{5.29}$$

which will converge inside the MT sphere at the origin and in the interstitial region, outside the neighbouring MT spheres but inside the sphere centred at the origin and passing through the nearest-neighbour sites, Fig.5.2.

To see this, we write (5.28) in the form

$$\chi_L^{\mathbf{k}}(E,\kappa,\mathbf{r}) = \chi_L(E,\kappa,\mathbf{r}) + \sum_{\mathbf{R}\neq\mathbf{0}} e^{i\mathbf{k}\cdot\mathbf{R}}\,\chi_L(E,\kappa,\mathbf{r}-\mathbf{R}) \tag{5.30}$$

and realise that the last term in (5.30) is the sum of the tails from the muffin-tin orbitals centred at all the sites in the crystal except the one at $\mathbf{R} = \mathbf{0}$. These tails, whether augmented (5.24) or normal Neumann, obey an expansion theorem of the form (5.4), and hence the sum of the tails in (5.30) may be written as a one-centre expansion, i.e.

$$\sum_{\mathbf{R}\neq\mathbf{0}} e^{i\mathbf{k}\cdot\mathbf{R}}\,\kappa N_\ell(\kappa,\mathbf{r}-\mathbf{R}) = \sum_{L'} J_{L'}(\kappa,\mathbf{r})B_{L'L}^{\mathbf{k}}(\kappa) \quad, \tag{5.31}$$

which immediately leads to (5.29). The region of convergence stated above follows from the fact that the expansion theorem (5.4) is valid inside the sphere passing through the nearest-neighbour sites while the tails are defined only outside their own MT spheres.

In (5.31) we introduced the KKR structures constants defined by

$$B_{L'L}^{\mathbf{k}} = 4\pi \sum_{L''} C_{LL'L''} \sum_{\mathbf{R}\neq\mathbf{0}} e^{i\mathbf{k}\cdot\mathbf{R}}\,\kappa n_{L''}^*(\kappa,\mathbf{R}) \quad, \tag{5.32}$$

which play the role of coefficients in the one-centre expansion of the MTO tails. These structure constants are independent of the potential, and the matrix $B_{\ell'm;\ell m}^{\mathbf{k}}$ is Hermitian, as may be seen by recalling the properties of spherical harmonics, i.e.

$$Y_\ell^m(-\hat{R}) = (-)^m Y_\ell^m(\hat{R})$$

$$Y_\ell^m(\hat{R})^* = (-)^m Y_\ell^{-m}(\hat{R}) \quad, \tag{5.33}$$

which lead to

$$C_{\ell m,\ell'm',\ell''m''} = (-)^m C_{\ell-m,\ell''m'',\ell'm'} \quad, \tag{5.34}$$

to be used in (5.32) together with the fact that a space lattice has inversion symmetry.

5.6 The LCMTO Secular Matrix

Instead of applying tail cancellation as in Sect.2.1 where we derived the KKR-ASA equations, one may use the linear combination of muffin-tin orbitals (5.27) directly in a variational procedure. This has the advantages that it leads to an eigenvalue problem and that it is possible to include non-muffin-tin perturbations to the potential. According to the Rayleigh-Ritz variational principle, one varies Ψ to make the energy functional stationary, i.e.

$$\delta <\Psi| H - E |\Psi> = 0 \quad , \tag{5.35}$$

where E is the Lagrange multiplier needed to ensure normalisation of Ψ. Equation (5.35) has solutions whenever

$$\det \{<\chi_{L'}^k| H - E |\chi_L^k>\} = 0 \quad . \tag{5.36}$$

Since the muffin-tin orbital is everywhere continuous and differentiable we may evaluate the integral over all space in (5.36) as a sum of integrals over all atomic polyhedra. After repeated use of the Bloch condition (5.28) and rearrangement of the lattice sums, we obtain the well-known result

$$N^{-1} <\chi_{L'}^k| H - E |\chi_L^k> = <\chi_{L'}^k| H - E |\chi_L^k>_0 \quad , \tag{5.37}$$

where the integral on the right-hand side extends only over the polyhedron at the origin.

The LCMTO secular matrix is now simply obtained by inserting the one-centre expansion (5.29) into the matrix (5.37). We find

$$<\chi_{L'}^k| H - E |\chi_L^k>_0 = <\chi_{L'}| H - E |\chi_L>_0$$

$$+ \sum_{L''} \left\{ <\chi_{L'}| H - E |J_{L''}>_0 B_{L''L}^k + <J_{L''}| H - E |\chi_L>_0 B_{L'L''}^k \right\}$$

$$+ \sum_{L''} \sum_{L'''} B_{L'L''}^k <J_{L''}| H - E |J_{L'''}>_0 B_{L'''L}^k \quad . \tag{5.38}$$

If the cellular potential, which has not been specified so far in this section, is spherically symmetric and the cells approximated by spheres (or $\kappa^2 = E - V_{MTZ}$), the cellular integrals in (5.38) become diagonal in L:

$$<\chi_{L'}| H - E |\chi_L>_0 \equiv \int_0 \int \int r^2 dr \, d\hat{\Omega} \chi_{\ell'}(r)[i^{\ell'} Y_{\ell'}^{m'}(\hat{r})]^*$$

$$\times [- \nabla^2 + V_{MT}(r) - \kappa^2] \chi_\ell(r) i^\ell Y_\ell^m(\hat{r})$$

$$= \int_0^\infty dr\, r\chi_{\ell'}(r) \left[-\frac{d^2}{dr^2} + \frac{\ell(\ell+1)}{r^2} + v_{MT}(r) - \kappa^2 \right] r\chi_\ell(r)$$

$$\times \int\int d\hat{\Omega} \left[i^{\ell'} Y_{\ell'}^{m'}(\hat{r}) \right]^* i^\ell Y_\ell^m(\hat{r})$$

$$= \int_0^\infty dr\, r\chi_\ell(r) \left[-\frac{d^2}{dr^2} + \frac{\ell(\ell+1)}{r^2} + v_{MT}(r) - \kappa^2 \right] r\chi_\ell(r)$$

$$\equiv \langle \chi_\ell | H - E | \chi_\ell \rangle_0 \quad . \tag{5.39}$$

Then the L"' summations vanish and the LCMTO secular matrix reduces to

$$\langle \chi_{L'}^k | H - E | \chi_L^k \rangle = \langle \chi_\ell | H - E | \chi_\ell \rangle_0 \delta_{L'L}$$

$$+ \{ \langle \chi_{\ell'} | H - E | J_\ell \rangle_0 + \langle J_\ell | H - E | \chi_\ell \rangle_0 \} B_{L'L}^k$$

$$+ \sum_{L''} B_{L'L''}^k \langle J_{\ell''} | H - E | J_{\ell''} \rangle_0 B_{L''L}^k \quad . \tag{5.40}$$

The LCMTO matrix (5.40), which forms the basis of the band-structure technique we are about to develop, is in a form closely related to the LCAO method. One clearly recognises the one-centre term, zeroth order in B, the two-centre terms, first order in B, and the three-centre or crystal-field term, second order in B. The convergence of the LCMTO method is similar to that of the KKR method and occurs when the phase shifts vanish for $\ell > \ell_{max}$. It may even be improved by extending the internal L" summation in the three-centre term beyond ℓ_{max}, thus including the tails of the higher-ℓ partial waves without increasing the dimension of the secular matrix.

To turn the LCMTO method into an efficient calculational technique, in the following we introduce the atomic-sphere approximation and parametrise the energy dependence of the one-, two-, and three-centre or overlap integrals appearing in (5.40) by means of the results in Sect.3.5. The resulting procedure constitutes the so-called linear muffin-tin orbital (LMTO) method.

5.7 The LMTO Method

The linear LCMTO equations, which we shall refer to as the LMTO equations, may easily be written in the basis of the energy-independent muffin-tin orbitals (5.25). Inspection of (5.40) shows that this involves seven different integrals, which may be evaluated by means of the matrix elements (3.48,49) when D' and D are substituted by $D\{n\}$ and $D\{j\}$, in the approximation that the muffin-tin radius used so far in this chapter is substituted by the atomic-sphere radius S. In writing (5.40) we have already assumed that the potential is spherically symmetric and that the cells are spheres. We use the notation $\Phi\{n\} = \Phi(D\{n\},S)$, $\Phi\{j\} = \Phi(D\{j\},S)$ and find:

$$\langle \chi_\ell | H - E_{\nu\ell} | \chi_\ell \rangle = \left(\frac{\kappa n_\ell(\kappa S)}{\Phi_\ell\{n\}}\right)^2 \omega_\ell\{n\}$$

$$\langle \chi_{\ell'} | H - E_{\nu\ell'} | j_{\ell'} \rangle = \frac{\kappa n_{\ell'}(\kappa S) j_{\ell'}(\kappa S)}{\Phi_{\ell'}\{n\}\Phi_{\ell'}\{j\}} \omega_{\ell'}\{j\}$$

$$\langle j_\ell | H - E_{\nu\ell} | \chi_\ell \rangle = \frac{\kappa n_\ell(\kappa S) j_\ell(\kappa S)}{\Phi_\ell\{n\}\Phi_\ell\{j\}} \omega_\ell\{n\}$$

$$\langle j_{\ell''} | H - E_{\nu\ell''} | j_{\ell''} \rangle = \left(\frac{j_{\ell''}(\kappa S)}{\Phi_{\ell''}\{j\}}\right)^2 \omega_{\ell''}\{j\}$$

$$\langle \chi_\ell | \chi_\ell \rangle = \left(\frac{\kappa n_\ell(\kappa S)}{\Phi_\ell\{n\}}\right)^2 (1 + \omega_\ell^2\{n\}\langle\dot{\phi}_\ell^2\rangle)$$

$$\langle \chi_{\ell'} | j_{\ell'} \rangle = \frac{\kappa n_{\ell'}(\kappa S) j_{\ell'}(\kappa S)}{\Phi_{\ell'}\{n\}\Phi_{\ell'}\{j\}} (1 + \omega_{\ell'}\{n\}\omega_{\ell'}\{j\}\langle\dot{\phi}_{\ell'}^2\rangle)$$

$$\langle j_{\ell''} | j_{\ell''} \rangle = \left(\frac{j_{\ell''}(\kappa S)}{\Phi_{\ell''}\{j\}}\right)^2 (1 + \omega_{\ell''}^2\{j\}\langle\dot{\phi}_{\ell''}^2\rangle) \quad . \tag{5.41}$$

To give expressions for the Hamiltonian and overlap matrices which are written in terms of the four potential parameters $\omega\{n\}$, $\omega\{j\}$, $\Phi\{n\}$, and $\Phi\{j\}$, we need the following two relations

$$D\{n\} - D\{j\} = \frac{1}{\kappa S n(\kappa S) j(\kappa S)} \quad ; \tag{5.42}$$

$$S\Phi\{n\}\Phi\{j\} = - \frac{\omega\{n\} - \omega\{j\}}{D\{n\} - D\{j\}} \quad . \tag{5.43}$$

The first of these, which connects the logarithmic derivatives of the spherical Bessel and Neumann functions, is a direct consequence of the Wronskian relation $nj' - jn' = S^2$ [5.4], while the latter may be derived quite generally in analogy to (4.17), to which it reduces when the two boundary conditions are $-\ell - 1$ and ℓ. In addition, we renormalise the structure constants

$$S^{\mathbf{k}}_{L'L}(\kappa) = \frac{B^{\mathbf{k}}_{L'L}(\kappa)}{S\kappa n_{\ell'}(\kappa S)\kappa n_{\ell}(\kappa S)/2} \tag{5.44}$$

so that the formalism given below also holds for the choice $\kappa^2 = 0$.

The LMTO secular matrix may now be written in the form $\underline{H} - E\underline{O}$, which corresponds to the generalised eigenvalue problem

$$\sum_{L} (H^{\mathbf{k}}_{L'L} - E^{j\mathbf{k}} O^{\mathbf{k}}_{L'L}) a^{j\mathbf{k}}_{L} = 0 \; , \tag{5.45}$$

and which may be solved by efficient numerical techniques [5.7] to give the eigenvalues $E^{j\mathbf{k}}$ and eigenvectors $a^{j\mathbf{k}}_{L}$. The Hamiltonian matrix is given by

$$H^{\mathbf{k}}_{L'L} = \left[\frac{\omega\{n\} + E_{\nu}(1 + \omega^2\{n\}<\dot{\phi}^2>)}{(S/2)\phi^2\{n\}} \right]_{\ell} \delta_{L'L}$$

$$+ \left\{ \left[\frac{\omega\{j\} + E_{\nu}(1 + \omega\{j\}\omega\{n\}<\dot{\phi}^2>)}{\omega\{j\} - \omega\{n\}} \right]_{\ell'} + [\ldots]_{\ell} - 1 \right\} S^{\mathbf{k}}_{L'L}(\kappa)$$

$$+ \sum_{L''} S^{\mathbf{k}}_{L'L''}(\kappa) \left[\frac{\omega\{j\} + E_{\nu}(1 + \omega^2\{j\}<\dot{\phi}^2>)}{2(D\{j\} - D\{n\})^2 S\phi^2\{j\}} \right]_{\ell''} S^{\mathbf{k}}_{L''L}(\kappa) \; ; \tag{5.46}$$

while the overlap matrix is

$$O^{\mathbf{k}}_{L'L} = \left[\frac{1 + \omega^2\{n\}<\dot{\phi}^2>}{(S/2)\phi^2\{n\}} \right]_{\ell} \delta_{L'L}$$

$$+ \left\{ \left[\frac{1 + \omega\{j\}\omega\{n\}<\dot{\phi}^2>}{\omega\{j\} - \omega\{n\}} \right]_{\ell'} + [\ldots]_{\ell} \right\} S^{\mathbf{k}}_{L'L}(\kappa)$$

$$+ \sum_{L''} S^{\mathbf{k}}_{L'L''}(\kappa) \left[\frac{1 + \omega^2\{j\}<\dot{\phi}^2>}{2(D\{j\} - D\{n\})^2 S\phi^2\{j\}} \right]_{\ell''} S^{\mathbf{k}}_{L''L}(\kappa) \; . \tag{5.47}$$

These expressions are used in the LMTO programme, Sect.9.4.2.

Although we prefer to use the LMTO method in conjunction with a cellular potential which is spherically symmetric inside the atomic sphere but not flat in any region, we shall briefly indicate here how the LMTO method may be used in connection with a muffin-tin potential. The simplest procedure is to fix κ^2 at zero and to integrate the Schrödinger equation out to the boundary of the atomic sphere. In this $\kappa^2 = 0$ limit the logarithmic derivatives of the Bessel and Neumann functions become, see (5.82),

$$D_\ell\{n\} = - \ell - 1 \quad ; \quad D_\ell\{j\} = \ell \tag{5.48}$$

and the LMTO matrices (5.46,47) are expressed entirely in terms of the canonical structure constants

$$S^k_{L'L} = \lim_{\kappa^2 \to 0} S^k_{L'L}(\kappa) \tag{5.49}$$

and the potential parameters in Chap.4 evaluated at the atomic sphere. This constitutes the LMTO method in the ASA with a muffin-tin potential.

A different procedure is to use S_{MT} for S in the structure constants (5.44,49) as well as in the crystal potential parameters. In this case, the muffin-tin orbitals involved are properly defined by (5.1,25), but the overlap integrals extend over the muffin-tin sphere only. The contribution to these integrals from the remainder of the atomic polyhedron vanishes however, provided that the cellular potential *is* a muffin-tin potential *and* $\kappa^2 =$ $E - V_{MTZ}$, because then both χ_ν and J_ν satisfy the proper Schrödinger equation in the interstitial region. We may therefore solve the eigenvalue problem (5.45) at a number of fixed κ values and obtain the energies $E^j(k,\kappa)$. A variational estimate of the band energy is subsequently obtained as the solution of the "self-consistency " equation

$$E^j(k,\kappa) = \kappa^2 + V_{MTZ} \quad . \tag{5.50}$$

If non-muffin-tin perturbations are added to the muffin-tin potential used in the definition of the muffin-tin orbitals, the constant intersitial potential V_{MTZ} may not be defined. In this case, the wave number κ may be regarded as a non-linear variational parameter and (5.5) should be substituted by

$$\frac{\partial E^j(k,\kappa)}{\partial(\kappa^2)} = 0 \tag{5.51}$$

[5.2,8]. A typical example where such a procedure is needed is in the application of the LMTO method to molecules. Furthermore, in this situation Bloch's theorem does not apply and k is therefore not a good quantum number. Instead, the k dependence should be substituted by a Q', Q dependence, where Q is a site index. Formally, the LMTO matrix for molecules may be obtained by substituting

$$\ell m \to Q\ell m \tag{5.52}$$

everywhere in (5.46,47). The technique is described in [5.2], and further developments and applications may be found in [5.9-13]. The corresponding KKR-ASA technique is described in Sect.8.1.

80

The LMTO method as defined in this section may be regarded as an LCAO formalism in which the muffin-tin potential, rather than the atomic potential, defines the set of basis functions used to construct the trial functions of the variational procedure. Consequently, all overlap integrals can be expressed in terms of the logarithmic derivative parameters, and the muffin-tin Hamiltonian can be solved to any accuracy.

6. The Atomic-Sphere Approximation (ASA)

"Atomic-sphere approximation" is a common label given by *O.K. Andersen* to the combination of essentially two approximations, one being that the kinetic energy κ^2 of the tail of the partial wave (5.10) may be fixed independently of E, the other that the atomic polyhedron of *Wigner* and *Seitz* may be approximated by an atomic sphere with radius S, defined in (1.8). In practice we include only s, p, d, (and f) orbitals, neglecting all higher partial waves, fix κ^2 at zero, and integrate the radial Schrödinger equation (5.6) out to the boundary of the atomic sphere. Here we first present the atomic-sphere approximation (ASA) and discuss its accuracy. Then we derive the canonical structure constants which may be used in connection with the LMTO matrices (5.46,47) in the limit $\kappa^2 \rightarrow 0$. Next, we prove that the LMTO-ASA and KKR-ASA methods are mathematically equivalent, and show how the electron density needed in a self-consistent calculation may be found from the LMTO eigenvectors. Finally, we describe a correction to the ASA which may be used if very accurate energy bands are needed.

The original material for the chapter may be found in the unpublished notes by *Andersen* [6.1] and in his paper [6.2].

6.1 The Kinetic Energy κ^2

One of the approximations included in the ASA is to fix κ^2 at some suitable value. The incentive to do so stems from the energy dependence implied by (5.3) of the structure constants (5.44) which, for instance, limits the computational speed of the KKR method. As shown in Sect.5.4, the solution to this problem is simply to disregard (5.3), and use a single characteristic κ^2 value. In terms of the potential this means that the constant value outside the sphere is $E - \kappa^2$ rather than V_{MTZ} (Fig.6.1). How good such an approximation will be depends on the wavelength $2\pi/\kappa$ of the partial wave in the interstitial region.

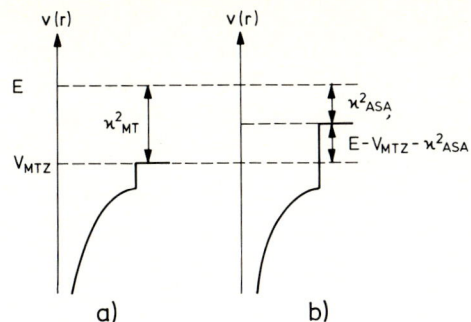

Fig.6.1. The kinetic energy κ^2 of the interstitial region in (a) the muffin-tin approximation, and (b) the atomic-sphere approximation

In many applications we are interested only in the states below the Fermi level, corresponding to an occupancy of approximately one nearly free electron. In that case, the bottom of the s band $B_s \approx V_{MTZ}$ and the minium wavelength can be estimated as follows:

$$2\pi/\kappa_F \approx 2\pi/k_F = 2\pi S(9\pi/4)^{-1/3} \approx 10(S_E - S_{MT}) \quad , \tag{6.1}$$

where $(S_E - S_{MT})$ is the radial extent of the interstitial region (Fig.5.1a). Hence, the wavelength of the partial wave in this interstitial region is large, and when we therefore consider matching the two separate wave functions at the muffin-tin sphere, Fig.5.1b, we realise that the tail will be insensitive to changes in κ when $\psi_\ell(E,S_{MT})$ and $\psi_\ell'(E,S_{MT})$ are given. Consequently, it may be a good approximation to disregard (5.3) and use partial waves or muffin-tin orbitals with a tail wave number κ independent of the energy E.

6.2 An Error Estimate

Following the procedure outlined for a muffin-tin potential in Sect.5.7, we should solve the LMTO equations as functions of κ^2 and find those eigenvalues which satisfy (5.50) or (5.51). However, we should like to be able to use the LMTO method in a way which involves only a single κ^2 value. In that case the tails of the orbitals are no longer solutions of the proper Schrödinger equation in the interstitial region, i.e. $\kappa^2 \neq E - V_{MTZ}$, and the contribution to the overlap integrals from that region does not vanish. We must therefore include the non-muffin-tin perturbation, Fig.6.1,

$$\Delta v(r) = (V_{MTZ} - E + \kappa^2)[\theta_{WS}(r) - \theta_{MT}(r)] \quad , \tag{6.2}$$

where $\theta_{WS}(r)$ and $\theta_{MT}(r)$ are step functions which select out the region between the muffin-tin sphere and the atomic polyhedron. Section 6.9 indicates how this correction may be evaluated to first order. In view of the analysis in the preceding section we expect that the LMTO method for a single κ^2 value and with the perturbation (6.2) will yield eigenvalues $E_i(\mathbf{k},\kappa)$, which in a large neighbourhood of their minima (5.51) are independent of κ^2.

Provided that the errors of the LMTO method have been taken into account to first order in $(V_{MTZ} - E + \kappa^2)$ by means of the correction technique described in Sect.6.9, the leading error will be of second order. This second-order error has been estimated by *Andersen* [6.2] who gives the following expression:

$$\Delta E \approx (E - V_{MTZ} - \kappa^2)^2 \phi^3(S)\dot{\phi}(S) \frac{\Omega_I}{4\pi}^2 \quad , \tag{6.3}$$

where Ω_I is the volume of the interstitial region.

The error ΔE will be smallest for low-lying bands, i.e. $E \approx V_{MTZ} \approx B_s$ ($\kappa^2 = 0$), but it will also be small for narrow bands where the probability $\sim S\phi^2(S)$ of finding an electron in the intersititial region is small. For d states in chromium we have $-S\phi^2 S\dot{\phi} = 0.04$ and $(E_\nu - V_{MTZ})S^2 \approx (E_\nu - B_s)S^2 = 3$ [Table 4.1, (4.13,17), and Fig.4.1]. With $\Omega_I = 0.3$ and $\kappa^2 = 0$, the error $\Delta E_d S^2$ is 0.03 which is one per cent of the bandwidth, $(A_d - B_d)S^2 = 3.6$. The errors in the s and p bands at E are 0.1 and 0.2 per cent of the respective bandwidths, corresponding to a maximum absolute error of 0.006 Ry. For closely packed structures, where Ω_I is small, one may therefore neglect the energy dependence of κ^2 within a range of about 10 S^{-2}, and work with structure constants $S^{\mathbf{k}}_{\ell'm',\ell m}$ which are independent of energy.

6.3 The Atomic Sphere and the ASA

The formalism presented so far is based on a muffin-tin geometry. The potential is assumed to be a muffin-tin potential, Fig.5.1, the partial waves (5.9) and the muffin-tin orbitals (5.13,25) are formed by matching wave functions at the muffin-tin sphere, and the LMTO overlap integrals extend only over the muffin-tin sphere. However, we have already substituted the atomic sphere for the atomic polyhedron as the cell used to divide space into suitable units.

We now take the last major step in the ASA: the atomic sphere of radius S defined by (1.8) is also substituted for the muffin-tin sphere.

In such a procedure the basic unit in space is the atomic sphere which by construction has the same volume as the atomic polyhedron. The interstitial region vanishes and any kinetic energy κ^2 consistent with (5.11) may be chosen. In the ASA we choose $\kappa^2 = 0$ with the computational and conceptual consequences described in Chap.2. Furthermore, we integrate the Schrödinger equation (1.17) out to the atomic radius, using either a muffin-tin potential (5.1) or a cellular potential which may be more accurate in the region between S_{MT} and S. In self-consistent calculations we prefer the lattice choice.

All integrals over the cell are replaced by integrals over the atomic sphere, whereby the contribution from the region between the muffin-tin and the atomic sphere will be taken exactly into account. The size of this contribution is approximately

$$<\phi|V_{MTZ} - E + \kappa^2|\phi>_I$$

$$= (V_{MTZ} - E + \kappa^2)\ \frac{1}{3}\ S^3 \phi^2(S)\ \frac{\Omega_I}{\Omega}\ , \tag{6.4}$$

where the fractional volume Ω_I/Ω included in the intersitital region is 0.26 and 0.32 for the close-packed fcc and bcc structures, respectively. It may be appreciable for wide s and p bands.

The continuous and differentiable matching of ψ_ℓ and χ_ℓ to a tail of fixed κ^2 is performed at S. When we therefore use the $\kappa^2 = 0$ analogue of the one-centre expansion (5.29) to derive the LMTO matrix, we treat the orbital from any nearest-neighbour site incorrectly in the region where its own atomic sphere overlaps that at the origin. However, the error corresponds to using a curvature associated with zero kinetic energy, instead of $E - V_{MTZ}$. For the orbital at the origin we do use the correct kinetic energy in the region between the muffin tin and the atomic sphere. Furthermore, when we include the correction (6.2) to first order we make only second-order errors of the kind discussed in Sect.6.2, and expect (6.3) to describe these errors quite well. The degree to which this is the case may be judged form Fig.5 in [6.2].

Finally, in the ASA we use s, p, d, (and f) orbitals only. Therefore, the Hamiltonian and overlap matrices are given by (5.46,47) with $\kappa^2 = 0$ inserted and truncated at $\ell = 2$ or 3. The LMTO eigenvalues then have small errors of order $(E - V_{MTZ})$ in addition to those of second order (6.3). In Sect.6.9 a technique is described that corrects for these and other errors of the pure ASA.

6.4 The Canonical Structure Constants

In the $\kappa^2 = 0$ limit used in the atomic-sphere approximation, the wave equation (5.7) used to construct the tail of the partial wave (5.10) turns into the Laplace equation. Hence, in the definition of the muffin-tin orbitals (5.13,25) the spherical Bessel and Neumann functions should be substituted by the harmonic functions $(r/S)^\ell$ and $(r/S)^{-\ell-1}$, respectively. By means of the small κr limits (5.8) of the spherical Bessel and Neumann functions, the expansion theorem (5.14) becomes

$$\left(\frac{S}{|r-R|}\right)^{\ell+1} i^\ell Y_\ell^m(\widehat{r-R})$$

$$= \sum_{L'} \left(\frac{r}{S}\right)^{\ell'} i^{\ell'} Y_{\ell'}^{m'}(\hat{r}) \; 4\pi \; \frac{(2\ell''-1)!!}{(2\ell-1)!!(2\ell'+1)!!} \; C_{LL'L''}$$

$$\times \left(\frac{S}{R}\right)^{\ell''+1} [i^{\ell''} Y_{\ell''}^{m''}(\hat{R})]^* \quad . \tag{6.5}$$

In the process of deriving (6.5) we encounter a term $\kappa^{\ell+\ell'-\ell''}$ on the right-hand side which survives in the limit only if $\ell'' = \ell + \ell'$. For this reason the summation on L'', which is already restricted to $m'' = m' - m$, is reduced to include only the term with $\ell'' = \ell + \ell'$.

Inspection of expansion (6.5) shows that it is invariant under a uniform change in the scale of the lattice. This is a direct consequence of the fact that the solutions of the Laplace equation are harmonic functions, and hence the structure constants given below are also invariant under uniform lattice scalings.

In the limit of vanishing κ^2 the Bloch sum (5.31) of MTO tails may be written as the one-centre expansion

$$\sum_{R\neq 0} e^{ik\cdot R}\left(\frac{S}{|r-R|}\right)^{\ell+1} i^\ell Y_\ell^m(\widehat{r-R})$$

$$= \sum_{L'} \frac{-1}{2(2\ell'+1)}\left(\frac{r}{S}\right)^{\ell'} i^{\ell'} Y_{\ell'}^{m'}(\hat{r}) S_{L'L}^k \quad , \tag{6.6}$$

where we have introduced the canonical structure constants in the ℓm representation, i.e.

$$S_{L'L}^k = g_{\ell'm';\ell m}$$

$$\times \sum_{R\neq0} e^{ik\cdot R} \left(\frac{S}{R}\right)^{\ell''+1} [\sqrt{4\pi} \, i^{\ell''} Y^{m''}_{\ell''}(\hat{R})]^* \; , \qquad (6.7)$$

and where the factor $[2(2\ell' + 1)]^{-1}$ appears explicitly in (6.6) to make the structure-constant matrix Hermitian. The factor $g_{\ell'm';\ell m}$ may be expressed in terms of the Gaunt coefficients (5.15) as

$$g_{\ell'm';\ell m} = \frac{-2(2\ell'' - 1)!!}{(2\ell' - 1)!!(2\ell - 1)!!} \sqrt{2\ell'' + 1} \; c^{\ell''}(\ell'm';\ell m) \qquad (6.8)$$

or calculated directly from

$$g_{\ell'm';\ell m}$$

$$= (-)^{m+1} \, 2\left[\frac{(2\ell' + 1)(2\ell + 1)}{2\ell'' + 1} \frac{(\ell'' + m'')!(\ell'' - m'')!}{(\ell' + m')!(\ell' - m')!(\ell + m)!(\ell - m)!}\right]^{\frac{1}{2}}$$

$$\qquad (6.9)$$

where, as above, $\ell'' = \ell' + \ell$, $m'' = m' - m$, and $(-1)!! = 1$. The connection to the original KKR structure constants [6.3,4]

$$B^{k}_{\ell',m;\ell m}(\kappa) = \sum_{\ell''} \sqrt{2\ell'' + 1} \; c^{\ell''}(\ell'm';\ell m)$$

$$\times \sum_{R\neq0} e^{ik\cdot R} \, \kappa n_{\ell''}(\kappa R)[\sqrt{4\pi} \, i^{\ell''} Y^{m''}_{\ell''}(\hat{R})]^* \qquad (6.10)$$

is provided by (5.49,44).

The structure constants $S^{k}_{L'L}$ introduced by *Andersen* [6.2,5] have a number of simple properties which the conventional KKR structure constants (6.10) lack. First of all, they are independent of the energy $E = \kappa^2 + V_{MTZ}$, which is extremely convenient from a calculational point of view. Secondly, by the specific choice $\kappa^2 = 0$, the structure constants have been made invariant under a uniform scaling of the lattice. Therefore they need only be calculated and stored once and for all for a given crystal structure. It is exactly these two properties, independence of volume and energy, which are the basis of the label *canonical* given to the S-structure constants.

As shown in [6.6] the LMTO-ASA Hamiltonian matrix may be transformed into the two-centre form [6.7] where the hopping integrals are products of potential parameters and the canonical structure constants. This result was already stated in Sect.2.5. A less accurate two-centre approximation based upon the KKR-ASA equations will be presented in Sect.8.1.2. The canonical structure constants which, after multiplication by the appropriate potential parameters, form the two-centre hopping integrals are listed in Table 6.1. The

Table 6.1. Canonical structure constants [6.8,9] in the two-centre notation of *Slater* and *Koster* [6.7]. The present, real structure constants are equal to those defined in (6.7,8.8) times $i^{\ell'-\ell}$, and $S(\ell'm',\ell m) = (-)^{\ell'+\ell} S(\ell m,\ell'm')$ where ℓm refers to the angular momentum. The vector from the first to the second orbital has a length R, and direction cosines ℓ, m, and n. The distance S, which also enters the definition of the potential functions, is arbitrary. The entries not given in the table may be found by cyclically permuting the coordinates and direction cosines

$S(s,s)$		$-2(S/R)$
$S(s,x)$	ℓ	$2\sqrt{3}(S/R)^2$
$S(x,x)$	$(3\ell^2 - 1)$	$6(S/R)^3$
$S(x,y)$	$3\ell m$	$6(S/R)^3$
$S(s,xy)$	$-\sqrt{3}\ell m$	$2\sqrt{5}(S/R)^3$
$S(s,x^2-y^2)$	$-\sqrt{3}(\ell^2-m^2)/2$	$2\sqrt{5}(S/R)^3$
$S(s,3z^2-r^2)$	$(1-3n^2)/2$	$2\sqrt{5}(S/R)^3$
$S(x,xy)$	$(1-5\ell^2)m$	$6\sqrt{5}(S/R)^4$
$S(x,x^2-y^2)$	$[1-5(\ell^2-m^2)/2]\ell$	$6\sqrt{5}(S/R)^4$
$S(x,yz)$	$-5\ell mn$	$6\sqrt{5}(S/R)^4$
$S(z,x^2-y^2)$	$-5n(\ell^2-m^2)/2$	$6\sqrt{5}(S/R)^4$
$S(x,3z^2-r^2)$	$(\sqrt{3}/2)(1-5n^2)\ell$	$6\sqrt{5}(S/R)^4$
$S(z,3z^2-r^2)$	$(\sqrt{3}/2)(3-5n^2)n$	$6\sqrt{5}(S/R)^4$
$S(xy,xy)$	$(-35\ell^2m^2-5n^2+4)$	$10(S/R)^5$
$S(x^2-y^2,x^2-y^2)$	$[-35(\ell^2-m^2)^2/4-5n^2+4]$	$10(S/R)^5$
$S(3z^2-r^2,3z^2-r^2)$	$(-3/4)(35n^4-30n^2+3)$	$10(S/R)^5$
$S(xy,x^2-y^2)$	$-35\ell m(\ell^2-m^2)/2$	$10(S/R)^5$
$S(zx,x^2-y^2)$	$-5[7(\ell^2-m^2)/2-1]\ell n$	$10(S/R)^5$
$S(yz,x^2-y^2)$	$5[7(m^2-\ell^2)/2-1]mn$	$10(S/R)^5$
$S(yz,zx)$	$-5(7n^2-1)\ell m$	$10(S/R)^5$
$S(x^2-y^2,3z^2-r^2)$	$(-\sqrt{3}/2)5(7n^2-1)(\ell^2-m^2)/2$	$10(S/R)^5$
$S(xy,3z^2-r^2)$	$(-\sqrt{3}/2)5(7n^2-1)\ell m$	$10(S/R)^5$

expressions given in the table are very simple but for crystals only the pd and dd interactions converge with sufficient speed in real space to be useful. In general an Ewald procedure (Sect.9.2.1) must be used to perform the lattice summations in (6.7).

In addition to the above simplifications, the poles which occur when $|\mathbf{k} + \mathbf{G}| = \kappa^2$ in the original KKR structure constants have been reduced to singularities at the reciprocal lattice points only. In practical terms this means that the ss and sp structure constants diverge while the pp terms are discontinuous at the centre of the Brillouin zone. In contrast, the $B_{L'L}^{\mathbf{k}}$'s become singular everywhere on the free-electron parabola. Finally, there is no sum on ℓ'' as in (6.10), the Gaunt coefficients are particularly simple when $\ell'' = \ell + \ell'$, and the Bessel and Neumann functions entering the KKR formalism become r^ℓ and $r^{-\ell-1}$, respectively.

6.5 Muffin-Tin Orbitals in the ASA

In the limit where κ^2 goes to zero the conventional phase-shift notation (5.10,11) becomes meaningless, and the muffin-tin orbitals (5.13,25) should be redefined. We have already given the $\kappa^2 = 0$ analogue of the energy-dependent muffin-tin orbital in Chap.2. Here we present the augmented, energy-independent muffin-tin orbital which may be used to derive the LMTO-ASA equations directly.

We redefine (5.25,26) in the form

$$\chi_L(\mathbf{r}) = i^\ell Y_\ell^m(\hat{r}) \begin{cases} \Phi_\ell(-\ell-1,r)/[\sqrt{S/2}\ \Phi_\ell(-\ell-1)] & r \leq S \\ \\ N(r/S)/\sqrt{S/2} & r \geq S \end{cases} \tag{6.11}$$

$$J_\ell(r) = \begin{cases} \Phi_\ell(\ell,r)/[\sqrt{S/2}\ \Phi_\ell(\ell)] & r \leq S \\ (r/S)^\ell/\sqrt{S/2} & r \geq S \ . \end{cases} \tag{6.12}$$

The normalisation of $\chi_\ell(r)$ and $J_\ell(r)$ is such that their amplitudes at the sphere boundary are $1/\sqrt{S/2}$ where the factor $\sqrt{S/2}$ is introduced for reasons of notation. The augmented spherical Neumann function $N(r/S)$ appearing in (6.11) may be defined by the $\kappa^2 = 0$ limit as (5.24)

$$N_\ell(|\mathbf{r} - \mathbf{R}|/S) i^\ell Y_\ell^m(\widehat{\mathbf{r} - \mathbf{R}})$$

$$= \begin{cases} -\sum_{L'} \dfrac{\Phi_{L'}(\ell',\mathbf{r} - \mathbf{R}')}{2(2\ell'+1)\Phi_\ell'(\ell')} g_{L'L}\left(\dfrac{S}{|\mathbf{R} - \mathbf{R}'|}\right)^{\ell''+1} \\ \qquad\qquad\qquad\qquad\qquad\qquad\qquad |r-R'| \leq S \\ \times\ [\sqrt{4\pi}\ i^{\ell''} Y_{\ell''}^{m''}(\widehat{\mathbf{R} - \mathbf{R}'})]^* \qquad \forall R' \neq R \qquad (6.13) \\ \\ (|\mathbf{r} - \mathbf{R}|/S)^{-\ell-1} \qquad\qquad\qquad \text{otherwise} \end{cases}$$

where $\ell'' = \ell' + \ell$, $m'' = m' - m$, and where we have used (5.15,6.8,12).

The energy-independent muffin-tin orbital $\chi_L(\mathbf{r} - \mathbf{R})$ centred at $\mathbf{R}$ is especially designed to have a tail orthogonal to the core states inside the sphere centred at $\mathbf{R}'$. Therefore the function $\Phi_{L'}(\ell',\mathbf{r} - \mathbf{R}')$ appears in the

definition (6.13) of the tail. In addition, the tail is constructed to obey
the expansion theorem (6.5) for spherical Bessel and Neumann functions in the
ASA. As a result, the Bloch sum of energy-independent muffin-tin orbitals

$$\chi_L^{\mathbf{k}}(\mathbf{r}) = \sum_{\mathbf{R}} e^{i\mathbf{k}\cdot\mathbf{R}} \chi_L(\mathbf{r} - \mathbf{R}) \tag{6.14}$$

may be written as the simple one-centre expansion

$$\chi_L^{\mathbf{k}}(\mathbf{r}) = \frac{\Phi_L(-\ell - 1, \mathbf{r})}{\sqrt{S/2}\ \Phi_\ell(-\ell - 1)} - \sum_{L'} \frac{\Phi_{L'}(\ell', \mathbf{r})}{2(2\ell' + 1)\sqrt{S/2}\ \Phi_{\ell'}(\ell')} S_{L'L}^{\mathbf{k}} \tag{6.15}$$

in analogy with (5.29). To see this one may use (6.14,11,13,7).

The one-centre expansion (6.15) is specialised to the case where $\mathbf{R'} = \mathbf{0}$,
and is valid inside the atomic sphere centred at the origin. It may be used
to derive the LMTO equations and with the normalisation implied by (6.11)
it is consistent with the secular matrices (5.46,47) in the ASA. In linear
methods in band theory [6.2] *Andersen* presented the one-centre expansion in
the form (6.15) and derived the LMTO formalism from that assumption. His LMTO
formalism is equivalent to that presented here apart from the normalising
factor $[\sqrt{S/2}\ \Phi_\ell(-\ell-1)]^{-1}$ appearing in the definition (6.11) of the energy-
dependent muffin-tin orbital.

6.6 Relation Between the LMTO and KKR Matrices

In Chap.2 we introduced the concept of canonical bands based upon the KKR-ASA
equations and used it to interpret energy bands calculated by the LMTO method.
We did this because the KKR-ASA and LMTO-ASA methods are mathematically equivalent, as proven below. Specifically, we show that in a range around E_ν so
narrow that the small parameter $\langle\dot\phi^2\rangle$ may be neglected the LMTO-ASA and KKR-
ASA equations will lead to the same eigenvalues.

To this end we define the three matrices

$$\underline{\underline{\Pi}} = \left[\frac{1}{\sqrt{S/2}\ \Phi(-)}\right]_\ell \delta_{L'L} + \left[\frac{\sqrt{S/2}\ \Phi(-)}{\omega(+) - \omega(-)}\right]_{\ell'} S_{L'L}^{\mathbf{k}}$$

$$\underline{\underline{\Omega}} = \left[\frac{\omega(-)}{\sqrt{S/2}\ \Phi(-)}\right]_\ell \delta_{L'L} + \left[\frac{\sqrt{S/2}\ \Phi(-)\omega(+)}{\omega(+) - \omega(-)}\right]_{\ell'} S_{L'L}^{\mathbf{k}}$$

$$\langle\dot\phi^2\rangle = \langle\dot\phi_\ell^2\rangle\delta_{L'L} \tag{6.16}$$

and write the LMTO matrices (5.46,47) in the form

$$\underline{H} = \underline{\Pi}^{\dagger}\underline{\Omega} + E_{\nu}\underline{\Pi}^{\dagger}\Pi + E_{\nu}\underline{\Omega}^{\dagger}<\dot{\phi}^2>\underline{\Omega}$$

$$\underline{O} = \underline{\Pi}^{\dagger}\underline{\Pi} + \underline{\Omega}^{\dagger}<\dot{\phi}^2>\underline{\Omega} \quad . \tag{6.17}$$

To see that the matrix multiplications in (6.17) actually give (5.46,47) one must use (4.17) in the three-centre term. It now follows that if $<\dot{\phi}^2>$ may be neglected, the LMTO matrix $<L'|H - E|L>$ can be expressed as the product

$$\underline{H} - E\underline{O} = \underline{\Pi}^{\dagger}(\underline{\Omega} - \varepsilon\underline{\Pi}) \quad , \tag{6.18}$$

where for simplicity $\varepsilon = E - E_{\nu}$ independent of ℓ.

The matrix of the KKR-ASA equations (2.8) may be written in the form

$$\underline{A} = P_{\ell}(D)\delta_{L'L} - S^{\mathbf{k}}_{L'L} \quad , \tag{6.19}$$

where the potential function P_{ℓ} is defined by (2.9). The latter may be para-metrised by means of (3.51,4.15), i.e.

$$P_{\ell}(E) = 2(2\ell + 1)\left[\frac{\phi(+)}{\phi(-)}\frac{\varepsilon - \omega(-)}{\varepsilon - \omega(+)}\right]_{\ell} \quad . \tag{6.20}$$

If we now define the matrix

$$\underline{A} = \left[\frac{\omega(+) - \varepsilon}{\omega(+) - \omega(-)}\right]_{\ell}\delta_{L'L} \tag{6.21}$$

we realise by means of (4.17) that

$$\underline{A}\underline{A} = \underline{\Omega} - \varepsilon\underline{\Pi} \quad . \quad . \tag{6.22}$$

Hence, the KKR-ASA matrix $\underline{A}$ is a factor of the LMTO-ASA matrix, and the LMTO and KKR methods are equivalent in the neighbourhood of E_{ν}, as we wished to prove.

6.7 Wave Functions and ℓ Character

Let us assume that we have solved the LMTO eigenvalue problem (5.45) by standard numerical techniques. We therefore know the $(2\ell + 1)$ eigenvalues $E^{j\mathbf{k}}$ at any given k vector, and the corresponding eigenvectors $a^{j\mathbf{k}}_{L}$ which sa-tisfy the orthonormality relation

$$\sum_{L'} \sum_{L} a^{j\mathbf{k}*}_{L} O^{\mathbf{k}}_{L'L} a^{j\mathbf{k}}_{L} = \delta_{j'j} \quad . \tag{6.23}$$

With this information we can construct the wave functions corresponding to the j'th eigenvalue at **k** from (5.27), i.e.

$$\psi^{j\mathbf{k}}(\mathbf{r}) = \sum_L \psi_L^{j\mathbf{k}}(\mathbf{r}) \quad , \tag{6.24}$$

where the ℓm-decomposed wave function is given by

$$\psi_L^{j\mathbf{k}}(\mathbf{r}) = a_L^{j\mathbf{k}} \chi_L^{\mathbf{k}}(\mathbf{r}) \quad . \tag{6.25}$$

The Bloch sum of muffin-tin orbitals $\chi_L^{\mathbf{k}}(\mathbf{r})$ is most conveniently expressed by the one-centre expansion (6.15) valid inside the sphere at the origin

$$\chi_L^{\mathbf{k}}(\mathbf{r}) = \frac{\Phi_L(- \ell -1,\mathbf{r})}{\sqrt{S/2}\, \Phi_\ell(-)} - \sum_{L'} \frac{\Phi_{L'}(\ell',\mathbf{r})}{2(2\ell' + 1)\sqrt{S/2}\, \Phi_{\ell'}(+)} S_{L'L}^{\mathbf{k}} \quad . \tag{6.26}$$

The trial function $\Phi_L(D,\mathbf{r})$ which enters (6.26) is defined by (3.44,47), i.e.

$$\Phi_L(D,\mathbf{r}) = i^\ell Y_\ell^m(\hat{r})\Phi_\ell(D,\mathbf{r})$$

$$\Phi_\ell(D,\mathbf{r}) = \phi_{\nu\ell}(\mathbf{r}) + \omega(D)\dot{\phi}_{\nu\ell}(\mathbf{r}) \quad . \tag{6.27}$$

By analogy we make the following definitions of the partial wave and its energy derivative, cf. (3.4,5),

$$\phi_{\nu L}(\mathbf{r}) = i^\ell Y_\ell^m(\hat{r})\phi_{\nu\ell}(\mathbf{r})$$

$$\dot{\phi}_{\nu L}(\mathbf{r}) = i^\ell Y_\ell^m(\hat{r})\dot{\phi}_{\nu\ell}(\mathbf{r}) \quad . \tag{6.28}$$

If we now combine (6.24-28) we may express the ℓm-decomposed wave function $\psi_L^{j\mathbf{k}}$ in the form

$$\psi_L^{j\mathbf{k}}(\mathbf{r}) = A_L^{j\mathbf{k}}\phi_{\nu L}(\mathbf{r}) + B_L^{j\mathbf{k}}\dot{\phi}_{\nu L}(\mathbf{r}) \quad , \tag{6.29}$$

where the coefficients A and B are given by

$$A_L^{j\mathbf{k}} = \left[a_L^{j\mathbf{k}} - \gamma_\ell \sum_{L'} a_L^{j\mathbf{k}}S_{LL'}^{\mathbf{k}}\right]/[\sqrt{S/2}\,\Phi_\ell(-)]$$

$$B_L^{j\mathbf{k}} = \left[a_L^{j\mathbf{k}}\omega_\ell(-) - \gamma_\ell \omega_\ell(+)\sum_{L'} a_L^{j\mathbf{k}}S_{LL'}^{\mathbf{k}}\right]/[\sqrt{S/2}\,\Phi_\ell(-)] \quad . \tag{6.30}$$

The potential parameter γ_ℓ is defined by (4.16), and to obtain an expression for $\chi_L^{\mathbf{k}}$ rather than $\chi_{L'}^{\mathbf{k}}$, we have at one stage interchanged the summation labels L and L'.

In the expression (6.29) for the ℓm-decomposed wave function, the coefficients $A_L^{j\mathbf{k}}$ and $B_L^{j\mathbf{k}}$ are completely specified by the band-structure calculation, as (6.30) shows. Furthermore, the partial wave $\phi_{\nu\ell}(r)$ and its energy deriva-tive $\dot{\phi}_{\nu\ell}(r)$ are determined by the numerical integration of the radial Schrö-dinger equation in the sphere, and correspond to the first two terms of a Taylor expansion for $\phi_\ell(E,r)$. As a result, the combined expression (6.29) is accurate to first order in $(E - E_\nu)$ and, since it is extremely convenient, one may use it to construct wave-function-dependent quantities.

The ℓm character of the j'th eigenvalue at a given k vector may be found from the following integral over the sphere:

$$
\begin{aligned}
C_{Lj}^{\mathbf{k}} &= <\psi_{Lj}^{\mathbf{k}} | \psi_{Lj}^{\mathbf{k}}> \\
&= <\phi_{\nu\ell}^2>|A_L^{j\mathbf{k}}|^2 + <\dot{\phi}_{\nu\ell}^2>|B_L^{j\mathbf{k}}|^2 \\
&\quad + <\phi_{\nu\ell}|\dot{\phi}_{\nu\ell}>(A_L^{j\mathbf{k}*}B_L^{j\mathbf{k}} + B_L^{j\mathbf{k}*}A_L^{j\mathbf{k}}) \quad .
\end{aligned}
\tag{6.31}
$$

Owing to the orthogonality of ϕ and $\dot{\phi}$ (3.8), the cross-terms in A and B vanish and the integral may be reduced to the simple expression

$$
C_{\ell m}^{j\mathbf{k}} = |A_{\ell m}^{j\mathbf{k}}|^2 + <\dot{\phi}_{\nu\ell}^2>|B_{\ell m}^{j\mathbf{k}}|^2 \quad ,
\tag{6.32}
$$

with the ℓm quantum numbers displayed explicitly. Similarly, for the ℓ charac-ter we find that

$$
C_\ell^{j\mathbf{k}} = \sum_m |A_L^{j\mathbf{k}}|^2 + <\dot{\phi}_{\nu\ell}^2>\sum_m |B_L^{j\mathbf{k}}|^2
\tag{6.33}
$$

which we shall use to construct ℓ-projected state densities.

As a last step we wish to evaluate the normalisation integral in the sphere

$$
<\psi^{j\mathbf{k}}|\psi^{j\mathbf{k}}> = \sum_L (|A_L^{j\mathbf{k}}|^2 + <\dot{\phi}_{\nu\ell}^2>|B_L^{j\mathbf{k}}|^2
\tag{6.34}
$$

and show that it is unity. To do this, we use the matrix notation of the pre-vious section, and write the A and B coefficients in (6.29) as column vectors. From (6.16,30) we find

$$
\mathbf{A} = \underline{\underline{\Pi}}\mathbf{a}
$$

$$
\mathbf{B} = \underline{\underline{\Omega}}\mathbf{a} \quad ,
\tag{6.35}
$$

where $\mathbf{a}$ is the LMTO eigenvector satisfying (6.23). According to (6.17,23,35), the normalisation integral (6.34) is now

$$
\begin{aligned}
<\psi^{j\mathbf{k}}|\psi^{j\mathbf{k}}> &= \mathbf{A}^\dagger \cdot \mathbf{A} + \mathbf{B}^\dagger<\dot{\phi}^2>\mathbf{B} \\
&= \mathbf{a}^\dagger\underline{\underline{\Pi}}^\dagger \cdot \underline{\underline{\Pi}}\mathbf{a} + \mathbf{a}^\dagger\underline{\underline{\Omega}}^\dagger<\dot{\phi}^2>\underline{\underline{\Omega}}\mathbf{a}
\end{aligned}
$$

93

$$= a^{\dagger} \cdot \mathbf{0} \cdot a$$

$$= 1 \tag{6.36}$$

as we wished to show. It now follows from (6.33,34) that

$$\sum_{\ell} c_{\ell}^{jk} = 1 \tag{6.37}$$

and hence all the ℓ characters at any given eigenvalue E^{jk} at any given point in k space add up to unity.

6.8 Projected State Density and Density of Electrons

With the properties of the ℓ character in mind one may obtain the ℓ-projected state density from the analogue of (2.37), i.e.

$$N_{\ell}(E) = (2\pi)^{-3} \Omega \sum_{j} \int_{BZ} dk \, c_{\ell}^{jk} \delta[E - E_{j}(k)] \quad . \tag{6.38}$$

In addition we realise from (6.31) that the spherically averaged electron density n(r) may be expressed as

$$4\pi n(r) = \sum_{\ell j} \left[\phi_{\nu\ell}^{2}(r)|A_{L}^{jk}|^{2} + \dot{\phi}_{\nu\ell}^{2}(r)|B_{L}^{jk}|^{2} \right]$$

$$+ \sum_{\ell j} \left\{ \phi_{\nu\ell}(r)\dot{\phi}_{\nu\ell}(r) \left[A_{L}^{jk*}B_{L}^{jk} + B_{L}^{jk*}A_{L}^{jk} \right] \right\} \quad . \tag{6.39}$$

If we instead solve the KKR-ASA equations (2.8) in the partial-wave representation (1.20) we obtain

$$4\pi n(r) = \sum_{\ell} \int^{E_{F}} 2N_{\ell}(E)\phi_{\ell}^{2}(E,r)dE \quad , \tag{6.40}$$

which by means of the Taylor series (3.10) may be approximated by an expansion in the moments of the projected state densities, i.e.

$$4\pi n(r) = \sum_{\ell} \left\{ \phi_{\nu\ell}^{2}(r)n_{\ell} + 2\phi_{\nu\ell}(r)\dot{\phi}_{\nu\ell}(r) \int^{E_{F}} 2N_{\ell}(E)(E - E_{\nu\ell})dE \right.$$

$$+ \left[\dot{\phi}_{\nu\ell}^{2}(r) + \ddot{\phi}_{\nu\ell}(r)\phi_{\nu\ell}(r) \right] \int^{E_{F}} 2N_{\ell}(E)(E - E_{\nu\ell})^{2}dE + \ldots \right\} \quad . \tag{6.41}$$

Here we used the definition

$$n_{\ell} = \int^{F_{F}} 2N_{\ell}(E)dE \tag{6.42}$$

94

of the number of ℓ electrons and the factor of 2 accounts for spin.

A comparison between the correct expression (6.39) and the moment expansion (6.41) shows that they are identical to first order in $(E - E_{\nu\ell})$. Since the ℓ characters add up to unity (6.31,37), the integrals $\int_0^S \phi\dot{\phi}r^2\,dr$ and $\int_0^S (\dot{\phi}^2 + \phi\ddot{\phi})r^2\,dr$ over the sphere are zero (3.8,9), and $\int_0^S \phi^2 r^2\,dr$ is unity (3.7), we find from both (6.39,41) that

$$4\pi \int_0^S n(r)r^2\,dr = \sum_\ell n_\ell \; . \tag{6.43}$$

Consequently, we need not truncate (6.41) after the first two terms, even though the projected state densities have been obtained by means of the linear method. Since (6.41), on account of the ϕ, $\ddot{\phi}$ term, is the more accurate expression, we prefer to use it instead of (6.39) to construct charge densities. It follows from the above that only the first term in (6.41) contributes net charge to the sphere, and hence the second and third terms represent the radial displacement of charge due to the broadening of the band around $E_{\nu\ell}$.

6.9 The Combined Correction Term

Although the errors introduced by the atomic-sphere approximation are unimportant for many applications, e.g. self-consistency procedures, there are cases where energy bands of high accuracy are needed, and where one should include the perturbation (6.2) in some form. Below, we derive an expression which accounts to first order for the differences between the sphere, atomic or muffin tin, and the atomic polyhedron, re-establishes the correct kinetic energy in the region between the sphere and the polyhedron, and corrects for the neglect of higher partial waves. The extra terms added to the LMTO matrices which accomplish these corrections are called the combined correction terms [6.2].

We evaluate the perturbation (6.2) over the interstitial region (I) as a difference between an integral over the cell (C), i.e. the polyhedron, and an integral over the sphere (S). Hence,

$$\langle \tilde{\chi}_L^k | \Delta v | \tilde{\chi}_L^k \rangle = (V_{MTZ} - E + \kappa^2)\langle \tilde{\chi}_L^k | \tilde{\chi}_L^k \rangle_C$$

$$- (V_{MTZ} - E + \kappa^2)\langle \tilde{\chi}_L^k | \tilde{\chi}_L^k \rangle_S \tag{6.44}$$

and we define $\tilde{\chi}_L^k$ as the Bloch sum (5.28) of the free-electron orbital

$$
\tilde{\chi}_\ell(\kappa,r) =
\begin{cases}
\dfrac{\kappa n_\ell(\kappa S)}{\tilde{\phi}_\ell\{n_\ell\}}\ \tilde{\Phi}_\ell(D\{n_\ell\},r) & r \leq S \\[4mm]
\kappa n_\ell(\kappa r) & r \geq S
\end{cases}
\tag{6.45}
$$

where the augmented Neumann function (5.24) is equal to n_ℓ because $\tilde{D}_\nu = D\{j_\ell\}$, and where

$$
\tilde{\Phi}_\ell(D,r) = \tilde{\phi}_\ell(r) + \tilde{\omega}_\ell(D)\dot{\tilde{\phi}}_\ell(r)
$$

$$
\tilde{\phi}_\ell(r) = j_\ell(\kappa r)/<j_\ell^2(\kappa r)>^{\frac{1}{2}} \quad .
\tag{6.46}
$$

With this definition, $\tilde{\chi}_L$ is equal to the proper orbital in the interstitial region only, while inside the spheres it is derived from a constant (pseudo) potential $v(r) = E - \kappa^2$. For that reason the integral over the sphere appearing in (6.44) is simply the LMTO overlap matrix (5.47) evaluated for the free-electron potential parameters from Sect.4.4 corresponding to $\tilde{D}_\nu = D\{j_\ell(\kappa S)\}$. Hence, the contribution from the second term in (6.44) may be included in the LMTO equations (5.45) by subtracting

$$
\frac{1 + \tilde{\omega}^2\{n\}<\dot{\tilde{\phi}}^2>}{S\tilde{\phi}^2\{n\}/2}
$$

$$
\frac{1 + \tilde{\omega}\{n\}\tilde{\omega}\{j\}<\dot{\tilde{\phi}}^2>}{\tilde{\omega}\{j\} - \tilde{\omega}\{n\}}
$$

$$
\frac{1 + \tilde{\omega}^2\{j\}<\dot{\tilde{\phi}}^2>}{2(D\{j\} - D\{n\})^2 S\tilde{\phi}^2\{j\}}
\tag{6.47}
$$

in the one-, two-, and three-centre terms, respectively, of the overlap matrix (5.47), and by subtracting $(V_{MTZ} - \kappa^2)$ times (6.47) in the Hamiltonian matrix (5.46). In the important case where $\kappa^2 = 0$, we have $\tilde{D}_\nu = \ell$ and the necessary standard parameters may be found in Table 4.5. In addition, $S^2\tilde{\omega}\{j\} = 0$, $S^3\tilde{\phi}^2\{j\} = 2\ell + 3$, and $D\{j\} - D\{n\} = 2\ell + 1$.

We proceed by evaluating the integral over the cell as a single sum in reciprocal space. To this end we note that the Fourier transform of a function, which inside the sphere equals a solution $\phi_L(\kappa^2,r)$ of the wave equation and is zero outside, may be written as

$$
<\mathbf{K}|\phi_L> = \int e^{i\mathbf{K}\cdot\mathbf{r}}\ \phi_\ell(\kappa r) i^\ell Y_\ell^m(\hat{r})\ d\mathbf{r}
$$

$$
= 4\pi Y_\ell^m(\mathbf{K}) \int_0^S j_\ell(Kr)\phi_\ell(\kappa r)r^2\ dr \quad ,
\tag{6.48}
$$

where we used the expansion $e^{i\mathbf{K} \cdot \mathbf{r}} = \sum_{\ell m} 4\pi i^{\ell} Y_{\ell}^{m} (\hat{K})^{*} j_{\ell}(Kr) Y_{\ell}^{m}(\hat{r})$ of a plane wave in spherical waves, and integrated over solid angle. In addition, we need Green's theorem (3.13) specialised to the case of spherically symmetric potentials, i.e.

$$\int_{0}^{S} \psi_1(r)[v_1(r) - E_1 - v_2(r) + E_2]\psi_2(r)r^2 \, dr$$

$$= S\psi_1(S)\psi_2(S)[D\{\psi_1(S)\} - D\{\psi_2(S)\}] \quad . \tag{6.49}$$

If one of the functions, say ψ_1, is $\dot{\phi}$, the analogue of (6.49) is

$$\int_{0}^{S} \dot{\phi}(r)[v_1(r) - E_1 - v_2(r) + E_2]\psi_2(r)r^2 \, dr$$

$$= \int_{0}^{S} \dot{\phi}(r)\psi_2(r)r^2 \, dr$$

$$+ S\dot{\phi}(S)\psi_2(S)[D\{\dot{\phi}(S)\} - D\{\psi(S)\}] \quad , \tag{6.50}$$

where the extra term arises because $\dot{\phi}$ satisfies (3.12) rather than the Schrödinger equation (3.11).

The Fourier transform of the free-electron muffin-tin orbital (6.45) is now found by taking the transforms of $\tilde{\phi}_{\ell}$, $\dot{\tilde{\phi}}_{\ell}$, and n_{ℓ} separately. For ϕ_{ℓ} we calculate

$$\langle \mathbf{K}, \phi_{\ell} \rangle = 4\pi Y_{\ell}^{m}(\hat{K}) \int_{0}^{S} j_{\ell}(Kr)\phi_{\ell}(\kappa r)r^2 \, dr \quad , \tag{6.51}$$

which is easily obtained from (6.49) by the substitutions $v_1 = v_2 = 0$, $E_1 = K^2$, $E_2 = \kappa^2$, $\psi_1 = j_{\ell}(Kr)$, $\psi_2 = \phi_{\ell}(\kappa r)$. Hence,

$$\langle \mathbf{K}, \phi_{\ell} \rangle = 4\pi Y_{\ell}^{m}(\hat{K}) \, S \, j_{\ell}(KS)\phi_{\ell}(\kappa S) \, \frac{D\{\phi_{\ell}(\kappa S)\} - D\{j_{\ell}(KS)\}}{K^2 - \kappa^2} \quad . \tag{6.52}$$

For $\dot{\phi}_{\ell}$ we find from (6.50)

$$\langle \mathbf{K}, \dot{\phi}_{\ell} \rangle = 4\pi Y_{\ell}^{m}(\hat{K}) \, S \, j_{\ell}(KS) \, \dot{\phi}_{\ell}(\kappa S) \, \frac{D\{\dot{\phi}_{\ell}(\kappa S)\} - D\{j_{\ell}(KS)\}}{K^2 - \kappa^2}$$

$$+ 4\pi Y_{\ell}^{m}(\hat{K}) \, S \, j_{\ell}(KS) \, \phi_{\ell}(\kappa S) \, \frac{D\{\phi_{\ell}(\kappa S)\} - D\{j_{\ell}(KS)\}}{(K^2 - \kappa^2)^2} \quad . \tag{6.53}$$

By analogy, the Fourier transform of the tail is

$$\langle K | n_\ell \rangle = -4\pi Y_\ell^m(\hat{K}) \, S \, j_\ell(KS) \, n_\ell(\kappa S) \, \frac{D\{n_\ell(\kappa S)\} - D\{j_\ell(KS)\}}{K^2 - \kappa^2} \quad . \tag{6.54}$$

If we now piece the individual terms together, with the appropriate amplitudes, we find

$$\langle K | \tilde{\chi} \rangle = 4\pi Y_\ell^m(\hat{K}) \, j_\ell(KS)\kappa n_\ell(\kappa S)$$

$$\times \, [D\{j_\ell(\kappa S)\} - D\{n_\ell(\kappa S)\}] \, \frac{D\{j_\ell(\kappa S)\} - D\{j_\ell(KS)\}}{(K^2 - \kappa^2)^2} \, S\phi_\ell^2(\kappa S) \quad . \tag{6.55}$$

Hence, with the definition

$$F_L(\kappa S, KS) = S^3 \phi_\ell^2(\kappa S)[D\{j_\ell(\kappa S)\} - D\{n_\ell(\kappa S)\}]$$

$$\times \, \frac{D\{j_\ell(\kappa S)\} - D\{j_\ell(KS)\}}{[(KS)^2 - (\kappa S)^2]^2} \, j_\ell(KS)Y_\ell^m(\hat{K}) \tag{6.56}$$

the Bloch sum of free-electron muffin-tin orbitals has the reciprocal lattice representation

$$\tilde{\chi}_L^k(\kappa, r) = \frac{4\pi S^3}{\Omega} \, \kappa n_\ell(\kappa S) \, \sum_G e^{iK \cdot r} \, F_L(\kappa S, KS) \quad , \tag{6.57}$$

where $K = k + G$.

The contribution to the LMTO matrix (5.45) from the integral over the cell is obtained by adding

$$\frac{1}{N} \, \frac{\langle \tilde{\chi}_{L'}^k | \tilde{\chi}_L^k \rangle}{S \kappa n_{\ell'}(\kappa S)\kappa n_\ell(\kappa S)/2}$$

$$= \frac{(4\pi)^2 2 S^5}{\Omega} \, \sum_G F_{L'}^*(\kappa S, KS) \, F_L(\kappa S, KS) \tag{6.58}$$

to the overlap matrix (5.47) and $(V_{MTZ} - \kappa^2)$ times (6.58) to the Hamiltonian matrix (5.46).

When $\kappa^2 = 0$, $S^3 \phi_\ell^2(\kappa S) = 1/m_\ell(0) = 2\ell + 3$ and $D\{j\} - D\{n\} = 2\ell + 1$, whereby (6.56) reduces to

$$F_L(0, KS) = (2\ell + 1)(2\ell + 3)(KS)^{-3} j_{\ell+1}(KS)Y_\ell^m(\hat{K}) \quad , \tag{6.59}$$

where we have used the well-known relation [6.10] between $j_{\ell+1}(\kappa r)$ and the derivative of $j_\ell(\kappa r)$.

With the combined correction term included in an LMTO calculation, one corrects for the errors of the ASA to first order in $E - V_{MTZ} - \kappa^2$, and one expects the error expression (6.3) to give a reasonably accurate description of the remaining errors, provided that Ω_I is the volume between the polyhedron and its inscribed sphere. How well this latter expectation is fulfilled may be judged from Fig.5 in [6.2].

When the corrections are included, the ℓ characters (6.33) no longer add up to unity (6.37) because, although we have restored the polyhedron, we still normalise the partial waves $\phi_\ell(r)$ over the sphere. The simplest way to correct this error is to renormalise C_ℓ^{jk} such that the sum $\sum_\ell C_\ell^{jk}$ is unity for each jk.

7. Ground-State Properties

The ability to calculate properties such as lattice parameter, compressibility, cohesive energy, and magnetic moment for crystalline solids in their ground state represents an important step towards the understanding of the behaviour of a variety of materials. Here we shall consider a method for calculating ground-state properties of metals and inorganic compounds based on the so-called (spin) density-functional formalism. It requires that we perform self-consistent electronic-structure calculations, and to this end we use the LMTO method in connection with the scaling principle outlined in Sect.2.5. The only input to such calculations is the atomic number (and the crystal structure), and when the self-consistency procedure is completed we may use the potential parameters together with canonical band theory to understand the calculated properties in basic physical terms.

The original material for the present chapter may be found in the papers by *Hohenberg* and *Kohn* [7.1], *Kohn* and *Sham* [7.2], *Andersen* [7.3], and *Mackintosh* and *Andersen* [7.4].

7.1 Cohesive Properties

The most fundamental ground-state property of a crystalline solid is its total energy, since by calculating total energies under changing external circumstances one should be able to answer such questions as: "What are the atomic equilibrium volume and the bulk modulus (compressibility)?", "Is there a magnetic moment?" However, owing to problems associated with the large core energies and the so-called double counting term, it proves much more convenient to calculate and use the electronic pressure as the basis for an interpretation of cohesive properties. This is especially so because, as shown below, the pressure is directly related to the sum of the one-electron energies.

At zero temperature the pressure P is the volume derivative of the total energy, i.e.

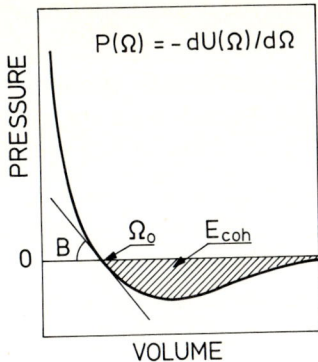

$$P(\Omega) = -dU(\Omega)/d\Omega$$

Fig.7.1. Equation of state, equilibrium volume Ω_0, bulk modulus B, and cohesive energy E_{coh}

Fig.7.2. ➤

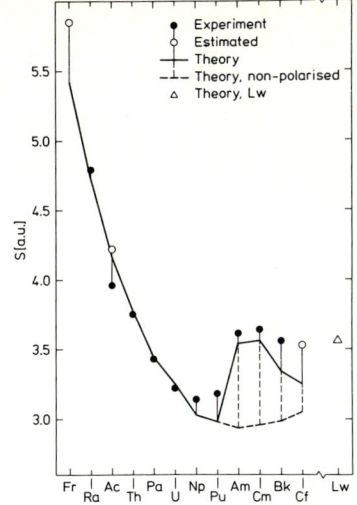

Fig.7.2. Calculated and measured equilibrium atomic raddi for the light actinide metals [7.6]. The decreasing radii in the beginning of the series, Ac-Pu, are caused by increasing occupation of bonding 5f states while the calculated anomaly between Pu and Am is due to the onset of spin polarisation

$$P = -d<\Phi|H|\Phi>/d\Omega \quad , \tag{7.1}$$

to which may be added the small contribution

$$P_z = (9/8)\gamma k_B \, \Theta_D/\Omega \tag{7.2}$$

from the zero-point motion of the nuclei as obtained in a Debye model [7.5]. In (7.2) γ is the Grüneisen parameter, k_B is Boltzmann's constant and Θ_D is the Debye temperature.

As we shall see, the pressure expression derived from (7.1) allows us to calculate the equation of state $P(\Omega)$ from the results of self-consistent energy-band calculations. In Fig.7.1 a typical equation of state is sketched and it is indicated that the equilibrium atomic volume Ω_0 is determined by

$$P(\Omega_0) = 0 \quad ; \tag{7.3}$$

the bulk modulus B by

$$B = -dP/d\ln \Omega|_{\Omega_0} \quad ; \tag{7.4}$$

and the cohesive energy E_{coh} per atom by

$$E_{coh} = - \int_{\Omega_0}^{\infty} P \, d\Omega \quad . \tag{7.5}$$

To exemplify the accuracy of such a procedure, Fig.7.2 shows the atomic radius for most of the actinide metals as calculated by *Skriver* et al. [7.6]. Similar plots for the 3d and 4d metals have been obtained by *Moruzzi* et al. [7.7], for 4d and 5d metals by *Glötzel* (reviewed in [7.4]) and for the 3d monoxides by *Andersen* et al. [7.8,9].

In the remainder of this chapter I present a pressure relation based on (7.1), in addition to an approximate expression which may be used in simple model calculations and which may form the basis for interpreting a complete calculation, such as presented in Fig.7.2, in terms of the volume derivatives of the potential parameters.

7.2 Density-Functional Theory

Density-functional formalism is based on two theorems by *Hohenberg* and *Kohn* [7.1] who considered the Hamiltonian

$$H = T + U + V$$

$$= \sum_i^M (-\nabla_i^2) + \frac{1}{2} \sum_{i \neq j}^M \frac{2}{r_{ij}} + \sum_i^M v_{ext}(\mathbf{r}_i) \tag{7.6}$$

of a system of M interacting electrons moving in some fixed external potential v_{ext}. Here T is the kinetic energy, U is the electron-electron Coulomb repulsion and V is the interaction with the external potential, which includes the electrostatic interaction with the fixed nuclei.

Firstly, *Hohenberg* and *Kohn* showed that the external potential is a unique functional of the electron density $n(\mathbf{r})$, and hence that the ground state Φ, and the energy functionals

$$<\Phi|H|\Phi> = F[n] + \int v_{ext}(\mathbf{r})n(\mathbf{r})d\mathbf{r}$$

$$F[n] = <\Phi|T + U|\Phi> \tag{7.7}$$

are unique functionals of $n(\mathbf{r})$. In addition, they separated out the classical Hartree contribution

$$F[n] \equiv 1/2 \iint \frac{2n(\mathbf{r})n(\mathbf{r}')}{|\mathbf{r} - \mathbf{r}'|} d\mathbf{r} \, d\mathbf{r}' + G[n] \tag{7.8}$$

by defining yet another functional G[n], which represents the kinetic energy plus the difference between the true interaction energy and that given by the Hartree interaction term. It follows that F[n] and G[n] are universal functionals of the electron density, valid for any external potential and any

number of electrons, and that the freedom to specify the external potential
may be used to generate different electron densities.

Hohenberg and *Kohn* showed secondly that the energy functional (7.7) assumes its minimum value, the ground-state energy, for the correct ground-
state density. Hence, if the universal functional $F[n] = <\Phi|T + U|\Phi>$ were
known it would be relatively simple to use this variational principle to
determine the ground-state energy and density for any specified external potential. Unfortunately, the functional is not known, and the full complexity
of the many-electron problem is associated with its determination.

In this situation it is useful to note that the theorems described above
apply equally well to the case of non-interacting electrons, i.e. to a system with the Hamiltonian

$$H_s = T + V$$
$$= \sum_i^M (-\nabla_i^2) + \sum_i^M v_s(\mathbf{r}_i) \quad . \tag{7.9}$$

The ground state Φ_s of this single-particle problem is simply a Slater determinant obtained by populating the lowest-lying one-electron orbitals defined by the Schrödinger equation

$$[- \nabla^2 + v_s(\mathbf{r})]\psi_j(\mathbf{k},\mathbf{r}) = E_j(\mathbf{k})\psi_j(\mathbf{k},\mathbf{r}) \quad , \tag{7.10}$$

so the density is given by

$$n(\mathbf{r}) = \sum_{j\mathbf{k}}^{\text{occ.}} |\psi_j(\mathbf{k},\mathbf{r})|^2 \quad . \tag{7.11}$$

Hence, guided by the success of the one-electron picture, *Kohn* and *Sham*
[7.2] considered a system of non-interacting electrons together with the
real system, and proceeded to determine the external potential $v_s(\mathbf{r})$ such
that (7.11) is also the ground-state density of the real system. We shall
follow their reasoning.

First we write G[n] in the form

$$G[n] = T_s[n] + E_{xc}[n] \quad , \tag{7.12}$$

where $T_s[n]$ is the kinetic energy

$$<\Phi_s|T|\Phi_s> = \sum_{j\mathbf{k}}^{\text{occ.}} \int \psi_j^*(\mathbf{k},\mathbf{r})(-\nabla^2)\psi_j(\mathbf{k},\mathbf{r})d\mathbf{r} \tag{7.13}$$

of the non-interacting electrons of density $n(\mathbf{r})$, and $E_{xc}[n]$ is the so-called
exchange-correlation energy functional. In total we have now isolated two
terms, the Hartree term in (7.8) and the kinetic energy in (7.12), which

play a decisive role in the single-electron picture, and which are presumably also the dominant terms in the interacting system. The remainder has been collected in $E_{xc}[n]$, which therefore describes the difference between the true kinetic energy and that of the non-interacting system, plus the difference between the true interaction energy and that included by the Hartree contribution.

Then we introduce the local-density approximation

$$E_{xc}[n] = \int \varepsilon_{xc}(n(\mathbf{r}))d\mathbf{r} \tag{7.14}$$

which is exact in the limit of slowly varying densities, and, on account of including the kinetic energy correctly, also in the limit of high densities. The exchange-correlation energy density $\varepsilon_{xc}(n)$ is obtained from a homogeneous electron gas of density n.

Collecting together the above assumptions, we write the energy functional in the form

$$<\Phi|H|\Phi> = <\Phi_s|T|\Phi_s>$$

$$+ \int \left\{ 1/2 \int \frac{2n(\mathbf{r}')}{|\mathbf{r} - \mathbf{r}'|} d\mathbf{r}' + v_{ext}(\mathbf{r}) + \varepsilon_{xc}(n(\mathbf{r})) \right\} n(\mathbf{r})d\mathbf{r} \tag{7.15}$$

and a minimalisation with respect to the density n(r) now leads to the effective single-particle Schrödinger equation

$$\left\{ -\nabla^2 + \int \frac{2n(\mathbf{r}')}{|\mathbf{r} - \mathbf{r}'|} d\mathbf{r}' + v_{ext}(\mathbf{r}) + v_{xc}(n(\mathbf{r})) \right\} \psi_j(\mathbf{k},\mathbf{r})$$

$$= E_j(\mathbf{k})\psi_j(\mathbf{k},\mathbf{r}) \quad . \tag{7.16}$$

A comparison with (7.10) shows that the non-interacting electrons must move in the effective potential

$$v_s(\mathbf{r}) = \int \frac{2n(\mathbf{r}')}{|\mathbf{r} - \mathbf{r}'|} d\mathbf{r}' + v_{ext}(\mathbf{r}) + v_{xc}(n(\mathbf{r})) \quad . \tag{7.17}$$

Here, the first term is the classical Hartree potential or electron-electron repulsion, the second term is the external potential which in most applications includes the Coulomb attraction from the nuclei, and the final term is the exchange-correlation potential. The latter is given by

$$v_{xc}(\mathbf{r}) = d[n\varepsilon_{xc}(n)]/dn$$

$$\equiv \mu_{xc}[n(\mathbf{r})] \quad , \tag{7.18}$$

where μ_{xc} is the exchange-correlation part of the chemical potential in a homogeneous electron gas of density n(r). The exchange only part of this potential is proportional to $n(\mathbf{r})^{1/3}$, and is just the Slater $X\alpha$ potential with $\alpha = 2/3$. Useful estimates of ε_{xc} and μ_{xc} have been given by *Hedin Lundqvist* [7.10].

7.2.1 Spin-Density-Functional Theory

Density-functional formalism may be extended to a spin-density formalism, in which the spin-up and spin-down densities $n^\uparrow(\mathbf{r})$ and $n^\downarrow(\mathbf{r})$ are the independent variables. In terms of these basic variables the electron and magnetisation densities are

$$n(\mathbf{r}) = n^\uparrow(\mathbf{r}) + n^\downarrow(\mathbf{r})$$
$$m(\mathbf{r}) = n^\uparrow(\mathbf{r}) - n^\downarrow(\mathbf{r}) \quad, \tag{7.19}$$

respectively. Allowing for spin polarisation, we minimise the energy functional

$$<\Phi|H|\Phi> = T_s[n^\uparrow] + T_s[n^\downarrow]$$
$$+ \int \left\{ 1/2 \int \frac{2n(\mathbf{r}')}{|\mathbf{r} - \mathbf{r}'|} \, d\mathbf{r}' + v_{ext}(\mathbf{r}) + \varepsilon_{xc}[n^\uparrow(\mathbf{r}), n^\downarrow(\mathbf{r})] \right\} n(\mathbf{r}) d\mathbf{r} \tag{7.20}$$

and obtain an effective one-electron Schrödinger equation of the form (7.10) for each spin direction. For spin-up we have specifically

$$[- \nabla^2 + v_s^\uparrow(\mathbf{r})]\psi_j^\uparrow(\mathbf{k},\mathbf{r}) = E_j^\uparrow(k)\psi_j^\uparrow(\mathbf{k},\mathbf{r}) \quad, \tag{7.21}$$

where the effective potential is given by

$$v_s^\uparrow(\mathbf{r}) = \int \frac{2n(\mathbf{r}')}{|\mathbf{r} - \mathbf{r}'|} \, d\mathbf{r}' + v_{exc}^\uparrow(\mathbf{r}) + v_{xc}^\uparrow[n^\uparrow(\mathbf{r}), n^\downarrow(\mathbf{r})]$$

$$v_{xc}^\uparrow(n^\uparrow, n^\downarrow) = d[n\varepsilon_{xc}(n^\uparrow, n^\downarrow)]/dn^\uparrow \quad. \tag{7.22}$$

Useful estimates of ε_{xc} and the corresponding potentials $v_{xc}^{\uparrow\downarrow}$ have been given by *von Barth* and *Hedin* [7.11], and by *Gunnarsson* and *Lundqvist* [7.12].

7.3 Self-Consistent Band-Structure Problem

We now specify the external potential included by the last term of (7.6) to be the Coulomb attraction

$$v_c(\mathbf{r}) = - \sum_{\mathbf{R}} \frac{2Z}{|\mathbf{r} - \mathbf{R}|} \tag{7.23}$$

from the nuclei of charge Z, and positioned at lattice vectors $\mathbf{R}$, and we add to (7.6) the nuclear-nuclear repulsion

$$V_n = 1/2 \sum_{R \neq R'} \sum \frac{2Z^2}{|R - R'|} \quad . \tag{7.24}$$

In the atomic-sphere approximation the electron density is spherically symmetric inside spheres of radius S, giving

$$n(r) = \sum_R n(|r - R|)\theta(|r - R| - S) \quad , \tag{7.25}$$

where $n(r)$ has the form (6.40,41). For the case of one atom per cell the number of electrons in the atomic sphere is Z. Hence, when we include (7.23, 24) in the energy functional (7.15), we find that by means of (7.25) the electrostatic interactions reduce to an interaction with the field $-2Z/r$ from the nucleus which is therefore the external potential $v_{ext}(r)$. Furthermore, since this interaction is restricted to the sphere we are led to minimise the energy functional (per atom)

$$<\Phi|H|\Phi> = <\Phi_s|T|\Phi_s>_{ASA}$$

$$+ \int_S \left\{ 1/2\ v_H(r) - 2Z/r + \varepsilon_{xc}(n(r)) \right\} n(r) dr \quad . \tag{7.26}$$

Here, the integrals extend over the atomic sphere of radius S, and we defined the Hartree potential $v_H(r)$ by

$$v_H(r) = \int_S \frac{2n(r')}{|r - r'|} dr' \quad . \tag{7.27}$$

It may be found from Poisson's equation (1.3).

The minimalisation of (7.26) results in a one-electron Schrödinger equation of the form (7.10), valid inside each atomic sphere and with an effective one-electron potential given by

$$v_s(r) = v_H(r) - 2Z/r + v_{xc}(n(r)) \quad . \tag{7.28}$$

Matching the solutions of (7.10,28) from sphere to sphere finally gives the electronic energy-band structure of the crystal in question.

The kinetic energy (7.13) of the non-interacting system may be obtained from (7.10,28) as

$$<\Phi_s|T|\Phi_s>_{ASA} = \sum_{jk}^{occ.} E_j(k) - \int_S v_s(r)n(r)dr \tag{7.29}$$

and the total energy (7.26) may then be written in the form

$$<\Phi|H|\Phi> = \sum_{jk}^{occ.} E_j(k) - \int_S v_s(r)n(r)dr$$

$$+ \int_S \left\{ 1/2 \; v_H(r) - 2Z/r + \varepsilon_{xc}(n(r)) \right\} n(r) d\mathbf{r} \quad , \tag{7.30}$$

i.e. as the sum of the one-electron energies corrected for double counting. In this connection it is important to realise that the effective one-electron potential (7.28) used to construct the electron density and the sum of the one-electron energies through (7.10,11), and which also enters the second term in the kinetic energy (7.29), is exactly equal to the variation in the last term of (7.30) *only* when a self-consistent solution is obtained, as described below. Hence, this last term should be evaluated by using the electron density (7.11), and one should not use (7.28) to combine the second and third terms in (7.30) for the total energy.

Since the effective one-electron potential given by (7.28) depends on the electron density we want to find, we must solve (7.10,11,18,27,28) self-consistently in analogy with the classical Hartree or Hartree-Fock methods. We begin by assuming an effective one-electron potential based, for instance, on a free-atom electron density renormalised to the Wigner-Seitz sphere, and solve the energy-band problem (7.28) by the LMTO method described in Chap.5. We then obtain an electron density by means of (6.41) and proceed to construct a new one-electron potential (7.28) from (7.27) and from parametrised versions of (7.18) or (7.22), as given for example by *von Barth* and *Hedin* [7.11]. Finally, the whole cycle is iterated to self-consistency, by which stage we have all the pieces needed to calculate ground-state properties of the electronic system.

7.4 Electronic Pressure Relation

With the current formalism one may calculate total energies. However, the total energy has a large contribution (thousands of Rydbergs) from the core-levels, and owing to the corrections for double counting and exchange correlation in (7.30) there is no simple relation to the sum of the one-electron energies. Nonetheless, structural energy differences, for instance, are of the order of mRy, and we know from explicit calculations [7.13-15] that most of the physics is contained in the sum of the one-electron energies for the valence states. Hence, when we calculate total energy differences, large cancellations will take place, and for that reason we prefer to calculate total energy derivatives rather than the total energy itself.

We now show that the change in total energy upon a virtual displacement of the atoms may be obtained from the difference in the sum of the one-electron energies, provided that the one-electron potential is frozen during the displacement. This fact was first noticed by *Pettifor* [7.16] who used the atomic-sphere approximation and the virial expression by *Liberman* [7.17] to obtain a pressure relation in the form (7.37) given below. Here (7.37) is derived by means of the canonical band theory presented in Chap.2 and the so-called force theorem proved by *Andersen* [7.4,8]. According to the latter the change in total energy is given by the change in the sum of the one-electron energies plus a surface term, and with one atom per cell we have in the atomic-sphere approximation

$$dU \equiv <\Phi|H|\Phi>$$

$$= \delta \int^{E_F} E\ 2N(E)dE$$

$$+ \left\{ \varepsilon_{xc}(n(S)) - v_{xc}(n(S)) \right\} 4\pi S^2 n(S) dR \quad . \tag{7.31}$$

Here, δ indicates the restricted variation with a frozen rather than self-consistently relaxed one-electron potential, $N(E)$ is the state density per spin for the valence electrons, and dR is the displacement. The use of a restricted variation ensures that the chemical shifts of the core levels and the double-counting term do not enter the force relation.

In terms of the number-of-states function $n(E)$ we obtain by partial integration

$$\delta \int^{E_F} E\ 2N(E)dE = \delta \left[E_F n - \int^{E_F} n(E)dE \right]$$

$$= - \int^{E_F} \delta n(E)dE \tag{7.32}$$

since the total number of electrons $n = n(E_F)$ is constant. According to canonical band theory we now regard the number-of-states function as a function of the vector $\mathbf{D}(E) = \{D_s(E), D_p(E), D_d(E), \ldots\}$ uniquely related to $\mathbf{P}(E)$, Sect.2.5, and find

$$\delta \int^{E_F} E\ 2N(E)dE = - \sum_\ell \int^{E_F} \frac{\partial n(\mathbf{D})}{\partial D_\ell} \delta D_\ell(E)dE$$

$$= - \sum_\ell \int^{E_F} 2N_\ell(E)\dot{D}_\ell(E)^{-1} \delta D_\ell(E)dE$$

$$= \sum_\ell \int^{E_F} 2N_\ell(E)S\phi_\ell^2(E,S)\delta D_\ell(E)dE \quad . \tag{7.33}$$

In the last two steps we used (2.42) with $\mathbf{D}$ substituted for $\mathbf{P}$, and inserted $\dot{D}(E)$ from (3.38).

The change in the sum of the one-electron energies (7.33) has now been written in terms of the projected state density, the partial wave evaluated at the sphere, and the change in boundary condition of the solutions in the sphere. This boundary condition was imposed by the KKR-ASA equations, Sect.

2.1, and its variation upon, for instance, uniform compression of the lattice may be found from the radial Schrödinger equation (1.17)

$$- \frac{\delta D_\ell(E)}{\delta \ln S} = [D_\ell(E) + \ell + 1][D_\ell(E) - \ell] + [E - v(S)]S^2$$

$$- \left\{ \varepsilon_{xc}(n(S)) - v_{xc}(n(S)) \right\} S^2 \quad . \tag{7.34}$$

Here the last term has been added in order to include the surface term of (7.31), as will become clear if one simply inserts this last term into (7.33) and uses (6.40). In actual calculations one often shifts the potential such that the electrostatic part is zero at the atomic radius. In that case $v(S) = v_{xc}(n(S))$ and (7.34) reduces to

$$- \frac{\delta D_\ell(E)}{\delta \ln S} = [D_\ell(E) + \ell + 1][D_\ell(E) - \ell] + [E - \varepsilon_{xc}]S^2 \tag{7.35}$$

which may be used in (7.33) to calculate the restricted variation in the sum of the one-electron energies.

We wish to calculate the electronic pressure at zero temperature (7.1) which is the change in total energy upon a uniform compression of the lattice. Since

$$d\langle\Phi|H|\Phi\rangle = -Pd\Omega = -3PV \, d\ln(S) \quad , \tag{7.36}$$

we find from (7.31,33,34)

$$3P\Omega = \sum_\ell \int^{E_F} 2N_\ell(E)S\phi_\ell^2(E)[D_\ell(E) + \ell + 1][D_\ell(E) - \ell]dE$$

$$+ \sum_\ell \int^{E_F} 2N_\ell(E)S\phi_\ell^2(E)[E - v(S) - \varepsilon_{xc} + v_{xc}]S^2 \, dE \quad . \tag{7.37}$$

This is the pressure formula of *Nieminen* and *Hodges* [7.18] and of *Pettifor* [7.16], who derived it from the virial expression given by *Liberman* [7.17].

Expression (7.37) may be cast in a form analogous to (6.41) which may be used in connection with the LMTO method. Our starting point is (3.27) with n = ν, i.e.

$$[D(E) - D_\nu]S\phi(E)\phi_\nu = -(E - E_\nu)\langle\phi(E)|\phi_\nu\rangle \quad , \tag{7.38}$$

where on the right-hand side we insert the Taylor expansion (3.10). We find

$$[D(E) - D_\nu]S\phi(E)\phi_\nu = -\varepsilon\langle\phi_\nu|\phi_\nu\rangle - \varepsilon^2\langle\dot{\phi}_\nu|\phi_\nu\rangle + o(\varepsilon^2)$$

$$= -\varepsilon + o(\varepsilon^2) \tag{7.39}$$

and hence

$$[D(E) + \ell + 1]S\phi(E) = (D_\nu + \ell + 1)S\phi(E) - \varepsilon/\phi_\nu$$

$$[D(E) - \ell]S\phi(E) = (D_\nu - \ell)S\phi(E) - \varepsilon/\phi_\nu \quad , \tag{7.40}$$

whereby

$$[D(E) + \ell + 1][D(E) - \ell]S\phi^2(E) = (D_\nu + \ell + 1)(D_\nu - \ell)S\phi^2(E)$$
$$- \epsilon(2D_\nu + 1)S\phi(E)/S\phi_\nu + \epsilon^2/S\phi_\nu^2 \quad . \quad (7.41)$$

We then expand $\phi(E)$ and $\phi^2(E)$ to second order in ϵ and fixed

$$S\phi^2(E)\left\{[D(E) + \ell + 1][D(E) - \ell] + [E - v(S) - \epsilon_{xc} + v_{xc}]S^2\right\}$$

$$= S\phi_\nu^2 X$$

$$+ \epsilon[2S\phi_\nu\dot\phi_\nu X - (2D_\nu + 1) + S^3\phi^2]$$

$$+ \epsilon^2\left[(S\dot\phi_\nu^2 + S\phi_\nu\ddot\phi_\nu)X - \frac{\dot\phi_\nu}{\phi_\nu}(D_\nu + 1) + \frac{1}{S\phi_\nu^2} + 2S^3\phi_\nu\dot\phi_\nu\right] + o(\epsilon^2)$$

$$\equiv A + \epsilon B + \epsilon^2 C + o(\epsilon^2)$$

$$X = \left\{(D_\nu + \ell + 1)(D_\nu - \ell) + [E_\nu - v(S) - \epsilon_{xc} + v_{xc}]S^2\right\} \quad . \quad (7.42)$$

The pressure formula may now be written as

$$3P\Omega = \sum_\ell \left[An_\ell + B \int^{E_F} 2N_\ell(E)(E - E_{\nu\ell})dE \right.$$

$$\left. + C \int^{E_F} 2N_\ell(E)(E - E_{\nu\ell})^2 dE\right] \quad , \quad (7.43)$$

where A, B, and C are defined by (7.42). Expression (7.43) is no more complicated than (6.41) for the electron density and, as it expresses the pressure in terms of the moments of the projected state densities, it is directly applicable in connection with our self-consistency procedure.

7.5 First-Order Pressure Relation

Expression (7.43) is the pressure relation we shall use in the complete calculations of pressure-volume curves. It will prove useful, however, to use an expression which is less accurate but more directly related to simple potential parameters such as band centres and masses in order to understand the more complete calculations. In the following we shall derive such an expression from canonical band theory.

According to the previous section we should calculate the change in the sum of the one-electron energies. To this end we write the one-electron energy in the canonical form

$$E = C[S] + \frac{E - C}{W}[n] \times W[S] \quad , \quad (7.44)$$

110

where the square brackets indicate that the centre C and the bandwidth W depend on the atomic volume only while the combination (E - C)/W depends solely on the occupation number n. The sum of the one-electron energies is therefore

$$\int^{E_F} E\, 2N(E)dE = C[S]n + W[S] \int^{E_F} \frac{E-C}{W}[n]dn \quad . \tag{7.45}$$

Since the integral on the right-hand side depends only on n, differentiation with respect to S may be performed directly, and for the ℓ partial pressure

$$3P_\ell\Omega = -\delta \int^{E_F} E\, 2N_\ell(E)dE/\delta\ln(S)$$

$$= n_\ell\left(-\frac{\delta C}{\delta\ln(S)}\right) + n_\ell(\bar{E}_\ell - C_\ell)\left(-\frac{\delta\ln(W_\ell)}{\delta\ln(S)}\right)$$

$$\bar{E}_\ell = \frac{1}{n_\ell} \int^{E_F} E\, 2N_\ell(E)dE \quad , \tag{7.46}$$

where $\bar{E}_\ell$ is the centre of gravity of the occupied part of the ℓ band. Similarly, we find by expansion around V_ℓ

$$3P_\ell\Omega = n_\ell\left(-\frac{\delta V_\ell}{\delta\ln(S)}\right) + n_\ell(\bar{E}_\ell - V_\ell)\frac{\delta\ln(\tau S^2)}{\delta\ln(S)} \tag{7.47}$$

which one may use for s and p bands, while (7.46) is more appropriate for d and f bands.

The volume derivatives of the potential parameters appearing in (7.46) may be found in Sect.4.5. To include the surface term of (7.31), the potential v in (4.44,48) should be replaced by the exchange-correlation energy density ε_{xc} as in (7.35), i.e. we imply that the electrostatic potential is zero at S. Hence, we have the relations

$$\frac{\delta C}{\delta\ln(S)} = -2\frac{C - \varepsilon_{xc}}{\mu}$$

$$\frac{\delta V}{\delta\ln(S)} = -(2\ell + 3)\frac{V - \varepsilon_{xc}}{\tau}$$

$$\frac{\delta\ln(\mu S^2)}{\delta\ln(S)} = -\frac{\delta\ln(W)}{\delta\ln(S)} = (2\ell + 1) + \frac{2}{\mu} - \frac{2(C - \varepsilon_{xc})S^2}{D_\nu^{\bullet} + \ell + 1}$$

$$\frac{\delta\ln(\tau S^2)}{\delta\ln(S)} = -(2\ell + 1) + \frac{2\ell + 3}{\tau} - \frac{2(V - \varepsilon_{xc})S^2}{D_\nu^{\bullet} - \ell} \tag{7.48}$$

which express the dependence of the band positions and bandwidths on the atomic volume.

7.6 Chromium: An Example

In Fig.7.3 we show some results from a calculation [7.19] of the properties
of antiferromagnetic chromium. The self-consistent energy bands were calculated as outlined in Sect.7.3, including the spin-polarised exchange-correlation terms mentioned in Sect.7.2.1. The pressure was calculated by means
of (7.43) for each spin and the bulk modulus was obtained by numerical differentiation. As indicated in the figure, a few per cent change in the atomic
radius produces complete agreement with experiment.

We shall now discuss the form of the equation of state, i.e. $P(\Omega)$, for
Cr in the light of the pressure expressions of the previous sections. To
this end, in Fig.7.4 the total pressure has been decomposed into its s, p,
and d contributions. We note that the s and p electrons provide repulsive,
i.e. positive, pressures while the d electrons provide an attractive or
bonding, i.e. negative, pressure. These facts may be understood if we write
(7.46) in terms of an average energy <E>, specified by a volume-independent
logarithmic derivative at the atomic sphere and the occupation number, i.e.

$$3P\Omega \sim \sum_{\ell} n_{\ell} \left(- \frac{\delta <E>_{\ell}}{\delta \ln(S)} \right) . \tag{7.49}$$

Now, the logarithmic derivative $D(E)$ is a decreasing function of energy, and
typical s and p orbitals have a negative curvature at the atomic sphere.
Therefore, when we decrease the volume we must increase the energy to keep
the logarithmic derivative at the sphere fixed, and the energy term in (7.49)
is positive. In contrast, d and f orbitals typically have a positive curvature at the atomic sphere and the corresponding pressures are negative. As
a result, the equation of state for chromium is a balance between a bonding
d contribution and repulsive s and p contributions.

The first-order expressions (7.46-48) provide a more useful decomposition
than (7.49). Therefore Fig.7.5 presents some relevant data, and for $S = 2.684$
a.u. data may also be found in Tables 4.1,2, since E_{ν} is $\bar{E}$ by choice. The
first-order estimates for the partial pressures are included in Fig.7.4 for
a few atomic radii. Judging from the figure, they provide accurate estimates,
the d contribution being only 10% wrong owing to errors of order higher than
$(\bar{E} - C)$, while the sp contributions are almost correct.

According to (7.46,47) the partial ℓ pressure may be decomposed into a
band-position term proportional to $\delta V/\delta \ln(S)$ or $\delta C/\delta \ln(S)$, and a band-broadening term proportional to $\delta \ln(\tau S^2)/\delta \ln(S)$ or $\delta \ln(\mu S^2)/\delta \ln(S)$. At large volumes
in chromium $V \simeq \mu_{xc} < \varepsilon_{xc}$ and $\tau \simeq 1$ for s and p states. The δV term therefore
provides an attractive pressure which is the correction due to exchange and
correlation to the repulsive bandwidth or kinetic energy term proportional to

112

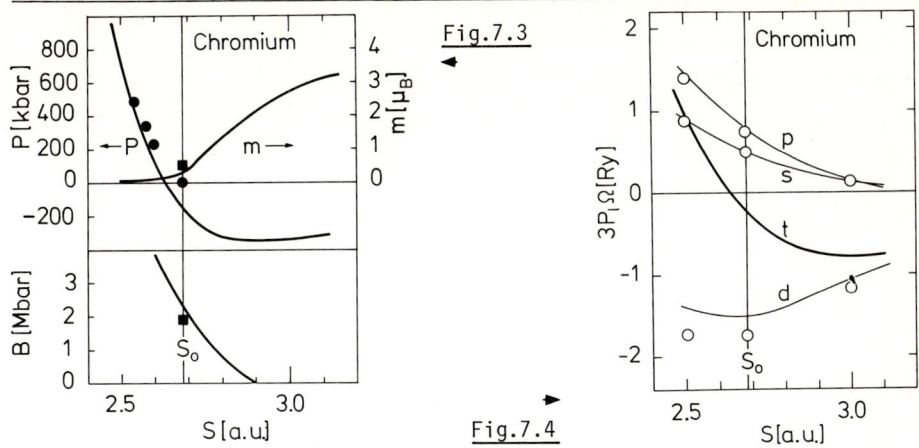

Fig.7.3

Fig.7.4

Fig.7.3. Calculated pressure P, bulk modulus B, and sublattice magnetisation m, as functions of atomic radius for commensurate antiferromagnetic chromium [7.19]. The filled circles show the measured equation of state, and the filled squares indicate the measured magnetisation, upper panel, and the measured bulk modulus, lower panel

Fig.7.4. The partial s, p, and d pressures, and the total pressure as obtained by the pressure relation (7.43) for antiferromagnetic chromium. The circles indicate the results of the first-order expressions (7.46-48)

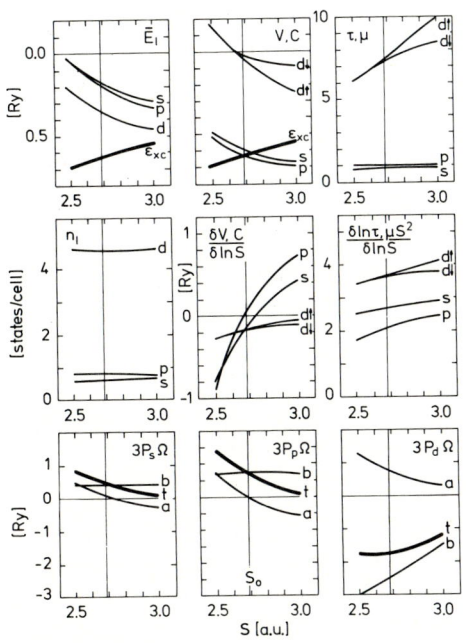

Fig.7.5. The volume dependence of the potential parameters and the occupation numbers which enter the first-order pressure expressions (7.46-48) as obtained in calculations on chromium. For s and p electrons the V, τ parameters corresponding to (7.47) are shown, while for d electrons we give the C, μ parameters corresponding to the central expression (7.46). In the lowest panels pressures calculated by the first-order expressions (7.46-48) are shown and decomposed into δV or δC terms indicated by a, and bandwidth terms indicated by b

($\bar{E}$ - V). As the volume decreases, the pseudopotential V increases above ε_{xc} owing to the relative increase in the core size. Hence, $\delta V/\delta \ln(S)$ changes sign near the equilibrium volume and the band-position term becomes repulsive. The bandwidth term on the other hand remains constant under compression because although $\bar{E}$ - V increases slightly the simultaneous increase in V - ε_{xc} makes $\delta \ln(\tau S^2)/\delta \ln(S)$ go down. In total, the s and p electrons provide strongly repulsive pressures which are responsible for the incompressibility, i.e. the large bulk modulus, of chromium.

For d electrons the centre C is well above ε_{xc} and the band-position term of (7.46) is therefore repulsive. As the volume decreases it becomes even more so because C and ε_{xc} move in opposite directions. The attraction is provided by the band-broadening term which is negative because $\bar{E}$ is below the centre C. This term becomes more bonding under compression simply because the d band widens. In total, the d contribution becomes more attractive as the volume decreases, until the relative increase in the core size forces the band position to increase rapidly and partly cancel the bonding effect.

Since the d band in chromium is approximately half-full, $\bar{E}$ is positioned essentially halfway between the bottom and the centre of the band, whereby the bonding contribution attains its maximum value relative to the neighbouring elements. This may be understood in a simple model where the state density is assumed to be rectangular and equal to $10/W_d$. In this case, $\bar{E} - C$ = $-W_d(10 - n_d)/20$, whence the band-broadening term of (7.46) gives rise to the well-known parabolic behaviour

$$3P_d\Omega = -n_d(10 - n_d) \frac{W_d}{20} \frac{\delta \ln(\mu S^2)}{\delta \ln(S)} \qquad (7.50)$$

which is reduced somewhat by the δC term. It is the d-band-broadening effect which is responsible for the parabolic behaviour of the lattice parameter in the d-transition metals. A similar f-band-broadening effect explains the parabolic trend in the first part of the actinide series, Fig.7.2 from francium to plutonium.

We have now discussed the equation of state in chromium within the framework of canonical band theory. The usefulness of this picture is not, however, restricted to a description of the more elaborate calculations because once we have understood the physics of the problem, we may change the external circumstances and use the first-order relations (7.46-48) to estimate the corresponding changes in the equation of state. An example of such use of the theory is given in [7.8] where the effect of antiferromagnetism in the 3d monoxides is judged from the ferromagnetic results.

8. Many Atoms per Cell

The LMTO method has the computational speed and flexibility needed to perform calculations of electron states in molecules and compounds. Therefore
in the present chapter we shall generalise the LMTO formalism purely within
the atomic-sphere approximation to include the case of many inequivalent
atoms per cell. The LMTO method is based on the variational principle in
conjunction with energy-independent muffin-tin orbitals but, in addition to
this approach, we have also considered the tail-cancellation principle which
led to the KKR-ASA condition (2.8). Since the latter has conceptual advantages, we apply the tail-cancellation principle to the simplest possible
case of more than one atom, namely the diatomic molecule. After that, we
turn to crystalline solids and generalise or sometimes rederive the important equations of LMTO formalism. Hence, in addition to giving the LMTO
equations for many atoms per cell, the present chapter may also serve as a
short and compact presentation of the crystal-structure-dependent part of
LMTO formalism. The potential-dependent part is treated in Chap.3. In the
final sections are listed the modifications needed to calculate ground-state
properties for materials with several atoms per cell.

   The present chapter is based on the papers by *Andersen* and co-workers
[8.1,2].

8.1 Molecules and Clusters

To introduce the subject of many atoms per cell, we apply the tail-cancellation theorem, Sect.2.1, to a collection of atoms. In the derivation it is
convenient to consider the simplest case, i.e. a diatomic molecule, but the
results will be valid for any molecule or cluster. Our starting point is the
energy-independent muffin-tin orbitals (2.1) in the atomic-sphere approximation, i.e.

$$\chi_{t\ell m}(\mathbf{r}) = i^{\ell}Y_{\ell}^{m}(\hat{r}) \begin{cases} \psi_{t\ell}(E,r) + P_{t\ell}(r/S)^{\ell} & r \leq S_t \\ \\ (r/S)^{-\ell-1} & r \geq S_t \end{cases} \tag{8.1}$$

In this approximation each individual atom of type t is surrounded by an atomic sphere of radius S_t, and the kinetic energy κ in the region outside the spheres is zero. Hence, the spherical Bessel and Neumann functions which enter the theory become polynomials in (r/S) where S may be taken as a common radius different from S_t. The requirement of continuity and differentiability at the individual radii S_t determines the normalisation of $\psi_{t\ell}$ and the function

$$P_{t\ell}(E) = \frac{D_{t\ell}(E) + \ell + 1}{D_{t\ell}(E) - \ell}\left(\frac{S}{S_t}\right)^{2\ell+1} \tag{8.2}$$

which reduces to (2.2) when $S = S_t$.

8.1.1 Tail Cancellation

We now place the orbitals at $\mathbf{Q}$ and $\mathbf{Q}'$, the positions of the two atoms, and write the total wave function as the linear combination

$$\Psi = \sum_L a_L \chi_{t\ell m}(E, \mathbf{r} - \mathbf{Q}) \quad , \tag{8.3}$$

where the index L is shorthand for $\mathbf{Q}\ell m$. Inside the sphere at $\mathbf{Q}' = \mathbf{0}$ the linear combination

$$\sum_{\ell'm'} a_L \cdot i^{\ell'} Y_{\ell'}^{m'}(\hat{r})[\psi_{t'\ell'}(E,r) + P_{t'\ell'}(E)(r/S)^{\ell'}] \tag{8.4}$$

will be a solution of the Schrödinger equation provided the terms

$$\sum_{\ell'm'} a_L \cdot i^{\ell'} Y_{\ell'}^{m'}(\hat{r}) P_{t'\ell'}(E)(r/S)^{\ell'} \tag{8.5}$$

are cancelled by the tails

$$\sum_{\ell m} a_L i^{\ell} Y_{\ell}^{m}(\widehat{\mathbf{r} - \mathbf{Q}})(|\mathbf{r} - \mathbf{Q}|/S)^{-\ell-1} \tag{8.6}$$

of the orbitals at the other site. When a similar cancellation has occurred inside the sphere at $\mathbf{Q}$, the expansion coefficients a_L can be determined such that the wave function (8.3) is a solution to the problem.

It follows from the addition theorem (5.14,6.5) that a Neumann function centred at $\mathbf{Q}$, i.e. $(|\mathbf{r} - \mathbf{Q}|/S)^{-\ell-1}$ in the ASA, may be expanded around $\mathbf{Q}'$ in terms of Bessel functions, i.e. $(r/S)^{\ell'}$, in the form

116

$$\left(\frac{r - Q}{S}\right)^{-\ell-1} i^\ell Y_\ell^m (\widehat{r - Q})$$

$$= \sum_{\ell'm'} \left(\frac{r}{S}\right)^{\ell'} i^{\ell'} Y_{\ell'}^{m'}(\hat{r}) \frac{-1}{2(2\ell' + 1)} S_{Q'\ell'm',Q\ell m} \quad , \tag{8.7}$$

where the expansion coefficients are essentially the structure constants

$$S_{Q'\ell'm',Q\ell m} = (1 - \delta_{Q'Q}) g_{\ell'm',\ell m} \left(\frac{S}{Q}\right)^{\ell''+1} [\sqrt{4\pi}\ i^{\ell''} Y_{\ell''}^{m''}(\hat{Q})]^* \tag{8.8}$$

evaluated for the common radius S. The factor $(1 - \delta_{Q'Q})$ is introduced because the tail centred at Q' does not participate in the cancellation inside its own sphere, and $g_{\ell'm',\ell m}$ is defined by (6.9), $\ell'' = \ell' + \ell$, and $m'' = m' - m$.

With the addition theorem (8.7) the cancellation required will take place if

$$\sum_L (P_{t\ell}(E)\delta_{L'L} - S_{L'L})a_L = 0 \tag{8.9}$$

for all values of $L' = Q'\ell'm'$. Here the potential function $P_{t\ell}(E)$ is defined by

$$P_{t\ell}(E) = 2(2\ell + 1) \frac{D_{t\ell}(E) + \ell + 1}{D_{t\ell}(E) - \ell} \left(\frac{S}{S_t}\right)^{2\ell+1} \tag{8.10}$$

and reduces to (2.9) if $S = S_t$. Equations (8.9) are linear in a_L, they are the analogues of the KKR-ASA equations (2.8) for the energy-band problem, and they have non-trivial solutions for those energies where the determinant of the coefficient matrix vanishes. Similarly to the KKR-ASA equations, they may be regarded as a specification of the boundary condition imposed on the solutions inside the atomic sphere by the presence of other atoms.

8.1.2 The Two-Centre Approximation

The potential function $P_{t\ell}(E)$ may be parametrised analogously to (3.41) in the form

$$P_{t\ell}(E) = \mu_{t\ell} S_t^2 (E - C_{t\ell})(S/S_t)^{2\ell+1} \quad , \tag{8.11}$$

neglecting terms of order higher than (E - C). Such a parametrisation is not very accurate for s and p states and the approximation derived from it may be useful only for d bands. If we introduce the bandwidth or overlap parameter

$$\Delta_{t\ell} = [\mu_{t\ell} S_t^2 (S/S_t)^{2\ell+1}]^{-1} \quad , \tag{8.12}$$

the linear equations (8.9) take the form

$$\sum_L (H_{L'L} - E\delta_{L'L})a_L = 0 \qquad (8.13)$$

with the effective one-electron Hamiltonian matrix

$$H_{L'L} = C_{t\ell}\delta_{L'L} + \Delta^{\frac{1}{2}}_{t'\ell'}S_{L'L}\Delta^{\frac{1}{2}}_{t\ell} \qquad (8.14)$$

which is seen to have the two-centre form [8.3]. A comparison will show that this two-centre approximation to first order is equal to the more accurate two-centre expression (2.32) derived from the LMTO-ASA equations. The structure constants which enter the effective transfer integrals $\Delta^{\frac{1}{2}}S\Delta^{\frac{1}{2}}$ have been given in Table 6.1 and they are found to be much simpler than those encountered in the LCAO method. Because the two-centre approximation (8.14) is so simple it may be used in model calculations to get insight into the properties of molecules and clusters. In accurate calculations one should use the LMTO method.

8.2 The LMTO Formalism

Returning now to our main topic of crystalline solids, we consider a primitive cell with h atoms centred at positions **q**. Some of these atoms may be of the same type, and we denote the number of type t atoms in the cell h_t. In the atomic-sphere approximation each atom is surrounded by a sphere of suitably chosen radius S_t subject to the constraint

$$\sum_t (4\pi/3)S_t^3 h_t = \Omega$$

$$\sum_t h_t = h \quad , \qquad (8.15)$$

whereby the collection of h spheres has the volume Ω of the primitive cell. In addition we define an average atomic radius S by

$$(4\pi/3)S^3 h = \Omega \quad , \qquad (8.16)$$

used to calculate the canonical structure constants.

8.2.1 Muffin-Tin Orbitals and One-Centre Expansion

The starting point is now the energy-independent muffin-tin orbitals (6.11, 12) in the atomic-sphere approximation, i.e.

$$\chi_{L\ell m}(\mathbf{r}) = i^\ell Y_\ell^m(\hat{\mathbf{r}})
\begin{cases}
\Phi_{t\ell}(-\ell-1,r)/[\sqrt{S_t/2}\Phi_{t\ell}(-)] & r \le S_t \\[2ex]
N_{t\ell}(r/S_t)/\sqrt{S_t/2} & r \ge S_t
\end{cases} \qquad (8.17)$$

$$J_\ell(r) = \begin{cases} \Phi_\ell(\ell, r)/[\sqrt{S_t/2}\,\Phi_{t\ell}(+)] & r \leq S_t \\[2mm] (r/S_t)^\ell/\sqrt{S_t/2} & r \geq S_t \end{cases} \qquad (8.18)$$

We note that the augmented spherical Neumann $N_{t\ell}(r/S_t)$ and Bessel $J_\ell(r)$ functions are in this case defined from a tail of radial dependence (r/S_t) rather than (r/S) as in the energy-dependent orbital (8.1). The addition theorem (6.13,8.7) which represents the expansion in the sphere at $\mathbf{q}'$ of the tail of the muffin-tin orbital centred at $\mathbf{q}$ must therefore be changed to include the correct radial dependences, i.e. (r/S_t). From (6.13,8.7) we find

$$N_{t\ell m}[(\mathbf{r} - \mathbf{q})/S_t] = - \sum_{\ell'm'} \frac{\Phi_{t'\ell'm}(\ell', \mathbf{r} - \mathbf{q}')}{2(2\ell' + 1)\Phi_{t'\ell'}(+)} \left(\frac{S_t}{S_{t'}}\right)^{\frac{1}{2}} S_{L'L} \qquad (8.19)$$

In this addition theorem the spherical harmonics $i^\ell Y_\ell^m$ are included in $N_{t\ell m}$ and $\Phi_{t,\ell'm'}$, L is shorthand for $t\ell qm$, and the factor $(S_t/S_{t'})^{\frac{1}{2}}$ is included to make the structure constants

$$S_{L'L} = \left(\frac{S_{t'}}{S}\right)^{\ell'+\frac{1}{2}} S_{\mathbf{q}'\ell'm', \mathbf{q}\ell m} \left(\frac{S_t}{S}\right)^{\ell+\frac{1}{2}} \qquad (8.20)$$

Hermitian. Further, $S_{\mathbf{q}'\ell'm', \mathbf{q}\ell m}$ is defined in (8.8).

By means of the addition theorem (8.19) the Bloch sum

$$\chi_L^{\mathbf{k}}(\mathbf{r}) = \sum_{\mathbf{R}} e^{i\mathbf{k}\cdot\mathbf{R}} \chi_{t\ell m}(\mathbf{r} - \mathbf{R} - \mathbf{q}) \qquad (8.21)$$

of energy-independent muffin-tin orbitals of type t centred at $\mathbf{q}$ is found to have the one-centre expansion

$$\chi_L^{\mathbf{k}}(\mathbf{r}) = \frac{\Phi_{t\ell m}(-\ell - 1, \mathbf{r} - \mathbf{q})}{\sqrt{S_t/2}\,\Phi_{t\ell}(-)} \delta_{\mathbf{q}'\mathbf{q}}$$

$$- \sum_{\ell'm'} \frac{\Phi_{t'\ell'm'}(\ell', \mathbf{r} - \mathbf{q}')}{2(2\ell' + 1)\sqrt{S_{t'}/2}\,\Phi_{t'\ell'}(+)} S_{L'L}^{\mathbf{k}} \qquad (8.22)$$

in the sphere of type t' centred at $\mathbf{q}'$. The structure constants $S_{L'L}^{\mathbf{k}}$ are given by (8.23,24), and the factor $(S_{t'})^{\frac{1}{2}}$ in (8.19) is taken together with the potential parameter Φ, as it is in the normalisation of the energy-independent muffin-tin orbitals (6.11,8.17,18).

8.2.2 Structure Constants and the LMTO Method

For a crystal with lattice vectors $\mathbf{R}$, the structure constants $S^{\mathbf{k}}_{L'L}$ in (8.22) may be obtained from (8.20) and the Bloch sum of the cluster structure constants (8.8), i.e.

$$S^{\mathbf{k}}_{L'L} = (S_{t'}/S)^{\ell'+\frac{1}{2}} \sum_{\mathbf{R}} e^{i\mathbf{k}\cdot\mathbf{R}} S_{q'\ell'm',(\mathbf{q}+\mathbf{R})\ell m}(S_t/S)^{\ell+\frac{1}{2}}$$

$$= (S_{t'}/S)^{\ell'+\frac{1}{2}} g_{\ell'm',\ell m} L^{\mathbf{k}}_{\mathbf{q}''\ell''m''}(S_t/S)^{\ell+\frac{1}{2}} \quad , \tag{8.23}$$

where the lattice sum

$$L^{\mathbf{k}}_{\mathbf{q}''\ell''m''} = \sum_{\mathbf{R}\neq\mathbf{q}''} e^{i\mathbf{k}\cdot\mathbf{R}} \left(\frac{S}{|\mathbf{R}-\mathbf{q}''|}\right)^{\ell''+1} [\sqrt{4\pi}\; i^{\ell''} Y^{m''}_{\ell''}(\widehat{\mathbf{R}-\mathbf{q}''})]^* \tag{8.24}$$

is evaluated for the average atomic radius S, $\mathbf{q}'' = \mathbf{q}' - \mathbf{q}$, $\ell'' = \ell' + \ell$, $m'' = m' - m$, and $g_{\ell'm',\ell m}$ is defined by (6.8,9).

The one-centre expansion (8.22) may be used to obtain the LMTO secular matrix in analogy with the procedure (5.37-40). Here the bracket $\langle||\rangle$ now indicates the sum of integrals in all the spheres in the cell, i.e. $\sum_{\mathbf{q}} \langle||\rangle_{\mathbf{q}}$, and the radial function Φ should be defined to vanish outside its own sphere. We then find that $\langle\Phi_L|H-E|\Phi_L\rangle$ is diagonal in $L = t\ell qm$, and that the cellular integrals in the analogue of (5.40) are those given by (5.41) with $t\ell$ substituted for ℓ. These integrals may be found from (3.48,49,8.18), and the LMTO matrix is now

$$\frac{\langle\chi^{\mathbf{k}}_{L'}|H-E|\chi^{\mathbf{k}}_{L}\rangle}{\sqrt{S_{t'}/2}\chi_{\ell'}(S)\sqrt{S_t/2}\chi_\ell(S)}$$

$$= \left[\frac{\omega(-) - \varepsilon[1 + \omega^2(-)\langle\dot{\phi}^2\rangle]}{(S/2)\Phi^2(-)}\right]_{t\ell} \delta_{L'L}$$

$$+ \left\{\left[\frac{\omega(+) - \varepsilon[1 + \omega(+)\omega(-)\langle\dot{\phi}^2\rangle]}{\omega(+) - \omega(-)}\right]_{t'\ell'} + [..]_{t\ell} - 1\right\} S^{\mathbf{k}}_{L'L}$$

$$+ \sum_{L''} S^{\mathbf{k}}_{L'L''} \left[\frac{\omega(+) - \varepsilon[1 + \omega^2(+)\langle\dot{\phi}^2\rangle]}{2(2\ell + 1)S\Phi^2(+)}\right]_{t''\ell''} S^{\mathbf{k}}_{L''L} \quad , \tag{8.25}$$

where $L = t\ell qm$ and $\varepsilon_{t\ell} = E - E_{vt\ell}$.

The correction to the atomic-sphere approximation has two contributions as described in Sect.6.9. The first of these is a subtraction of $(V_{MTZ}-E+\kappa^2)$

times the matrix elements (6.47) in the LMTO matrix. In the present case the matrix elements (6.47) should be indexed $t\ell$. The second contribution is the addition of $(V_{MTZ} - E + \kappa^2)$ times the reciprocal lattice sum (6.58), and here we obtain

$$\frac{<\tilde{\chi}_L^k | \tilde{\chi}_L^k>}{\sqrt{S_{t'}/2}\ \tilde{\chi}_{\ell'}(S)\sqrt{S_t/2}\ \tilde{\chi}_\ell(S_t)}$$

$$= \frac{2(4\pi)^2}{\Omega} S_{t'}^{5/2} \sum_{\mathbf{G}} F_L^*(KS_{t'})F_L(KS_t)S_t^{5/2} \quad , \tag{8.26}$$

where

$$F_L(KS_t) = (2\ell + 1)(2\ell + 3)(KS_t)^{-3} j_{\ell+1}(KS_t) Y_\ell^m(\hat{K}) e^{-i\mathbf{K}\cdot\mathbf{q}} \quad . \tag{8.27}$$

The remainder of the LMTO formalism may now be taken over by analogy. This requires essentially the substitutions $\ell \to t\ell$ and $m \to qm$, but in addition $\mathbf{r} \to \mathbf{r} - \mathbf{q}$ in appropriate places, cf. (8.22). For the $t\ell$ character (6.33) we find

$$c_{t\ell j}^k = \sum_{qm} |A_{Lj}^k|^2 + <\dot{\phi}_{\nu t\ell}^2> \sum_{qm} |B_{Lj}^k|^2 \quad , \tag{8.28}$$

for the ℓ-state density projected into all the spheres of type t

$$N_{t\ell}(E) = (2\pi)^{-3}\Omega \sum_j \int_{BZ} d\mathbf{k}\ c_{t\ell j}^k \delta[E - E_j(\mathbf{k})] \quad , \tag{8.29}$$

and for the electron density inside a sphere of type t

$$4\pi n_t(r) = h_t^{-1} \sum_\ell \int^{E_F} 2N_{t\ell}(E)\phi_{t\ell}^2(E,r)dE$$

$$= h_t^{-1} \sum_\ell \left\{ \phi_{\nu t\ell}^2(r)\ n_{t\ell} + 2\phi_{\nu t\ell}\ \dot{\phi}_{\nu t\ell}(r) \int^{E_F} 2N_{t\ell}(E)(E - E_{\nu t\ell}) \right. \tag{8.30}$$

$$\left. + [\dot{\phi}_{\nu t\ell}^2(r) + \phi_{\nu t\ell}(r)\ddot{\phi}_{\nu t\ell}(r)] \int^{E_F} 2N_{t\ell}(E)(E - E_{\nu t\ell})^2 dE + \dots \right\} \quad .$$

8.3 Total Energy and Self-Consistent Energy Bands

With several inequivalent atoms in the cell we specify the interactions included in (7.6) to be the Coulomb attraction

$$v_c(\mathbf{r}) = - \sum_R \sum_q \frac{2Z_t}{|\mathbf{r} - \mathbf{R} - \mathbf{q}|} \tag{8.31}$$

from the nuclei of charge Z_t positioned at $\mathbf{q}$ in the cell at $\mathbf{R}$. In addition we also include the nuclear-nuclear repulsion

$$V_n = 1/2 \sum_{\mathbf{R}} \sum_{\mathbf{qq'}}' \frac{2Z_{t'}Z_t}{|\mathbf{q} - \mathbf{q'} - \mathbf{R}|} \quad, \tag{8.32}$$

where the prime indicates that the term with $\mathbf{q} = \mathbf{q'}$ should be excluded. The charge density we write

$$n(\mathbf{r}) = \sum_{\mathbf{R}} \sum_{\mathbf{q}} n_t(|\mathbf{r} - \mathbf{R} - \mathbf{q}|)\theta(|\mathbf{r} - \mathbf{R} - \mathbf{q}| - S_t) \tag{8.33}$$

with $n_t(r)$ given by (8.30). The atomic sphere no longer needs to be neutral, rather it has the charge (ionicity)

$$z_t = Z_t - \int_t n_t(r)d\mathbf{r} \quad, \tag{8.34}$$

where the integration extends over the sphere of type t positioned at $\mathbf{q}$.

In the atomic-sphere approximation the electrostatic interactions of (7.15), i.e. the Hartree term and (8.31,32), reduce to the interaction with the field $-2Z_t/r$ inside the sphere. In addition, the spheres interact via the Madelung term

$$1/2 \sum_{\mathbf{q}} \sum_{\mathbf{q'}} 2z_{t'}z_t(-S_{q'q}/2S) \tag{8.35}$$

which we have written in terms of the ss structure constants

$$S_{q'q} = -2 \sum_{\mathbf{R}} S/|\mathbf{R} + \mathbf{q} - \mathbf{q'}| \tag{8.36}$$

in the limit of zero k vector. The prime on the summation (8.32) can be removed because of charge neutrality in the cell and the structure constant, or in this case Madelung constant, may be calculated by the technique described in Sect.9.2.1, if the divergent $\mathbf{G} = \mathbf{0}$ term in (9.3) is neglected.

The energy functional (per cell) equivalent to (7.26) is now given by

$$<\Phi|H|\Phi> = <\Phi_s|T|\Phi_s>_{ASA} + 1/2 \sum_{\mathbf{q'q}} z_{t'}z_t(S_{q'q}/S)$$

$$+ \sum_{\mathbf{q}} \int_t \left\{1/2\, v_H(r) - 2Z_t/r + \varepsilon_{xc}(n_t(r))\right\}n_t(r)d\mathbf{r} \quad ; \tag{8.37}$$

the Hartree potential is

$$v_H(r) = \int_t \frac{2n_t(r')}{|\mathbf{r} - \mathbf{r'}|} d\mathbf{r'} \quad ; \tag{8.38}$$

122

the effective one-electron potential is

$$v_t(r) = v_H(r) - 2Z_t/r + \sum_{q'} z_{t'}(S_{q'q}/S) + v_{xc}(n_t(r)) \quad ; \tag{8.39}$$

the one-electron Schrödinger equation inside the type t sphere at **q** is

$$[- \nabla^2 + v_t(r)]\psi_j(\mathbf{r},\mathbf{k}) = E_j(\mathbf{k})\psi_j(\mathbf{r},\mathbf{k}) \quad ; \tag{8.40}$$

and the kinetic energy of the non-interacting system is

$$<\Phi_s|T|\Phi_s> = \sum_j^{occ.} E_j(\mathbf{k}) - \sum_q \sum_t \int v_t(r)dr \quad . \tag{8.41}$$

As before in Sect.7.3, (8.30,39,40) must be solved-self-consistently using for instance the LMTO technique described in Sect.8.2. Owing to the simplifying assumptions of the atomic-sphere approximation, the only added complication is the Madelung contribution from point charges z_q positioned at the centres of the atomic spheres.

9. Computer Programmes

At this stage it should be clear that the linear theory outlined in the preceding chapters may be applied at many levels of approximation. In the most favourable cases one may obtain energy levels simply by means of a pocket calculator, and this will often suffice for an overall picture of the band structure one wants to study. On the other hand, for most applications one must resort to calculations on a large-scale electronic computer, and to that end we now present a package of computer programmes, LMTOPACK, based on the linear muffin-tin orbital method.

This LMTOPACK is a collection of Fortran routines which may be used to calculate the electronic structure of a given crystalline material from knowledge of the crystal symmetry and the atomic numbers of the constituents involved. The backbone of the package is formed by four programmes, STR, LMTO, DDNS, and SCFC, which are used to determine structure constants, eigen-values, state densities, and ground-state properties, respectively. In addition, the package contains several utility programmes which allow the user to interpret and display in various fashions the electronic structure of the material considered. Table 9.1 contains the names and functions of all the programmes included.

Here we describe in detail some of the most important computer programme contained in LMTOPACK. All the source programmes have been collected on a magnetic tape which together with a listing may be obtained from the author on request.

9.1 The Self-Consistency Loop

Before discussing the usage of the individual programmes indicated by an asterisk in Table 9.1, I shall briefly describe a typical application of these programme in a self-consistent calculation of the ground-state proper-ties for a monoatomic metal. The structure of such a calculation is indi-cated in Fig.9.1.

Table 9.1. Source programmes included in LMTOPACK. An asterisk indicates that the programme is listed and described in the following sections

Name		Function
CANON		Calculates canonical and unhybridised bands.
COR	*	Generates correction-term structure constants.
DDNS	*	Calculates ℓ-projected and total state densities plus ℓ-projected and total number of states.
FSAR		Calculates extremal areas and effective masses on the Fermi surface.
JDNS		Calculates joint density of states.
LMTO	*	Calculates energy bands.
PLTBND		Plots energy bands.
PLTDOS		Plots state densities.
READB		Reads and prints energy-band data sets.
RHFS		Generates atomic and frozen-core charge densities.
RLMTO		Calculates energy bands including spin-orbit coupling.
SCFC	*	Calculates self-consistent potential parameters and ground-state properties.
STR	*	Generates structure constants.
TDNS		Calculates total state densities only.

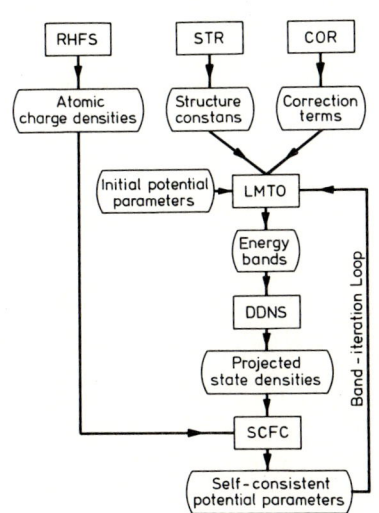

Fig.9.1. Illustration of the setup for self-consistent LMTO calculations utilizing the programmes listed in Table 9.1

At the outset one must generate data sets with structure constants and correction terms by means of STR and COR. These data depend only upon the crystal structure. In addition, one should generate a data set with atomic charge densities used to construct frozen-core charge densities. These data depend only upon the atomic number. None of the three data sets mentioned above should be recalculated during the iterations to self-consistency.

The first step in the calculation is to generate a suitable band structure by means of LMTO, and for that purpose we need a set of potential parameters, which may be obtained in several ways. The simplest possibility is to copy the appropriate values from Chap.10, and in that case the calculated band structure will be born self-consistent. The calculation is therefore completed, unless we want the energy bands for an atomic radius different from the equilibrium radius. A second possibility requires previously done calculations on a related metal, with the projected state densities stored. Execution of SCFC, which utilises the energy-scaling discussed in Sect.2.6, will then generate the necessary potential parameters. Since the potential construction in the first scaling iteration of SCFC is based upon the input atomic charge density, a third possibility is to feed SCFC any state density and use the potential parameters generated in this initial scaling iteration.

Having obtained initial potential parameters one may proceed to calculate energy bands by means of LMTO, projected state densities by means of DDNS, and finally pressure, etc. by means of SCFC. The scaling iteration in SCFC produces a new set of potential parameters which may subsequently be fed into LMTO, thus closing the band-iteration loop. Often the scaling iterations are so efficient that the calculated pressure, for instance, will converge to within 10^{-4} times the bulk modulus in only a few band iterations.

It should be evident that SCFC is the most important programme in the LMTO package. It is in SCFC that ground-state properties are calculated, and it is here that one may change the atomic radius and atomic number, and introduce spin polarisation. Also SCFC is computed rapidly, and typically it uses 20 s of Burroughs 7800 time. In comparison, LMTO (500 points in the Brillouin zone) uses 2 min and DDNS 1 min.

In summary, the self-consistency procedure has two iteration loops. One is based upon the energy-*scaling* principle of Sect.2.6 and implemented in SCFC, and the other is based upon *band* calculations and therefore requires consecutive execution of LMTO, DDNS and SCFC. At the end of the scaling iterations, and hence also at the end of a band iteration, one may compare the calculated ground-state properties with previous band iterations. If convergence is obtained one may stop at this point. If not, one may start a new band iteration using the potential parameters from the last scaling iteration.

9.2 Structure Constant Programme STR

The STR may be used to calculate the canonical structure constants defined by (6.7-9) or (8.23,24). In a typical application the programme is executed once for a given crystal structure. It produces and stores on disk or tape a set of structure-constant matrices distributed on a suitable grid in the irreducible wedge of the Brillouin zone. Whenever that particular crystal structure is encountered, the structure constant matrices may be retrieved and used to set up the LMTO eigenvalue problem which, in turn, leads to the energy bands of the material considered.

The basic input to STR is the translational vectors spanning the unit cell of the crystal, and the basis vectors giving the positions of the individual atoms in the cell. With this information STR may in principle be used to calculate canonical structure constants of any crystal structure, the only limitation being that central processor time grows rapidly as the number of atoms per cell is increased.

If one wants to calculate more than just energy levels, e.g. state densities, one must specify a grid in reciprocal space, and the use of STR is then restricted to the most commonly occurring Brillouin zones. It is, however, a simple matter to remove this restriction by adding a section in subroutine MESH (and in the corresponding routine TGEN in DDNS) specifying the grid needed.

The basic output from STR is the canonical structure constants used in CANON and LMTO to calculate band structures. In addition, STR produces a file with real and reciprocal space vectors which is used by the combined correction term programme COR. This file may also be read by STR next time the same crystal structure is encountered, thus saving the time used to generate these vectors.

9.2.1 Lattice Summations in STR

Calculating the structure constants (6.7,8.23) involves the time-consuming evaluation of the lattice sum

$$L = \sum_{R \neq q} e^{ik \cdot R} \left(\frac{S}{|R-q|}\right)^{\ell+1} [\sqrt{4\pi} \, i^\ell Y_\ell^m (\widehat{R-q})]^* \quad . \tag{9.1}$$

For $\ell \gtrsim 4$ the summation may be performed directly in real space, but for $\ell \lesssim 4$ the long-range part must be evaluated separately by Fourier transformation. In the standard technique, from Ewald, the terms in the reciprocal sum behave like $e^{-|k+G|^2/\lambda^2}$, where λ is a splitting parameter. The terms in real space, however, are more complicated functions, and we therefore prefer a simpler approach devised by *Andersen* [9.1,2] and based on the following Fourier transform:

$$\int e^{i\boldsymbol{\alpha}\cdot\boldsymbol{\beta}}[\alpha^{-\ell-1} - G_\ell^N(\alpha)][i^\ell Y_\ell^m(\alpha)]^* d\boldsymbol{\alpha}$$

$$= \frac{4\pi}{(2\ell - 1)!!} \frac{\beta^{\ell-2}}{(1 + \beta^2)^{N+1}} Y_\ell^m(\hat{\beta})^* \quad . \tag{9.2}$$

Here, N is an arbitrary splitting parameter and the functions G^N will be specified below.

The separation of L into real and reciprocal lattice sums is performed as follows:

$$(\lambda S)^{-\ell-1} L = \sum_R e^{i\mathbf{k}\cdot\mathbf{R}}[\alpha^{-\ell-1} - G_\ell^N(\alpha)][\sqrt{4\pi}\ i^\ell Y_\ell^m(\hat{\alpha})]^*$$

$$+ \sum_{R \neq q} e^{i\mathbf{k}\cdot\mathbf{R}}\ G_\ell^N(\alpha)[\sqrt{4\pi}\ i^\ell Y_\ell^m(\hat{\alpha})]^*$$

$$- \delta_{T,0}\ \lim_{\alpha\to 0}\ [\alpha^{-\ell-1} - G_\ell^N(\alpha)][\sqrt{4\pi}\ i^\ell Y_\ell^m(\hat{\alpha})]^*$$

$$= \frac{(4\pi)^{3/2}}{(2\ell - 1)!!\lambda^3\Omega}\ e^{i\mathbf{k}\cdot\mathbf{q}} \sum_G e^{i\mathbf{G}\cdot\mathbf{q}} \frac{\beta^{\ell-2}}{(1 + \beta^2)^{N+1}}\ Y_\ell^m(\hat{\beta})^*$$

$$+ \sum_{R \neq q} e^{i\mathbf{k}\cdot\mathbf{R}}\ G_\ell^N(\alpha)[\sqrt{4\pi}\ i^\ell Y_\ell^m(\hat{\alpha})]^*$$

$$- \delta_{q,0}\delta_{\ell,0}\ \lim_{\alpha\to 0}\ [\alpha^{-\ell-1} - G_\ell^N(\alpha)] \quad . \tag{9.3}$$

Here, $\alpha \equiv \lambda(R - q)$ and $\beta \equiv \lambda^{-1}(G + k)$, and the volume of the primitive cell is Ω.

The functions G_ℓ^N are

$$G_\ell^N = [(2\ell - 1)!!]^{-1} \sum_{n=0}^N h_{\ell,n} \quad , \tag{9.4}$$

where $h_{\ell,n}$ may be found by recursion

$$(n + 1)h_{\ell,n+1} = (\alpha^2/4n)h_{\ell,n-1} - (\ell - n + 1/2)h_{\ell,n}$$

$$h_{\ell+1,n} = h_{\ell-1,n} + (2\ell - 2n + 1)\alpha^{-1}h_{\ell,n} \tag{9.5}$$

using

$$h_{\ell,\ell} = \frac{\alpha^{\ell-1} e^{-\alpha}}{2^{\ell} \ell!}$$

$$h_{\ell,\ell+1} = \frac{\alpha}{2(\ell + 1)} h_{\ell,\ell}$$

$$h_{\ell,\ell-1} = 2\ell(\alpha^{-2} + \alpha^{-1}) h_{\ell,\ell} \quad . \tag{9.6}$$

Specifically, we obtain

$$G_0^3 = \alpha^{-1} e^{-\alpha}\left(1 + \frac{11}{16}\alpha\left(1 + \frac{3}{11}\alpha\left(1 + \frac{1}{9}\alpha\right)\right)\right)$$

$$G_1^3 = \alpha^{-2} e^{-\alpha}\left(1 + \alpha\left(1 + \frac{1}{2}\alpha\left(1 + \frac{7}{24}\alpha\left(1 + \frac{1}{7}\alpha\right)\right)\right)\right)$$

$$G_2^4 = \alpha^{-3} e^{-\alpha}\left(1 + \alpha\left(1 + \frac{1}{2}\alpha\left(1 + \frac{1}{3}\alpha\left(1 + \frac{1}{4}\alpha\left(1 + \frac{3}{16}\alpha\left(1 + \frac{1}{9}\alpha\right)...\right)\right)\right)\right)\right)$$

$$G_3^4 = \alpha^{-4} e^{-\alpha}\left(1 + \alpha\left(1 + \frac{1}{2}\alpha\left(1 + \frac{1}{3}\alpha\left(1 + \frac{1}{4}\alpha\left(1 + \frac{1}{5}\alpha\left(1 + \frac{1}{6}\alpha\left(1 + \frac{1}{8}\alpha\right)...\right)\right)\right)\right)\right)\right)$$

$$G_4^5 = \alpha^{-5} e^{-\alpha}\left(1 + \alpha\left(1 + \frac{1}{2}\alpha\left(1 + \frac{1}{3}\alpha\left(1 + \frac{1}{4}\alpha\left(1 + \frac{1}{5}\alpha\left(1 + \frac{1}{6}\alpha\left(1 + \frac{1}{7}\alpha\left(1 + \frac{1}{8}\alpha\right.\right.\right.\right.\right.\right.\right.\right.\right.$$

$$\left(1 + \frac{1}{10}\alpha\right)...\right) \quad . \tag{9.7}$$

Since

$$e^{\alpha} = 1 + \alpha\left(1 + \frac{1}{2}\alpha\left(1 + \frac{1}{3}\alpha\left(1 + ... \right.\right.\right. \quad , \tag{9.8}$$

it is seen that

$$\alpha^{-\ell-1} - G_{\ell}^{N}(\alpha) \rightarrow \text{const} \cdot \alpha^{\ell} \qquad \text{for} \qquad \alpha \rightarrow 0 \tag{9.9}$$

as demanded by the Fourier transform (9.2).

With (9.4-6) and using λ of the order of ten times the inverse lattice constant together with N specified by (9.7), the structure constants may be evaluated with an accuracy of a few per cent by summation over 3 shells in the direct lattice plus less than ten points of the reciprocal lattice. For an accuracy exceeding one part per thousand the summations must include 6-8 shells of both lattices.

```
C    ***********************************************************************
C    *                                                                     *
C    *     CALCULATION OF CANONICAL STRUCTURE CONSTANTS                     *
C    *     VERSION DEC. 1982                                                *
C    *                                                                     *
C    *     H.L.SKRIVER                                                      *
C    *     RISOE NATIONAL LABORATORY                                        *
C    *     DK-4000 ROSKILDE                                                 *
C    *     DENMARK                                                          *
C    *                                                                     *
C    ***********************************************************************
C    *                                                                     *
C    *     FILES AND THEIR ATTRIBUTES                                       *
C    *                                                                     *
C    *     FILE 1 :UNFORMATTED,TYPE=SEQUENTIAL,RECORDLENGTH=VARIABLE,       *
C    *             NAME=VEC/XXX                                             *
C    *     FILE 5 :FORMATTED,TYPE=SEQUENTIAL,RECORDLENGTH=80,               *
C    *             NAME=DATA/STR/XXX                                        *
C    *     FILE 6 :PRINTER                                                  *
C    *     FILE 9 :UNFORMATTED,TYPE=SEQUENTIAL,RECORDLENGTH=VARIABLE,       *
C    *             NAME=STR/XXX/A                                           *
C    *                                                                     *
C    ***********************************************************************
C    *                                                                     *
C    *     INPUT  (FILE 5) <MAIN>                                           *
C    *     -----                                                            *
C    *                                                                     *
C    *     ALAMDA :CONVERGENCE PARAMETER LAMBDA,SECT.(9.2.1)                *
C    *     BOA    :B OVER A LATTICE PARAMETER                               *
C    *     BSX                                                              *
C    *     BSY    :PRIMITIVE VECTORS IN REAL SPACE,UNITS OF A               *
C    *     BSZ                                                              *
C    *     COA    :C OVER A LATTICE PARAMETER                               *
C    *     GMAX   :RADIUS OF CONVERGENCE SPHERE IN RECIP. SPACE IN          *
C    *             UNITS OF 1/A                                             *
C    *     ICNVRG :=0:NO CONVERGENCE TEST                                   *
C    *             =1:CONVERGENCE TEST                                      *
C    *     LAT    :BRILLOUIN ZONE, SEE <MESH>                               *
C    *     MODE   :=0:AUTOMATIC GENERATION OF K-POINTS,<MESH>               *
C    *             =1:K-POINTS FROM INPUT                                   *
C    *     NGLN   :CONVERGENCE PARAMETER N,SECT.(9.2.1)                     *
C    *     NL     :NUMBER OF L-QUANTUM NUMBERS,LMAX=NL-1                     *
C    *     NOVCGN :=0:REAL AND RECIP. VECTORS FROM INPUT,FILE 1             *
C    *             =1:REAL AND RECIP. VECTORS GENERATED,<VECGEN>            *
C    *     NOWRT  :=0:SHORT PRINTOUT                                        *
C    *             =1:PRINT STRUCTURE CONSTANT MATRIX                       *
C    *     NPOINT :=NPX:NUMBER OF K-POINTS TO BE READ,MODE=1                *
C    *     NPX                                                              *
C    *     NPY    :NUMBERS FOR K-MESH CONSTRUCTION,SEE <MESH>               *
C    *     NPZ                                                              *
C    *     NQ     :NUMBER OF ATOMS IN THE PRIMITIVE CELL                    *
C    *     OKX                                                              *
C    *     OKY    :K-VECTOR IN UNITS OF PI/A,MODE=1. HEXAGONAL              *
C    *     OKZ     COORDINATES IF LAT=4                                     *
C    *     POINT  :SYMMETRY LABEL,MODE=1                                    *
C    *     QX                                                               *
C    *     QY     :BASIS VECTORS IN REAL SPACE IN UNITS OF A                *
C    *     QZ                                                               *
C    *     RMAX   :RADIUS OF CONVERGENCE SPHERE IN REAL SPACE IN UNITS      *
C    *             OF A                                                     *
C    *     TXT    :60 CHARACTERS OF TEXT FOR IDENTIFICATION                 *
```

```
C     *                                                                        *
C     *     INPUT   (FILE 1) <VECGEN>                                          *
C     *     -----                                                             *
C     *                                                                        *
C     *              INPUT WILL BE READ FROM FILE 1 IF NOVCGN=0,SEE           *
C     *              OUTPUT (FILE 1) FOR A SPECIFICATION                      *
C     *                                                                        *
C     **************************************************************************
C     *                                                                        *
C     *     OUTPUT (FILE 1) <VECGEN>                                          *
C     *     ------                                                            *
C     *                                                                        *
C     *              A SPECIFICATION OF OUTPUT NOT IN THE LISTS BELOW         *
C     *              CAN BE FOUND IN INPUT (FILE 5)                           *
C     *                                                                        *
C     *     AKX                                                               *
C     *     AKY      :G-VECTORS                                               *
C     *     AKZ                                                               *
C     *     ASX                                                               *
C     *     ASY      :R-VECTORS                                               *
C     *     ASZ                                                               *
C     *     DR       :LENGTH OF R-VECTORS                                     *
C     *     NG       :NUMBER OF G-VECTORS                                     *
C     *     NR       :NUMBER OF R-VECTORS                                     *
C     *     NRO      :NUMBER OF R-VECTORS AROUND (0,0,0)                      *
C     *     VOL      :VOLUME OF PRIMITIVE CELL SPANNED BY (BSX,BSY,BSZ)       *
C     *              IN UNITS OF A**3                                         *
C     *                                                                        *
C     *     OUTPUT (FILE 9) <MAIN>                                            *
C     *     ------                                                            *
C     *                                                                        *
C     *     DUM1     :NOT USED                                                *
C     *     KKX                                                               *
C     *     KKY      :K-VECTOR IN UNITS OF (DKX,DKY,DKZ)*PI/A                 *
C     *     KKZ                                                               *
C     *     LMQ      :DIMENSION OF STRUCTURE CONSTANT MATRIX,NQ*NLM          *
C     *     NLM      :SUM OF 2*L+1, =NL**2                                    *
C     *     NP       :COUNTER                                                 *
C     *     NPT      :NUMBER OF K-POINTS IN IRREDUCIBLE ZONE,MODE=0          *
C     *              NUMBER OF K-POINTS READ,MODE=1                          *
C     *     PKX                                                               *
C     *     PKY      :K-VECTOR IN UNITS OF PI/A,RECTANGULAR COORD.           *
C     *     PKZ                                                               *
C     *     SRL      :LOWER TRIANGLE OF STRUCTURE CONSTANT MATRIX,           *
C     *              REAL PART                                                *
C     *     SIL      :LOWER TRIANGLE OF STRUCTURE CONSTANT MATRIX,           *
C     *              IMAGINARY PART                                           *
C     *     SWS      :WIGNER-SEITZ RADIUS IN UNITS OF A                      *
C     *     WW       :SYMMETRY WEIGHTS,NOT USED                              *
C     *                                                                        *
C     **************************************************************************
C
      IMPLICIT REAL*8 (A-H,O-Z)
      COMPLEX*16 CIM,CCIM,SPHRM,DCMPLX
      COMMON/B1/CIM,CCIM,C(16,16),FAC(13),DAC(13),NGLN(10),
     1ALAMDA,TWOPI,FPL3OM,SWS,VOL,SOO,SWSL,FPILAM,RMAX,GMAX,ICNVRG,
     2NUMVR,NUMVG,NQ,NRO,NL2,NL2P,KLMO,NL,NLM,LMQ,NOWRT,IDIM
      COMMON/B2/SRL(2916),SIL(2916)
      COMMON/MSH/DKX,DKY,DKZ,DHX,BOA,COA,KX(2500),KY(2500),KZ(2500),
     1W(2500)
      COMMON/VEC/BSX(3),BSY(3),BSZ(3),BKX(3),BKY(3),BKZ(3),QX(6),QY(6),
     1QZ(6),ASX(500),ASY(500),ASZ(500),AKX(500),AKY(300),AKZ(300),
     2DR(500),DG(300),
     3USX(500),USY(500),USZ(500),UKX(300),UKY(300),UKZ(300)
      INTEGER*4 TLAT(8)/' SC',' FCC',' BCC',' HCP','  ST',' BCT','AL-U'
     1,'TRIG'/
      INTEGER TXT(15),POINT
      DATA POINT/'    '/,IDUM1/0/,DUM1/0.D0/
```

131

```
      1 FORMAT(15A4)
      2 FORMAT(3E15.7)
      3 FORMAT(2I5,5X,10I1,E15.7)
      4 FORMAT(1H ,//,11X,'BASIS VECTORS',//)
      5 FORMAT(1H1,10X,15A4,/)
      6 FORMAT(1H ,10X,'RMAX',F10.5,' GMAX',F10.5,/
     1,10X,' NL',I4,' NQ',I4,' NLM',I4,' LMQ',I4,/,10X,' LAMDA',F10.5,
     2' NGLN ',10I1,/,10X,' NOVCGN',I4,' NOWRT',I4,' ICNVRG',I4,/)
      7 FORMAT(4I5)
      8 FORMAT(3F10.6,A4)
      9 FORMAT(1H ,//,11X,'POINT NUMBER',I5,5X,'PKX,PKY,PKZ',3F10.4,
     15X,'W',F10.4,5X,A4)
     10 FORMAT(1H ,//,11X,'VOL',F15.7,'  SWS',F15.7,'  SOO',F15.7)
     11 FORMAT(1H ,11X,'(',F10.5,',',F10.5,',',F10.5,' )')
     12 FORMAT(1H ,//,11X,'PRIMITIVE VECTORS IN UNIT CELL OF ',A4,//,11X,
     1'BOA',F15.10,'  COA',F15.10,//)
     13 FORMAT(2F10.5,3I5)
     14 FORMAT(1H ,//,11X,'*MATRICES GENERATED AT AN EARLIER RUN*',//)
     15 FORMAT(I5,2E15.8)
     16 FORMAT(1H ,//,11X,'MODE,NPX,NPY,NPZ,NPT',5I5)
C
      IDIM=54
      CIM=DCMPLX(0.D00,1.D00)
      CCIM=DCMPLX(0.D00,-1.D00)
      PI=4.D00*DATAN(1.D00)
      TWOPI=2.D00*PI
      FOURPI=4.D00*PI
C
      READ(5,1) TXT
C
      TXT(15)=' NEW'
C
      READ(5,13) RMAX,GMAX,NOVCGN,NOWRT,ICNVRG
      READ(5,3) NL,NQ,NGLN,ALAMDA
      READ(5,15) LAT,BOA,COA
C
      NLM=NL*NL
      LMQ=NLM*NQ
      NL2=2*NL-1
C
C     IF LPP.GT.5 ONLY REAL SPACE SUMMATIONS IN <SMTRX>
C
      NL2P=MIN0(5,NL2)
      L=NL2-1
      KLM0=L*(L+1)+L+1
C
      CALL FACTOR(4*(NL-1))
      CALL GAUNT
      CALL LIMA0(NGLN(1),S00)
C
      WRITE(6,5) TXT
      WRITE(6,6) RMAX,GMAX,NL,NQ,NLM,LMQ,ALAMDA,NGLN,NOVCGN,
     1            NOWRT,ICNVRG
C
C     READ PRIMITIVE VECTORS BSX,BSY,BSZ OF REAL SPACE
C
      DO 20 I=1,3
   20 READ(5,2) BSX(I),BSY(I),BSZ(I)
C
C     READ THE NQ BASIS VECTORS QX,QY,QZ IN THE PRIMITIVE  CELL
C
      DO 21 I=1,NQ
   21 READ(5,2) QX(I),QY(I),QZ(I)
      WRITE(6,12) TLAT(LAT),BOA,COA
      DO 30 I=1,3
   30 WRITE(6,11) BSX(I),BSY(I),BSZ(I)
      WRITE(6,4)
      DO 25 I=1,NQ
   25 WRITE(6,11) QX(I),QY(I),QZ(I)
```

```
C
C         GENERATE THE NUMVR AND NUMVG SHORTEST VECTORS
C         OF REAL AND RECIPROCAL SPACE
C
          CALL VECGEN(NUMR,NUMG,NOVCGN,LAT)
C
          SWS=(3.D00*VOL/FOURPI/NQ)**(1.D00/3.D00)
          SWSL=SWS*ALAMDA
          FPL3OM=FOURPI/ALAMDA**3.D00/VOL
          FPILAM=FOURPI*ALAMDA
          WRITE(6,10) VOL,SWS,S00
          READ(5,7) MODE,NPX,NPY,NPZ
          IF(MODE.EQ.1) GO TO 28
C
C         GENERATE MESH IN K-SPACE
C
          CALL MESH(LAT,NPX,NPY,NPZ,NPT)
C
          WRITE(6,16) MODE,NPX,NPY,NPZ,NPT
          WRITE(9) TXT,NL,NQ,NLM,LMQ,SWS
          WRITE(9) NPX,NPY,NPZ,LAT,NPT,BOA,COA,DUM1
C
          DO 23 NP=1,NPT
          KKX=KX(NP)
          KKY=KY(NP)
          KKZ=KZ(NP)
          CALL KTRNSF(LAT,KKX,KKY,KKZ,PKX,PKY,PKZ)
          IF(NOWRT.GT.0) WRITE(6,9) NP,PKX,PKY,PKZ,W(NP),POINT
          WW=W(NP)
          WRITE(9) NP,PKX,PKY,PKZ,WW,POINT,KKX,KKY,KKZ
          PKXP=PKX*PI
          PKYP=PKY*PI
          PKZP=PKZ*PI
C
          CALL SMTRX(PKXP,PKYP,PKZP)
C
          J1=-IDIM
          DO 24 JJ=1,LMQ
          J1=J1+IDIM
          I1=J1+JJ
          I2=J1+LMQ
          WRITE(9) (SRL(LIN),LIN=I1,I2)
       24 WRITE(9) (SIL(LIN),LIN=I1,I2)
       23 CONTINUE
          CLOSE (9,DISP=CRUNCH)
          STOP
C
C         READ K-VECTOR IN UNITS OF PIA
C
       28 WW=0.D0
          NPOINT=NPX
          WRITE(6,16) MODE,NPX,NPY,NPZ,NPOINT
          WRITE(9) TXT,NL,NQ,NLM,LMQ,SWS
          WRITE(9) NPX,NPY,NPZ,LAT,NPOINT,BOA,COA,DUM1
          KKX=0
          KKY=0
          KKZ=0
          FX=PI
          FY=PI/BOA
          FZ=PI/COA
          FHX=0.D0
          IF(LAT.NE.4) GO TO 27
          FX=2.D0/3.D0*PI
          FY=PI/DSQRT(3.D0)
          FHX=-0.5D0*FX
```

```
C
   27 DO 29 I=1,NPOINT
      READ(5,8) OKX,OKY,OKZ,POINT
      WRITE(6,9) I,OKX,OKY,OKZ,WW,POINT
      WRITE(9) I,OKX,OKY,OKZ,WW,POINT,KKX,KKY,KKZ
      PKXP=OKX*FX+FHX*OKY
      PKYP=OKY*FY
      PKZP=OKZ*FZ
C
      CALL SMTRX(PKXP,PKYP,PKZP)
C
      J1=-IDIM
      DO 26 J=1,LMQ
      J1=J1+IDIM
      I1=J1+J
      I2=J1+LMQ
      WRITE(9) (SRL(LIN),LIN=I1,I2)
   26 WRITE(9) (SIL(LIN),LIN=I1,I2)
   29 CONTINUE
      CLOSE (9,DISP=CRUNCH)
      STOP
      END
      BLOCK DATA
C     ****************************************************************
C     *                                                              *
C     *    INSERT LM-DATA                                            *
C     *                                                              *
C     ****************************************************************
      COMMON/LDATA/LL(49),MM(49)
      DATA LL/0,3*1,5*2,7*3,9*4,11*5,13*6/
      DATA MM/0,-1,0,1,-2,-1,0,1,2,-3,-2,-1,0,1,2,3,-4,-3,-2,-1,0,1,2,3,
     14,-5,-4,-3,-2,-1,0,1,2,3,4,5,-6,-5,-4,-3,-2,-1,0,1,2,3,4,5,6/
      END
      SUBROUTINE VECGEN(NUMR,NUMG,NOVCGN,LAT)
C     ****************************************************************
C     *                                                              *
C     *    GENERATE VECTORS OF DIRECT AND RECIPROCAL SPACE FROM      *
C     *    BASIC TRANSLATION VECTORS (BSX,BSY,BSZ)                   *
C     *                                                              *
C     ****************************************************************
      IMPLICIT REAL*8 (A-H,O-Z)
      COMPLEX*16 CIM,CCIM
      COMMON/B1/CIM,CCIM,CC(16,16),FAC(13),DAC(13),NGLN(10),
     1ALAMDA,TWOPI,FPL3OM,SWS,VOL,SOO,SWSL,FPILAM,RMAX,GMAX,ICNVRG,
     2NUMVR,NUMVG,NQ,NRO,NL2,NL2P,KLMO,NL,NLM,LMQ,NOWRT,IDIM
      COMMON/MSH/DKX,DKY,DKZ,DHX,BOA,COA,KX(2000),KY(2000),KZ(2000),
     1W(2000)
      COMMON/VEC/BSX(3),BSY(3),BSZ(3),BKX(3),BKY(3),BKZ(3),QX(6),QY(6),
     1QZ(6),ASX(500),ASY(500),ASZ(500),AKX(500),AKY(300),AKZ(300),
     2DR(500),DG(300),
     3USX(500),USY(500),USZ(500),UKX(300),UKY(300),UKZ(300)
      DIMENSION CSX(500),CSY(500),CSZ(500),D(500),DD(3),DK(3)
      INTEGER NSR(50),NSG(50)
    1 FORMAT(1H ,//,15X,'SHELL NUMBER',I5,' WITH',I5,' POINTS',//)
    2 FORMAT(1H ,10X,I5,4F10.6)
    3 FORMAT(1H ,10X,'(',F10.5,',',F10.5,',',F10.5,')',/)
    4 FORMAT(1H1,14X,'RESULT FROM VECGEN FOR REAL SPACE VECTORS',//,
     114X,'NO',5X,'SX',8X,'SY',8X,'SZ',8X,'D',/)
    5 FORMAT(1H1,10X,'RESULT FROM VECGEN FOR RECIPROCAL SPACE VECTORS',
     1//,14X,'NO',5X,'KX',8X,'KY',8X,'KZ',8X,'D',/)
    6 FORMAT(1H ,//,11X,'PRIMITIVE VECTORS OF RECIPROCAL SPACE',/)
    7 FORMAT(1H ,//,11X,'R1',F10.4,' RA',F10.4,/,11X,'G1',F10.4,
     1' GA',F10.4,/)
    8 FORMAT(1H ,//,11X,'VALUES GENERATED AT PREVIOUS RUN OF VECGEN , (R
     1MAX,NUMVR,GMAX,NUMVG) =',//,11X,F10.5,I5,5X,F10.5,I5,//)
    9 FORMAT(1H ,'*** NR EXCEEDS DIMENSIONS OF ASX,ASY,ASZ',I5,3F10.4)
   10 FORMAT(1H ,'*** NG EXCEEDS DIMENSIONS OF AKX,AKY,AKZ',I5,3F10.4)
   11 FORMAT(1H ,10X,'NUMR',I4,' NUMG',I4,//)
```

134

```
C
      IF(NOVCGN.EQ.0) GO TO 32
      WRITE(6,6)
C
C     PRIMITIVE VECTORS (BKX,BKY,BKZ) OF RECIPROCAL SPACE
C
      DO 20 I=1,3
      I1=1+MOD(I,3)
      I2=1+MOD(I1,3)
      CALL CROSS(BSX(I1),BSY(I1),BSZ(I1),BSX(I2),BSY(I2),BSZ(I2),
     1BKX(I),BKY(I),BKZ(I))
   20 CONTINUE
      VOL=DABS(BSX(1)*BKX(1)+BSY(1)*BKY(1)+BSZ(1)*BKZ(1))
C
      DO 21 I=1,3
      BKX(I)=BKX(I)/VOL*TWOPI
      BKY(I)=BKY(I)/VOL*TWOPI
      BKZ(I)=BKZ(I)/VOL*TWOPI
   21 WRITE(6,3) BKX(I),BKY(I),BKZ(I)
C
C     CALCULATE RADII RA,GA OF SPHERES HOLDING ALL VECTORS USED IN LAT-
C     TICE SUMS. R1 IS LONGEST BASIS VECTOR. G1 IS LONGEST VECTOR IN
C     BRILLOUIN ZONE. MUST BE RECONSIDERED IN ANY NEW APPLICATIONS
C
      R1=1.E-06
      DO 35 IQ=1,NQ
      PQX=QX(IQ)
      PQY=QY(IQ)
      PQZ=QZ(IQ)
      DO 35 JQ=IQ,NQ
      X=PQX-QX(JQ)
      Y=PQY-QY(JQ)
      Z=PQZ-QZ(JQ)
      DQ=DSQRT(X*X+Y*Y+Z*Z)
      IF(DQ.LT.R1) GO TO 35
      R1=DQ
   35 CONTINUE
      R1=R1*1.001
      RA=RMAX+R1
      B1=BKX(1)+BKX(2)+BKX(3)
      B2=BKY(1)+BKY(2)+BKY(3)
      B3=BKZ(1)+BKZ(2)+BKZ(3)
      G1=DSQRT(B1*B1+B2*B2+B3*B3)/2.D00
      GA=GMAX+G1
      IF(LAT.EQ.3) GA=GMAX+6.2832
      IF(LAT.EQ.8) GA=GMAX+6.52
      IF(LAT.EQ.6) GA=GMAX+DSQRT(0.5+(1./COA)**2)*6.29
      WRITE(6,7) R1,RA,G1,GA
C
      DO 36 I=1,3
      DD(I)=DSQRT(BSX(I)**2+BSY(I)**2+BSZ(I)**2)
   36 DK(I)=DSQRT(BKX(I)**2+BKY(I)**2+BKZ(I)**2)
      DDM=DMAX1(DD(1),DD(2),DD(3))
      DKM=DMAX1(DK(1),DK(2),DK(3))
      DDM=TWOPI/DDM
      DKM=TWOPI/DKM
      NUMR=2*(IFIX(RA/DKM)+1)+1
      NUMG=2*(IFIX(GA/DDM)+1)+1
      NUMRH=NUMR/2+1
      NUMGH=NUMG/2+1
      WRITE(6,11) NUMR,NUMG
C
C     REAL SPACE
C
      WRITE(6,4)
      NR=0
      NRO=0
      DO 22 L=1,NUMR
```

```
      A=L-NUMRH
      DO 22 M=1,NUMR
      B=M-NUMRH
      DO 22 N=1,NUMR
      C=N-NUMRH
      SX=A*BSX(1)+B*BSX(2)+C*BSX(3)
      SY=A*BSY(1)+B*BSY(2)+C*BSY(3)
      SZ=A*BSZ(1)+B*BSZ(2)+C*BSZ(3)
      DX=DSQRT(SX*SX+SY*SY+SZ*SZ)
      IF(DX.GT.RA) GO TO 22
      IF(DX.LE.RMAX) NRO=NRO+1
      NR=NR+1
      IF(NR.GT.500) GO TO 33
      D(NR)=DX
      CSX(NR)=SX
      CSY(NR)=SY
      CSZ(NR)=SZ
   22 CONTINUE
C
C     SORT VECTORS IN ORDER OF INCREASING D
C
      DA=1.D-06
      NSH=0
      NSHL=-1
      DO 23 K=1,NR
      AMIN=1000.
      DO 24 N=1,NR
      IF(D(N)-AMIN)25,24,24
   25 AMIN=D(N)
      N1=N
   24 CONTINUE
      NSHL=NSHL+1
      ASX(K)=CSX(N1)
      ASY(K)=CSY(N1)
      ASZ(K)=CSZ(N1)
      DB=D(N1)
      DR(K)=DB
      IF(DB.GT.DA+1.D-06) GO TO 26
      WRITE(6,2)K, ASX(K),ASY(K),ASZ(K),DB
      GO TO 23
   26 NSH=NSH+1
      WRITE(6,1) NSH,NSHL
      NSR(NSH)=NSHL
      WRITE(6,2)K, ASX(K),ASY(K),ASZ(K),DB
      NSHL=0
      DA=DB
   23 D(N1)=1000.
      NSH=NSH+1
      NSHL=NSHL+1
      NSR(NSH)=NSHL
      WRITE(6,1) NSH,NSHL
      NUMVR=NR
      NSHLR=NSH
C
C     RECIPROCAL SPACE
C
      WRITE(6,5)
      NG=0
      DO 27 L=1,NUMG
      A=L-NUMGH
      DO 27 M=1,NUMG
      B=M-NUMGH
      DO 27 N=1,NUMG
      C=N NUMGH
      GX=A*BKX(1)+B*BKX(2)+C*BKX(3)
      GY=A*BKY(1)+B*BKY(2)+C*BKY(3)
      GZ=A*BKZ(1)+B*BKZ(2)+C*BKZ(3)
```

```
       DX=DSQRT(GX*GX+GY*GY+GZ*GZ)
       IF(DX.GT.GA) GO TO 27
       NG=NG+1
       IF(NG.GT.300) GO TO 34
       D(NG)=DX
       CSX(NG)=GX
       CSY(NG)=GY
       CSZ(NG)=GZ
   27 CONTINUE
C
C      SORT VECTORS IN ORDER OF INCREASING D
C
       DA=1.E-06
       NSH=0
       NSHL=-1
       DO 28 K=1,NG
       AMIN=1000.
       DO 29 N=1,NG
       IF(D(N)-AMIN)30,29,29
   30 AMIN=D(N)
       N1=N
   29 CONTINUE
       NSHL=NSHL+1
       AKX(K)=CSX(N1)
       AKY(K)=CSY(N1)
       AKZ(K)=CSZ(N1)
       DB=D(N1)
       IF(DB.GT.DA+1.D-07) GO TO 31
       WRITE(6,2) K,AKX(K),AKY(K),AKZ(K),DB
       GO TO 28
   31 NSH=NSH+1
       WRITE(6,1) NSH,NSHL
       WRITE(6,2)K, AKX(K),AKY(K),AKZ(K),DB
       NSG(NSH)=NSHL
       NSHL=0
       DA=DB
   28 D(N1)=1000.
       NSH=NSH+1
       NSHL=NSHL+1
       NSG(NSH)=NSHL
       WRITE(6,1) NSH,NSHL
       NUMVG=NG
       NSHLG=NSH
C
C      STORE VECTORS ON FILE 1
C
       WRITE(1) BKX,BKY,BKZ,VOL,RMAX,GMAX,NR,NG,NRO
       WRITE(1)(ASX(N),N=1,NR)
       WRITE(1)(ASY(N),N=1,NR)
       WRITE(1)(ASZ(N),N=1,NR)
       WRITE(1)(DR(N),N=1,NR)
       WRITE(1)(AKX(N),N=1,NG)
       WRITE(1)(AKY(N),N=1,NG)
       WRITE(1)(AKZ(N),N=1,NG)
       RETURN
   32 CONTINUE
C
C      READ VECTORS GENERATED IN AN EARLIER RUN
C
       READ(1) BKX,BKY,BKZ,VOL,RMAX,GMAX,NUMVR,NUMVG,NRO
       READ(1)(ASX(N),N=1,NUMVR)
       READ(1)(ASY(N),N=1,NUMVR)
       READ(1)(ASZ(N),N=1,NUMVR)
       READ(1)(DR(N),N=1,NUMVR)
       READ(1)(AKX(N),N=1,NUMVG)
       READ(1)(AKY(N),N=1,NUMVG)
       READ(1)(AKZ(N),N=1,NUMVG)
```

```
      WRITE(6,8) RMAX,NUMVR,GMAX,NUMVG
      RETURN
   33 WRITE(6,9) NR,DX,RA,RMAX
      STOP
   34 WRITE(6,10)NG,DX,GA,GMAX
      STOP
      END
      SUBROUTINE SMTRX(PKXP,PKYP,PKZP)
C     ****************************************************************
C     *                                                              *
C     *    CALCULATION OF THE STRUCTURE CONSTANT MATRIX              *
C     *                                                              *
C     ****************************************************************
      IMPLICIT REAL*8 (A-H,O-Z)
      COMPLEX*16 SUMR,SUMG,SR,CIM,CCIM,CIL,DCMPLX,YLM,CDEXP,CXPRDK,
     1SRC
      REAL*8 DREAL,DIMAG
      COMMON/B1/CIM,CCIM,C(256),FAC(13),DAC(13),NGLN(10),
     1ALAMDA,TWOPI,FPL3OM,SWS,VOL,SOO,SWSL,FPILAM,RMAX,GMAX,ICNVRG,
     2NUMVR,NUMVG,NQ,NRO,NL2,NL2P,KLMO,NL,NLM,LMQ,NOWRT,IDIM
      COMMON/B2/SRKLP(2916),SIKLP(2916)
      COMMON/LDATA/LL(49),MM(49)
      COMMON/VEC/BSX(3),BSY(3),BSZ(3),BKX(3),BKY(3),BKZ(3),QX(6),QY(6),
     1QZ(6),ASX(500),ASY(500),ASZ(500),AKX(500),AKY(300),AKZ(300),
     2DR(500),DG(300),
     3USX(500),USY(500),USZ(500),UKX(300),UKY(300),UKZ(300)
      COMMON/YLMD/YLM(49)
      DIMENSION GLNS(500,7),FBLN(500,7),BDOTQ(500),DRQ(500),
     1SR(49),CXPRDK(500),SUMR(49),SUMG(49)
  100 FORMAT(1H ,9(1PE13.5))
  101 FORMAT(1H ,///)
  102 FORMAT(1H0,10X,'QQ DIAGONAL BLOCK',/)
  103 FORMAT(1H ,10(1PE11.3))
  104 FORMAT(1H0,10X,'QP Q OFF DIAGONAL BLOCK , QP,Q =',2I5,/)
  105 FORMAT(1H0,25X,'CONVERGENCE TEST')
  106 FORMAT(1H0,'RECIPROCAL SPACE SUMS',/)
  107 FORMAT(1H0,'REAL SPACE SUMS',/)
C
C     QQ DIAGONAL BLOCKS
C
      PKX=PKXP
      PKY=PKYP
      PKZ=PKZP
      CXPRDK(1)=DCMPLX(1.D00,0.D00)
      DO 22 I=2,NRO
      DRI=DR(I)
      USX(I)=ASX(I)/DRI
      USY(I)=ASY(I)/DRI
      USZ(I)=ASZ(I)/DRI
      PDR=PKX*ASX(I)+PKY*ASY(I)+PKZ*ASZ(I)
      CXPRDK(I)=CDEXP(PDR*CIM)
      ALFA=ALAMDA*DRI
C
C     CONVERGENCE FUNCTION, REAL SPACE
C
      DO 25 LP=1,NL2P
      L=LP-1
      NG=NGLN(LP)
   25 GLNS(I,LP)=GLN(ALFA,L,NG)
   22 CONTINUE
      DO 35 I=1,NUMVG
      X=AKX(I)+PKX
      Y=AKY(I)+PKY
      Z=AKZ(I)+PKZ
      D=DSQRT(X*X+Y*Y+Z*Z)
      IF(D.GT.GMAX) GO TO 35
      BETA=D/ALAMDA
```

```
      UKX(I)=X/D
      UKY(I)=Y/D
      UKZ(I)=Z/D
C
C
      CONVERGENCE FUNCTION, RECIP. SPACE
C
      DO 33 LP=1,NL2P
      L=LP-1
      NG=NGLN(LP)
   33 FBLN(I,LP)=BETA**(L-2)/(1.D00+BETA*BETA)**(NG+1)
   35 DG(I)=D
C
C
      EVALUATION OF LATTICE SUMS FOR QPP .EQ. 0
C
      DO 44 KLM=1,KLM0
      SUMR(KLM)=DCMPLX(0.D00,0.D00)
   44 SUMG(KLM)=DCMPLX(0.D00,0.D00)
C
C     REAL SPACE
C
      DO 49 I=2,NRO
      CALL YLMRK(USX(I),USY(I),USZ(I),NL2)
      KLM=0
      DO 26 LP=1,NL2P
      L=LP-1
      NM=L*2+1
      DO 26 M=1,NM
      KLM=KLM+1
   26 SUMR(KLM)=SUMR(KLM)+GLNS(I,LP)*YLM(KLM)*CXPRDK(I)
      IF(NL2.LE.5) GO TO 49
C
C     IF L IS LARGER THAN 4, REAL SPACE SUMMATIONS ONLY
C
      DO 50 LP=6,NL2
      NM=LP*2-1
      SORO=(SWS/DR(I))**LP
      DO 50 M=1,NM
      KLM=KLM+1
   50 SUMR(KLM)=SUMR(KLM)+SORO*YLM(KLM)*CXPRDK(I)
   49 CONTINUE
C.
C     RECIP. SPACE
C
      DO 34 I=1,NUMVG
      IF(DG(I).GT.GMAX) GO TO 34
      CALL YLMRK(UKX(I),UKY(I),UKZ(I),NL2P)
      KLM=0
      DO 28 LP=1,NL2P
      L=LP-1
      NM=L*2+1
      DO 28 M=1,NM
      KLM=KLM+1
   28 SUMG(KLM)=SUMG(KLM)+FBLN(I,LP)*YLM(KLM)
   34 CONTINUE
C
C     SET UP STRUCTURE CONSTANT MATRIX FROM LATTICE SUMS
C
      CIL=CIM
      KLM=0
      DO 51 LP=1,NL2
      L=LP-1
      NM=L*2+1
      CIL=CIL*CCIM
      IF(LP.GT.5) GO TO 52
      AR=SWSL**LP
      AG=FPL30M*AR/DAC(LP)
      DO 45 M=1,NM
      KLM=KLM+1
```

139

```
   45 SR(KLM)=AR*SUMR(KLM)*CIL+AG*SUMG(KLM)
      GO TO 51
   52 DO 53 M=1,NM
      KLM=KLM+1
   53 SR(KLM)=SUMR(KLM)*CIL
   51 CONTINUE
      SR(1)=SR(1)-S00*SWSL
C
C     INSERT INTO LOWER TRIANGLE
C
      J1=-IDIM
      J2=-16
      DO 27 JLM=1,NLM
      J2=J2+16
      J1=J1+IDIM
      L=LL(JLM)
      M=MM(JLM)
      DO 27 ILM=JLM,NLM
      LIN=J1+ILM
      LINC=J2+ILM
      LPP=L+LL(ILM)
      MPP=MM(ILM)-M
      KLM=LPP*LPP+LPP+MPP+1
      SRC=SR(KLM)*C(LINC)
      SRKLP(LIN)=REAL(SRC)
      SIKLP(LIN)=AIMAG(SRC)
   27 CONTINUE
C
      IF(ICNVRG.EQ.0) GO TO 58
C
C     PRINT INDIVIDUAL CONTRIBUTIONS TO THE LATTICE SUMS. USED
C     FOR TEST OF THE CHOICE OF SPLITTING PARAMETERS <LAMDA>
C     AND <NGLN>
C
      WRITE(6,105)
      KLM=0
      DO 59 LP=1,NL2P
      NM=LP*2-1
      AR=SWSL**LP
      AG=FPL30M*AR/DAC(LP)
      DO 59 M=1,NM
      KLM=KLM+1
      SUMR(KLM)=SUMR(KLM)*AR
   59 SUMG(KLM)=SUMG(KLM)*AG
      WRITE(6,107)
      WRITE(6,103) (SUMR(K),K=1,KLM)
      WRITE(6,106)
      WRITE(6,103) (SUMG(K),K=1,KLM)
   58 IF(NOWRT.EQ.0) GO TO 30
C
C     PRINT STRUCTURE CONSTANT MATRIX
C
      WRITE(6,102)
      I1=-IDIM
      DO 42 ILM=1,NLM
      I1=I1+IDIM
      I2=I1+ILM
   42 WRITE(6,100) (SRKLP(LIN),LIN=ILM,I2,IDIM)
      WRITE(6,101)
      I1=-IDIM
      DO 43 ILM=1,NLM
      I1=I1+IDIM
      I2=I1+ILM
   43 WRITE(6,100) (SIKLP(LIN),LIN=ILM,I2,IDIM)
      WRITE(6,101)
C
   30 CONTINUE
      IF(NQ.EQ.1) RETURN
```

140

```
C
C      REPEAT FIRST DIAGONAL BLOCK
C
       JQQ=0
       DO 24 JQ=2,NQ
       JQQ=JQQ+(1+IDIM)*NLM
       J1=-IDIM
       DO 24 JLM=1,NLM
       J1=J1+IDIM
       DO 24 ILM=JLM,NLM
       LIN=J1+ILM
       LINQ=JQQ+LIN
       SRKLP(LINQ)=SRKLP(LIN)
   24 SIKLP(LINQ)=SIKLP(LIN)
       NROP=NRO+1
C
       DO 29 I=NROP,NUMVR
       PRD=PKX*ASX(I)+PKY*ASY(I)+PKZ*ASZ(I)
   29 CXPRDK(I)=CDEXP(CIM*PRD)
C
C      QQP OFF DIAGONAL BLOCKS
C
       NQM=NQ-1
       JQQ=-NLM*IDIM
       DO 20 JQ=1,NQM
       JQP=JQ+1
       JQQ=JQQ+NLM*IDIM
       QPX=QX(JQ)
       QPY=QY(JQ)
       QPZ=QZ(JQ)
       DO 20 IQ=JQP,NQ
       IQQ=(IQ-1)*NLM+JQQ
       QPPX=QX(IQ)-QPX
       QPPY=QY(IQ)-QPY
       QPPZ=QZ(IQ)-QPZ
       DO 36 I=1,NUMVR
       X=ASX(I)-QPPX
       Y=ASY(I)-QPPY
       Z=ASZ(I)-QPPZ
       D=DSQRT(X*X+Y*Y+Z*Z)
       DRQ(I)=D
       IF(D.GT.RMAX) GO TO 36
       ALFA=ALAMDA*D
       USX(I)=X/D
       USY(I)=Y/D
       USZ(I)=Z/D
C
C      CONVERGENCE FUNCTION, REAL SPACE
C
       DO 37 LP=1,NL2P
       L=LP-1
       NG=NGLN(LP)
   37 GLNS(I,LP)=GLN(ALFA,L,NG)
   36 CONTINUE
       DO 38 I=1,NUMVG
       D=DG(I)
       IF(D.GT.GMAX) GO TO 38
       BDOTQ(I)=(UKX(I)*QPPX+UKY(I)*QPPY+UKZ(I)*QPPZ)*D
       BETA=D/ALAMDA
C
C      CONVERGENCE FUNCTION, RECIP. SPACE
C
       DO 39 LP=1,NL2P
       L=LP-1
       NG=NGLN(LP)
   39 FBLN(I,LP)=BETA**(L-2)/(1.D00+BETA*BETA)**(NG+1)
   38 CONTINUE
```

```
C
C       EVALUATION OF LATTICE SUMS FOR QPP .NE. 0
C
        DO 46 KLM=1,KLMO
        SUMR(KLM)=DCMPLX(0.D00,0.D00)
     46 SUMG(KLM)=DCMPLX(0.D00,0.D00)
C
C       REAL SPACE
C
        DO 40 I=1,NUMVR
        IF(DRQ(I).GT.RMAX) GO TO 40
        CALL YLMRK(USX(I),USY(I),USZ(I),NL2)
        KLM=0
        DO 21 LP=1,NL2P
        L=LP-1
        NM=L*2+1
        DO 21 M=1,NM
        KLM=KLM+1
     21 SUMR(KLM)=SUMR(KLM)+GLNS(I,LP)*YLM(KLM)*CXPRDK(I)
        IF(NL2.LE.5) GO TO 40
C
C       IF L IS LARGER THAN 4, REAL SPACE SUMMATIONS ONLY
C
        DO 54 LP=6,NL2
        SORO=(SWS/DRQ(I))**LP
        NM=LP*2-1
        DO 54 M=1,NM
        KLM=KLM+1
     54 SUMR(KLM)=SUMR(KLM)+SORO*YLM(KLM)*CXPRDK(I)
     40 CONTINUE
C
C       RECIP. SPACE
C
        DO 41 I=1,NUMVG
        IF(DG(I).GT.GMAX) GO TO 41
        CALL YLMRK(UKX(I),UKY(I),UKZ(I),NL2P)
        KLM=0
        DO 47 LP=1,NL2P
        L=LP-1
        NM=L*2+1
        DO 47 M=1,NM
        KLM=KLM+1
     47 SUMG(KLM)=SUMG(KLM)+FBLN(I,LP)*YLM(KLM)*CDEXP(CIM*BDOTQ(I))
     41 CONTINUE
C
C       SET UP STRUCTURE CONSTANT MATRIX FROM THE LATTICE SUMS
C
        CIL=CIM
        KLM=0
        DO 56 LP=1,NL2
        L=LP-1
        NM=L*2+1
        CIL=CIL*CCIM
        IF(LP.GT.5) GO TO 55
        AR=SWSL**LP
        AG=FPL3OM*AR/DAC(LP)
        DO 48 M=1,NM
        KLM=KLM+1
     48 SR(KLM)=AR*SUMR(KLM)*CIL+AG*SUMG(KLM)
        GO TO 56
     55 DO 57 M=1,NM
        KLM=KLM+1
     57 SR(KLM)=SUMR(KLM)*CIL
     56 CONTINUE
C
C       INSERT INTO LOWER TRIANGLE
```

142

```
C
      J1=-IDIM
      J2=-16
      DO 23 JLM=1,NLM
      J1=J1+IDIM
      J2=J2+16
      L=LL(JLM)
      M=MM(JLM)
      DO 23 ILM=1,NLM
      LIN=J1+ILM
      LINC=J2+ILM
      LINQ=IQQ+LIN
      LPP=L+LL(ILM)
      MPP=MM(ILM)-M
      KLM=LPP*LPP+LPP+MPP+1
      SRC=SR(KLM)*C(LINC)
      SRKLP(LINQ)=REAL(SRC)
   23 SIKLP(LINQ)=AIMAG(SRC)
C
      IF(ICNVRG.EQ.0) GO TO 60
C
C     TEST OF SPLITTING PARAMETERS
C
      KLM=0
      DO 61 LP=1,NL2P
      NM=LP*2-1
      AR=SWSL**LP
      AG=FPL3OM*AR/DAC(LP)
      DO 61 M=1,NM
      KLM=KLM+1
      SUMR(KLM)=SUMR(KLM)*AR
   61 SUMG(KLM)=SUMG(KLM)*AG
      WRITE(6,107)
      WRITE(6,103) (SUMR(K),K=1,KLM)
      WRITE(6,106)
      WRITE(6,103) (SUMG(K),K=1,KLM)
   60 IF(NOWRT.EQ.0) GO TO 20
C
C     PRINT STRUCTURE CONSTANT MATRIX
C
      WRITE(6,104) IQ,JQ
      DO 31 ILM=1,NLM
      J1=IQQ+ILM
      J2=J1+(NLM-1)*IDIM
   31 WRITE(6,100) (SRKLP(LINQ),LINQ=J1,J2,IDIM)
      WRITE(6,101)
      DO 32 ILM=1,NLM
      J1=IQQ+ILM
      J2=J1+(NLM-1)*IDIM
   32 WRITE(6,100) (SIKLP(LINQ),LINQ=J1,J2,IDIM)
   20 CONTINUE
      RETURN
      END
      SUBROUTINE CROSS(AX,AY,AZ,BX,BY,BZ,CX,CY,CZ)
C     ************************************************************
C     *                                                          *
C     *    CROSS PRODUCT (CX,CY,CZ)=(AX,AY,AZ)*(BX,BY,BZ)        *
C     *                                                          *
C     ************************************************************
      IMPLICIT REAL*8 (A-H,O-Z)
C
      CX=AY*BZ-BY*AZ
      CY=BX*AZ-AX*BZ
      CZ=AX*BY-BX*AY
      RETURN
      END
      SUBROUTINE FACTOR(N)
```

143

```
C     *******************************************************************
C     *                                                                 *
C     *     CALCULATION OF FACTORIALS                                   *
C     *                                                                 *
C     *     FAC(N)=1*2*3*...*(N-1)                                      *
C     *                                                                 *
C     *     DAC(N)=1*1*3*5*...*(2*(N-1)-1)                              *
C     *                                                                 *
C     *******************************************************************
      IMPLICIT REAL*8 (A-H,O-Z)
      COMPLEX*16 CIM,CCIM
      COMMON/B1/CIM,CCIM,C(16,16),FAC(13),DAC(13),NGLN(10),
     1ALAMDA,TWOPI,FPL3OM,SWS,VOL,SOO,SWSL,FPILAM,RMAX,GMAX,ICNVRG,
     2NUMVR,NUMVG,NQ,NRO,NL2,NL2P,KLMO,NL,NLM,LMQ,NOWRT,IDIM
C
      NP=N+1
      FAC(1)=1.DO
      FAC(2)=1.DO
      DO 10 I=3,NP
   10 FAC(I)=FAC(I-1)*(I-1)
      DAC(1)=1.DO
      DAC(2)=1.DO
      DO 11 I=3,NP
   11 DAC(I)=DAC(I-1)*(2*I-3)
      RETURN
      END
      FUNCTION GLN(ALFA,LN,NN)
C     *******************************************************************
C     *                                                                 *
C     *     CALCULATION OF CONVERGENCE FUNCTIONS                        *
C     *                                                                 *
C     *******************************************************************
      IMPLICIT REAL*8 (A-H,O-Z)
      COMPLEX*16 CIM,CCIM
      COMMON/B1/CIM,CCIM,C(16,16),FAC(13),DAC(13),NGLN(10),
     1ALAMDA,TWOPI,FPL3OM,SWS,VOL,SOO,SWSL,FPILAM,RMAX,GMAX,ICNVRG,
     2NUMVR,NUMVG,NQ,NRO,NL2,NL2P,KLMO,NL,NLM,LMQ,NOWRT,IDIM
C
      ALFASF=ALFA*ALFA/4.D00
      EXA=DEXP(-ALFA)
      H00=EXA/ALFA
      H01=0.5D00*EXA
      H1=H00
      H2=H01
C
C     FOR L = 0
C
      GLN=H1+H2
      DO 20 NP=2,NN
      N=NP-1
      H3=(ALFASF/N*H1-(0.5D00-N)*H2)/NP
      GLN=GLN+H3
      H1=H2
   20 H2=H3
      IF(LN.EQ.0) RETURN
C
C     FOR L = 1
C
      H11=0.5D00*EXA
      H10=(1.D00+ALFA)/ALFA/ALFA*EXA
      H1=H10
      H2=H11
      GLN=H1+H2
      DO 21 NP=2,NN
      N=NP-1
      H3=(ALFASF/N*H1-(1.5D00-N)*H2)/NP
```

144

```
      GLN=GLN+H3
      H1=H2
   21 H2=H3
      IF(LN.EQ.1) RETURN
C
C     FOR L .GT. 1
C
      L=1
   23 L=L+1
      LP=L+1
      LM=L-1
      H21=H01+(2*LM-1)*H11/ALFA
      H20=H00+(2*LM+1)*H10/ALFA
      H1=H20
      H2=H21
      GLN=H1+H2
      DO 22 NP=2,NN
      N=NP-1
      XLN=L-N
      XLN=XLN+0.5D00
      H3=(ALFASF/N*H1-XLN*H2)/NP
      GLN=GLN+H3
      H1=H2
   22 H2=H3
      GLN=GLN/DAC(LP)
      IF(L.EQ.LN) RETURN
      H00=H10
      H10=H20
      H01=H11
      H11=H21
      GO TO 23
      END
      SUBROUTINE LIMA0(NN,FALFB)
C     ************************************************************************
C     *                                                                      *
C     *   CALCULATES ALPHA**-1-GLN(ALPHA) FOR L=0 AND ALPHA=0                *
C     *                                                                      *
C     ************************************************************************
      IMPLICIT REAL*8 (A-H,O-Z)
C
      NUM=0
      ALFA=1.E-03
      FALFA=10.
   10 ALFAS=ALFA*ALFA/4.D00
      EXA=DEXP(-ALFA)
      H1=EXA/ALFA
      H2=0.5D00*EXA
      GON=H1+H2
      DO 11 NP=2,NN
      N=NP-1
      H3=((ALFAS/N)*H1-(0.5-N)*H2)/NP
      GON=GON+H3
      H1=H2
   11 H2=H3
      FALFB=(1.-ALFA*GON)/ALFA
      DIF=FALFA-FALFB
      NUM=NUM+1
      IF(DIF.LT.1.E-05) RETURN
      IF(NUM.GT.10) WRITE(6,1)
    1 FORMAT(1H1,10X,'**TOO MANY ITERATIONS IN LIMA0**',/)
      FALFA=FALFB
      ALFA=ALFA-0.1E-04
      GO TO 10
      END
      SUBROUTINE YLMRK(XX,YY,ZZ,NLMAX)
```

```
C     ***************************************************************
C     *                                                             *
C     *    CALCULATES SPHERICAL HARMONICS                           *
C     *    YLM=CST*SQRT(4PI)*COMPLEX CONJUGATE(YLM)                  *
C     *    CST IS CALCULATED IN <GAUNT>                             *
C     *                                                             *
C     ***************************************************************
      IMPLICIT REAL*8(A-H,O-Z)
      COMPLEX*16 YLM,DCMPLX,CIM,CCIM
      COMMON/B1/CIM,CCIM,CC(16,16),FAC(13),DAC(13),NGLN(10),
     1ALAMDA,TWOPI,FPL3OM,SWS,VOL,SOO,SWSL,FPILAM,RMAX,GMAX,ICNVRG,
     2NUMVR,NUMVG,NQ,NRO,NL2,NL2P,KLMO,NL,NLM,LMQ,NOWRT,IDIM
      COMMON/YLMD/YLM(49)
      COMMON/LDATA/LL(49),MM(49)
      DIMENSION PLM(28),COSMP(7),SINMP(7)
C
C     CALCULATE LEGENDRE POLYNOMIALS BY RECURSION
C
      NLP=NLMAX
      P=DSQRT(XX*XX+YY*YY)
      X=ZZ
      Y=P
      XA=DABS(X)
      IF(XA.GT.1.D-06) GO TO 10
C
C     ABS(X) = 0
C
      DO 11 LP=1,NLP
      L=LP-1
      LA=L*(L+1)/2+1
      TA=2.D0**L
      DO 11 MP=1,LP
      M=MP-1
      K=L+M
      IF(K-2*(K/2).EQ.0) GO TO 12
      J=LA+M
      PLM(J)=0.D0
      GO TO 11
   12 IA=K+1
      IB=K/2+1
      JC=(L-M)/2
      IC=JC+1
      J=LA+M
      PLM(J)=(((-1)**JC)*FAC(IA))/(TA*FAC(IB)*FAC(IC))
   11 CONTINUE
      GO TO 32
   10 IF(XA.LT.0.999999D0) GO TO 20
C
C     ABS(X) = 1
C
      PLM(1)=1.D0
      PLM(2)=X
      DO 13 LP=3,NLP
      L=LP-1
      J=L*(L+1)/2+1
      L2=2*L-1
      K=J-L
      M=J-L2
   13 PLM(J)=(L2*X*PLM(K)-(L-1)*PLM(M))/L
      DO 14 LP=2,NLP
      L=LP-1
      LA=L*(L+1)/2
      DO 14 MP=2,LP
      J=LA+MP
   14 PLM(J)=0.D0
      GO TO 32
```

```
C
C       0 < ABS(X) < 1
C
   20 PLM(1)=1.D0
      PLM(2)=X
      PLM(3)=Y
      PLM(5)=3.D0*Y*X
      DO 21 LP=3,NLP
      L=LP-1
      J=L*(L+1)/2+1
      L2=2*L-1
      K=J-L
      M=J-L2
   21 PLM(J)=(L2*X*PLM(K)-(L-1)*PLM(M))/L
      DO 22 LP=4,NLP
      L=LP-1
      J=L*(L+1)/2+2
      L2=2*L-1
      K=J-L
      M=J-L2
   22 PLM(J)=(L2*X*PLM(K)-L*PLM(M))/(L-1)
      DO 23 LP=3,NLP
      L=LP-1
      NM=L*2+1
      LA=L*(L+1)/2
      DO 23 MP=3,LP
      M=MP-1
      J=LA+MP
      K=J-1
      N=K-1
      A=(M-1)*2.D0*X/Y
      B=(L+M-1)*(L-M+2)
   23 PLM(J)=A*PLM(K)-B*PLM(N)
   32 CONTINUE
C
C       FORM SPHERICAL HARMONICS
C
      IF(P.GT.1.D-06) GO TO 34
      COSPHI=1.D0
      SINPHI=0.D0
      GO TO 35
   34 COSPHI=XX/P
      SINPHI=YY/P
   35 COSMP(1)=1.D0
      SINMP(1)=0.D0
      DO 33 MP=2,NLP
      COSMP(MP)=COSMP(MP-1)*COSPHI-SINMP(MP-1)*SINPHI
   33 SINMP(MP)=SINMP(MP-1)*COSPHI+COSMP(MP-1)*SINPHI
      KLM=0
      DO 36 LP=1,NLP
      L=LP-1
      NM=L*2+1
      DO 36 MP=1,NM
      KLM=KLM+1
      M=MM(KLM)
      MA=IABS(M)+1
      MB=MA-1
      LB=L*(L+1)/2+MA
      IF(M.LE.0) GO TO 37
      YLM(KLM)=PLM(LB)*DCMPLX(COSMP(MA),-SINMP(MA))
      GO TO 36
   37 YLM(KLM)=PLM(LB)*DCMPLX(COSMP(MA),SINMP(MA))
   36 CONTINUE
      RETURN
      END
      SUBROUTINE GAUNT
```

```
C     ******************************************************************
C     *                                                                *
C     *      CALCULATION OF GAUNT COEFFICIENTS                         *
C     *                                                                *
C     ******************************************************************
      IMPLICIT REAL*8(A-H,O-Z)
      COMPLEX*16 CIM,CCIM
      COMMON/B1/CIM,CCIM,CC(16,16),FAC(13),DAC(13),NGLN(10),
     1ALAMDA,TWOPI,FPL3OM,SWS,VOL,SOO,SWSL,FPILAM,RMAX,GMAX,ICNVRG,
     2NUMVR,NUMVG,NQ,NRO,NL2,NL2P,KLMO,NL,NLM,LMQ,NOWRT,IDIM
      COMMON/LDATA/LL(49),MM(49)
      DIMENSION CYLM(28)
C
C     CALCULATE CONSTANTS FOR YLM
C
      DO 20 LP=1,NL2
      L=LP-1
      LA=L*(L+1)/2
      NM=L*2+1
      DO 20 MP=1,LP
      M=MP-1
      L1=L-M+1
      L2=L+MP
      KLM=LA+MP
      ARG=NM*FAC(L1)/FAC(L2)
   20 CYLM(KLM)=DSQRT(ARG)*(-1)**M
C
C     CALCULATE CYLM*GAUNT COEFFICIENTS
C
      DO 21 ILM=1,NLM
      LP=LL(ILM)
      MP=MM(ILM)
      DO 21 JLM=1,NLM
      L=LL(JLM)
      M=MM(JLM)
      LPP=LP+L
      MPP=MP-M
      MPPA=IABS(MPP)
      KLM=LPP*(LPP+1)/2+MPPA+1
      ISIGN=1
      IF(MPP.LT.0) ISIGN=(-1)**MPPA
      LPM=LPP+MPP+1
      LIM=LPP-MPP+1
      LM1P=LP+MP+1
      LM1M=LP-MP+1
      LM2P=L+M+1
      LM2M=L-M+1
      SETN=FAC(LPM)*FAC(LIM)
      SETD=FAC(LM1P)*FAC(LM1M)*FAC(LM2P)*FAC(LM2M)
      SET3=(2*LP+1)*(2*L+1)
      SET4=(2*LPP+1)
      SETF=SET3*SETN/SET4/SETD
   21 CC(ILM,JLM)=((-1)**(M+1))*ISIGN*CYLM(KLM)*DSQRT(SETF)*2.0D0
      RETURN
      END
      SUBROUTINE KTRNSF(LAT,KKX,KKY,KKZ,PKX,PKY,PKZ)
C     ******************************************************************
C     *                                                                *
C     *      TRANSFORMATION FROM SYMMETRY TO RECTANGULAR COORDINATES   *
C     *                                                                *
C     ******************************************************************
      IMPLICIT REAL*8(A-H,O-Z)
      COMMON/MSH/DKX,DKY,DKZ,DHX,ROA,COA,KX(2500),KY(2500),KZ(2500),
     1W(2500)
C
```

```
          GO TO (21,21,21,21,21,21,21,22),LAT
       21 PKX=DKX*KKX+DHX*KKY
          PKY=DKY*KKY
          PKZ=DKZ*KKZ
          DK=PKX*PKX+PKY*PKY+PKZ*PKZ
          IF(DK.LT.1.D-08) GO TO 20
          RETURN
       22 QKX=DKX*KKX+DHX*KKY
          QKY=DKY*KKY
          QKZ=DKZ*KKZ-1.5D0/DSQRT(3.D0)
          DK=QKX*QKX+QKY*QKY+QKZ*QKZ
          IF(DK.LT.1.D-08) GO TO 20
          SQ2=DSQRT(2.D0)
          SQ3=DSQRT(3.D0)
          SQ6=DSQRT(6.D0)
          PKX=QKX/SQ2+QKY/SQ6+QKZ/SQ3
          PKY=-QKX/SQ2+QKY/SQ6+QKZ/SQ3
          PKZ=-2.D0*QKY/SQ6+QKZ/SQ3
          RETURN
C
C         AVOID (0,0,0)
C
       20 PKX=0.D0
          PKY=0.D0
          PKZ=0.01D0
          RETURN
          END
          SUBROUTINE MESH(LAT,NPX,NPY,NPZ,NPT)
C    ********************************************************************
C    *                                                                  *
C    *     CONSTRUCTION OF MESH IN K-SPACE                              *
C    *                                                                  *
C    *     LAT=1  SIMPLE CUBIC                                          *
C    *                                                                  *
C    *     LAT=2  FACE CENTRED CUBIC                                    *
C    *                                                                  *
C    *     LAT=3  BODY CENTRED CUBIC                                    *
C    *                                                                  *
C    *     LAT=4  HEXAGONAL CLOSE PACKED                                *
C    *                                                                  *
C    *     LAT=5  SIMPLE TETRAGONAL                                     *
C    *                                                                  *
C    *     LAT=6  BODY CENTRED TETRAGONAL                               *
C    *                                                                  *
C    *     LAT=7  ALPHA URANIUM                                         *
C    *                                                                  *
C    *     LAT=8  TRIGONAL ZONE FOR ANTIFERROMAGNETIC OXIDES            *
C    *                                                                  *
C    ********************************************************************
          IMPLICIT REAL*8(A-H,O-Z)
          INTEGER X,Y,Z
          COMMON/MSH/DKX,DKY,DKZ,DHX,BOA,COA,KX(2500),KY(2500),KZ(2500),
         1WW(2500)
        1 FORMAT(1H ,10X,'NUMBER OF POINTS ON MESH EXCEEDS DIMENSION OF ARRA
         1YS',2I5)
C
          NDIM=2500
          NPXM=NPX-1
          NPYM=NPY-1
          NPZM=NPZ-1
          GO TO (20,21,22,23,24,25,26,27),LAT
       20 NP=0
C
C         SIMPLE CUBIC IRREDUCIBLE ZONE DEFINED BY
C         0 .LE. KZ .LE. KX .LE. KY .LE. PI/A
C
```

```
      DKX=1.D0/NPYM
      DKY=DKX
      DKZ=DKX
      DHX=0.D0
      DO 30 J=1,NPY
      Y=J-1
      DO 30 I=1,J
      X=I-1
      DO 30 K=1,I
      Z=K-1
      NP=NP+1
      IF(NP.GT.NDIM) GO TO 999
      KX(NP)=X
      KY(NP)=Y
      KZ(NP)=Z
      W=48.D0
      IF(X.EQ.Y.OR.X.EQ.Z.OR.Y.EQ.Z) W=W/2.D0
      IF(X.EQ.Y.AND.Y.EQ.Z) W=W/3.D0
      IF(Z.EQ.0) W=W/2.D0
      IF(X.EQ.0) W=W/2.D0
      IF(X.EQ.NPYM) W=W/2.D0
      IF(Y.EQ.NPYM) W=W/2.D0
      IF(X+Y+Z.EQ.3*NPYM) W=1.D0
      IF(X+Y+Z.EQ.0) W=1.D0
      WW(NP)=W
   30 CONTINUE
      NPT=NP
      RETURN
   21 NP=0
C
C     FCC IRREDUCIBLE ZONE DEFINED BY
C     0 .LE. KZ .LE. KX .LE. KY .LE. 2PI/A
C     KX + KY + KZ .LE. 3PI/A
C
      IF(NPYM.NE.4*(NPYM/4)) GO TO 998
      DKX=2.D0/NPYM
      DKY=DKX
      DKZ=DKX
      DHX=0.D0
      NPX=NPY
      NPH=NPY/2+1
      NPTH=(NPYM/2)*3
      DO 31 I=1,NPX
      Y=I-1
      M1=NPX-I+NPH
      NQY=MINO(I,M1)
      DO 31 J=1,NQY
      X=J-1
      M5=M1-J+1
      NPZ=MINO(J,M5)
      DO 31 K=1,NPZ
      Z=K-1
      NP=NP+1
      IF(NP.GT.NDIM) GO TO 999
      KX(NP)=X
      KY(NP)=Y
      KZ(NP)=Z
      W=48.D0
      IF(X.EQ.Y.OR.X.EQ.Z.OR.Y.EQ.Z) W=W/2.D0
      IF(X.EQ.Y.AND.Y.EQ.Z) W=W/3.D0
      IF(Z.EQ.0) W=W/2.D0
      IF(X.EQ.0) W=W/2.D0
      IF(Y.EQ.NPYM) W=W/2.D0
      IF(X+Y+Z.EQ.NPTH) W=W/2.D0
      IF(X+Y+Z.EQ.0) W=1.D0
      WW(NP)=W
```

```
   31 CONTINUE
      NPT=NP
      RETURN
   22 NP=0
C
C     BCC IRREDUCIBLE ZONE DEFINED BY
C     0 .LE. KZ .LE. KY .LE. KX .LE. 2PI/A
C     KX + KY .LE. 2PI/A
C
      IF(NPYM.NE.2*(NPYM/2)) GO TO 998
      DKX=2.D0/NPYM
      DKY=DKX
      DKZ=DKX
      DHX=0.D0
      NPH=NPY/2+1
      NPYH=NPYM/2
      DO 32 I=1,NPY
      X=I-1
      JM=MIN0(I,NPY-I+1)
      DO 32 J=1,JM
      Y=J-1
      DO 32 K=1,J
      Z=K-1
      NP=NP+1
      IF(NP.GT.NDIM) GO TO 999
      KX(NP)=X
      KY(NP)=Y
      KZ(NP)=Z
      W=48.D0
      IF(X.EQ.Y.OR.X.EQ.Z.OR.Y.EQ.Z) W=W/2.D0
      IF(X.EQ.Y.AND.Y.EQ.Z) W=W/3.D0
      IF(Z.EQ.0) W=W/2.D0
      IF(Y.EQ.0) W=W/2.D0
      IF(X+Y.EQ.NPYM) W=W/2.D0
      IF(Y.EQ.Z.AND.X+Y.EQ.NPYM) W=8.D0
      IF(Z.EQ.NPYH) W=2.D0
      IF(X.EQ.NPYM) W=1.D0
      IF(X+Y+Z.EQ.0) W=1.D0
      WW(NP)=W
   32 CONTINUE
      NPT=NP
      RETURN
   23 NP=0
C
C     HEXAGONAL CLOSE PACKED IRREDUCIBLE ZONE DEFINED BY
C     0 .LE. 2KY .LE. KX .LE. 4/3 PI/A
C     0 .LE. KZ .LE. A/C PI/A
C
      IF(NPYM.NE.2*(NPYM/2)) GO TO 998
      NPYH=NPYM/2
      NPH=NPY/2+1
      NPZM=NPZ-1
      AOC=1.D0/COA
      DKX=4.D0/3.D0/NPYM
      DKY=DSQRT(3.D0)/2.D0*DKX
      DKZ=AOC/NPZM
      DHX=-0.5D0*DKX
      DO 33 K=1,NPZ
      Z=K-1
      DO 33 J=1,NPH
      Y=J-1
      IM=2*Y+1
      DO 33 I=IM,NPY
      X=I-1
      NP=NP+1
      IF(NP.GT.NDIM) GO TO 999
      KX(NP)=X
```

```
      KY(NP)=Y
      KZ(NP)=Z
      W=24.D0
      IF(Z.EQ.0) W=W/2.D0
      IF(Y.EQ.0) W=W/2.D0
      IF(X.EQ.0) W=W/3.D0
      IF(X.EQ.2*Y) W=W/2.D0
      IF(Z.EQ.NPZM) W=W/2.D0
      IF(X.EQ.NPYM.AND.Y.EQ.0) W=W/3.D0
      IF(X.EQ.NPYM.AND.Y.NE.0) W=W/2.D0
      IF(X+Y+Z.EQ.0) W=1.D0
      IF(X+Y.EQ.0.AND.Z.EQ.NPZM) W=1.D0
      WW(NP)=W
   33 CONTINUE
      NPT=NP
      RETURN
   24 NP=0
C
C     SIMPLE TETRAGONAL IRREDUCIBLE ZONE DEFINED BY
C     0 .LE. KY .LE. KX .LE. PI/A
C     0 .LE. KZ .LE. A/C PI/A
C
      AOC=1.D0/COA
      DKX=1.D0/NPYM
      DKY=DKX
      DKZ=AOC/NPZM
      DHX=0.D0
      DO 34 I=1,NPY
      X=I-1
      DO 34 J=1,I
      Y=J-1
      DO 34 K=1,NPZ
      Z=K-1
      NP=NP+1
      IF(NP.GT.NDIM) GO TO 999
      KX(NP)=X
      KY(NP)=Y
      KZ(NP)=Z
      WW(NP)=0.
   34 CONTINUE
      NPT=NP
      RETURN
   25 NP=0
C
C     BODY CENTRED  TETRAGONAL IRREDUCIBLE ZONE DEFINED BY
C     0 .LE. KY .LE. KX .LE. PI/A
C     0 .LE. KZ .LE. 2*A/C PI/A
C
      AOC=1.D0/COA
      DKX=1.D0/NPYM
      DKY=DKX
      DKZ=2.D0*AOC/NPZM
      DHX=0.D0
      DO 35 I=1,NPY
      X=I-1
      DO 35 J=1,I
      Y=J-1
      DO 35 K=1,NPZ
      Z=K-1
      NP=NP+1
      IF(NP.GT.NDIM) GO TO 999
      KX(NP)=X
      KY(NP)=Y
      KZ(NP)=Z
      W=16.D0
      IF(X.EQ.0) W=W/2.D0
      IF(Y.EQ.0) W=W/2.D0
```

```
      IF(Z.EQ.0) W=W/2.D0
      IF(X.EQ.Y) W=W/2.D0
      IF(X.EQ.NPYM) W=W/2.D0
      IF(Y.EQ.NPYM) W=W/2.D0
      IF(Z.EQ.NPZM) W=W/2.D0
      WW(NP)=W
   35 CONTINUE
      NPT=NP
      RETURN
   26 NP=0
C
C     ALPHA URANIUM
C
      AOC=1.D0/COA
      DKX=1.D0/NPXM
      DKY=2.D0/BOA/NPYM
      DKZ=AOC/NPZM
      DHX=0.D0
      DO 36 I=1,NPX
      X=I-1
      DO 36 J=1,NPY
      Y=J-1
      DO 36 K=1,NPZ
      Z=K-1
      NP=NP+1
      IF(NP.GT.NDIM) GO TO 999
      KX(NP)=X
      KY(NP)=Y
      KZ(NP)=Z
      WW(NP)=0.D0
   36 CONTINUE
      NPT=NP
      RETURN
C
C     TRIGONAL ZONE FOR ANTIFERROMAGNETIC OXIDES
C
C     0. < KX < 4.*SQRT(2.)/3.
C     0. < KY < SQRT(2./3.)    (PI/A)
C     -1.5/SQRT(3.) < KZ < 1.5/SQRT(3.)
C
   27 NP=0
      IF(NPYM.NE.2*(NPYM/2)) GO TO 998
      NPYH=NPYM/2
      NPH=NPY/2+1
      NPZM=NPZ-1
      DKX=4.D0*DSQRT(2.D0)/3.D0/NPYM
      DKY=DSQRT(3.D0)/2.D0*DKX
      DKZ=3.D0/DSQRT(3.D0)/NPZM
      DHX=-0.5D0*DKX
      DO 37 K=1,NPZ
      Z=K-1
      DO 37 J=1,NPH
      Y=J-1
      IM=2*Y+1
      DO 37 I=IM,NPY
      X=I-1
      NP=NP+1
      IF(NP.GT.NDIM) GO TO 999
      KX(NP)=X
      KY(NP)=Y
      KZ(NP)=Z
      WW(NP)=0
   37 CONTINUE
      NPT=NP
      RETURN
  998 WRITE(6,2) NPY
    2 FORMAT(1H ,'*** WRONG NPY , NPY=',I4,'  ***')
      STOP
```

```
  999 WRITE(6,1) NP,NDIM
      STOP
      END
C
```

9.2.3 Execution of STR

Let us assume that the programme STR has been successfully compiled, and that
the user wants to calculate and store the canonical structure constants for
the bcc structure. Let us further assume that the data files needed have been
created according to the attributes given in the listing, Sect.9.2.2. The
user is now faced with the problem of generating the input data necessary to
make the programme run. To help him choose the correct value of the various
variables, Table 9.2 lists input data for four different cases explained in
detail below.

   The first line of the data is a text meant to describe the structure constants being calculated, and which will be transmitted to and printed by all
later programmes. The second line contains data for five variables. Here RMAX
and GMAX are cut-off radii and determine the number of real and reciprocal
lattice vectors included in the lattice sums (9.3). It should be realised that
these lattice sums are q and k dependent, and that RMAX and GMAX are radii
of spheres centred at q and k. The subroutine VECGEN therefore calculates
two (larger) radii RA and GA, which are used to select out all vectors that

Table 9.2. Input for STR

```
STRUCTURE CONSTANTS FOR BCC LATTICE   STRUCTURE CONSTANTS FOR CSCL LATTICE
2.5       17.8        1     1     0    2.25      14.1         1     0     0
    3    1     3445555555    7.            3    2    3445555555        5.
    3    1.0              1.0              1    1.0              1.0
0.5            0.5            -0.5     1.0            0.0            0.0
-0.5           0.5            0.5      0.0            1.0            0.0
0.5           -0.5            0.5      0.0            0.0            1.0
0.            0.             0.        0.            0.             0.
    0    0    17    0                  0.5            0.5            0.5
                                            0    0    11    0


STRUCTURE CONSTANTS FOR FCC LATTICE   STRUCTURE CONSTANTS FOR AUCU3 LATTICE
1.9       26.         0     0     0    2.5       16.4         1     0     0
    4    1     3445555555    7.            3    4    3445555555        5.
    2    1.0              1.0              1    1.0              1.0
0.5            0.5            0.0      1.0            0.             0.
0.0            0.5            0.5      0.             1.             0.
0.5            0.0            0.5      0.             0.             1.
0.            0.             0.        0.            0.             0.
    1    2    13    0                  0.             0.5            0.5
1.5            1.5            0.0   K  0.5            0.             0.5
1.0            2.0            0.0   W  0.5            0.5            0.
                                            0    0    11    0
```

154

may possibly enter into the sums. It should also be mentioned that while the procedure for calculating RA is general, this is not the case for GA. Thus, if the programme is changed to include new Brillouin zones, the procedure for obtaining GA must be reconsidered.

The values of RMAX and GMAX are compromises between good convergence and short running time: the larger RMAX and GMAX, the better the convergence and the slower the programme. Suitable values may be established by examining the structure constants generated in a few test runs with the print option NOWRT = 1. In these test runs and in the first production run NOVCGN must be set to 1. In subsequent runs NOVCGN may be 0, and the programme will then read lattice vectors from the file VEC/XXX. ICNVRG = 1 is used to determine the best value of the splitting parameter λ, as explained below. Normally ICNVRG = 0.

The third line contains data for four variables: NL = 3, and NQ = 1 means that ℓ_{max} = 2, i.e. we include s, p, and d electrons, and that the crystal structure has one atom per primitive cell. The choice of NGLN's should in principle be determined by convergence tests, but in practice the values given in Table 9.2 have proven to be adequate. The best value of ALAMDA, λ, may be determined in test runs with ICNVRG = 1. In this mode the programme will print the individual contributions to the lattice sums, and good convergence is usually obtained when ALAMDA is such that the real- and reciprocal-space sums do not differ by more than a few orders of magnitude.

The fourth line contains input for three variables. LAT selects the Brillouin zone to be used in the calculation according to the table in subroutine MESH. BOA and COA should be set equal to the actual lattice parameters b/a and c/a.

The next three lines contain the translational vectors spanning the unit cell in real space. Since computer time increases as NQ to some power, and since the primitive cell contains the smallest number of atoms per cell, the natural choice of translational vectors is the primitive vectors. In special case one may, however, make other choices as long as LAT is selected accordingly. The next NQ lines contain the basis vectors giving the positions of the individual atoms in the unit cell.

The next line contains data for four variables. With MODE = 0 the programme will generate structure constants distributed on a grid in k space selected by LAT. MODE = 1 is typically used to test convergence, and where one wants structure constants at symmetry points and symmetry lines only. NPX, NPY, and NPZ are used to establish the grid in k space, and designate the number of points along the x, y, and z axes (hexagonal coordinates when LAT = 4). For historical reasons NPY is the important parameter, and (NPX,

NPY, NPZ) = (0,17,0) leads to 285 k points in the irreducible zone of the bcc structure.

An edited output corresponding to the bcc data in Table 9.2 is given in Sect.9.2.4. In a production run one would normally suppress the printing of the real and reciprocal space vectors and of the structure constants, i.e. NOVCGN = 0 (if one has a suitable set of vectors in VEC/XXX) and NOWRT = 0.

If MODE = 1 NPX lines should follow with coordinates specifying k vectors at symmetry points and symmetry lines or at general points in the zone. Note that the centre of the zone Γ, because of the diverging s-structure constants, should be specified as $(0.,0.,0.01)\pi/a$. The data set for the fcc structure in Table 9.2 is an example of the MODE = 1 mode of operation. In this case NOVCGN = 0, assuming that STR has been executed once with the appropriate choice of RMAX and GMAX, and NL = 4 showing that we include f electrons.

The data sets for the CsCl and $AuCu_3$ lattices are examples of cases with more than one basis vector. In addition, NL = 3 and therefore only s, p, and d states will be included in the subsequent band calculations. The choice (NPX, NPY, NPZ) = (0, 11, 0) leads to a grid of 286 k points in the irreducible zone of the simple cubic (LAT = 1) Brillouin zone.

9.2.4 Sample Output from STR

STRUCTURE CONSTANTS FOR BCC LATTICE

```
RMAX    2.50000 GMAX   17.80000
NL    3 NQ    1 NLM    9 LMQ    9
LAMDA    7.00000 NGLN 3445555555
NOVCGN    1 NOWRT    1 ICNVRG    0
```

PRIMITIVE VECTORS IN UNIT CELL OF   BCC

BOA   1.0000000000   COA   1.0000000000

```
(    0.50000,    0.50000,   -0.50000 )
(   -0.50000,    0.50000,    0.50000 )
(    0.50000,   -0.50000,    0.50000 )
```

BASIS VECTORS

```
(    0.00000,    0.00000,    0.00000 )
```

PRIMITIVE VECTORS OF RECIPROCAL SPACE

```
(    6.28319,    6.28319,    0.00000)

(    0.00000,    6.28319,    6.28319)

(    6.28319,    0.00000,    6.28319)
```
156

```
R1     0.0000 RA     2.5000
G1    10.8828 GA    24.0832

NUMR   9 NUMG    9


RESULT FROM VECGEN FOR REAL SPACE VECTORS

NO     SX         SY         SZ         D

 1  0.000000   0.000000   0.000000   0.000000


SHELL NUMBER    1 WITH    1 POINTS

   2 -0.500000 -0.500000 -0.500000   0.866025
   3 -0.500000 -0.500000  0.500000   0.866025
   4  0.500000 -0.500000 -0.500000   0.866025
   5 -0.500000  0.500000 -0.500000   0.866025
   6  0.500000 -0.500000  0.500000   0.866025
   7 -0.500000  0.500000  0.500000   0.866025
   8  0.500000  0.500000 -0.500000   0.866025
   9  0.500000  0.500000  0.500000   0.866025


   SHELL NUMBER    2 WITH    8 POINTS


  10  0.000000 -1.000000  0.000000   1.000000
  11 -1.000000  0.000000  0.000000   1.000000
  12  0.000000  0.000000 -1.000000   1.000000

       <PRINTOUT SHORTENED>

RESULT FROM VECGEN FOR RECIPROCAL SPACE VECTORS

   NO     KX         KY         KZ         D

   1  0.000000   0.000000   0.000000   0.000000


   SHELL NUMBER    1 WITH    1 POINTS


   2 -6.283185 -6.283185  0.000000   8.885766
   3  0.000000 -6.283185  6.283185   8.885766
   4 -6.283185  0.000000  6.283185   8.885766
   5  0.000000 -6.283185 -6.283185   8.885766
   6  6.283185 -6.283185  0.000000   8.885766
   7 -6.283185  0.000000 -6.283185   8.885766
   8  6.283185  0.000000  6.283185   8.885766
   9 -6.283185  6.283185  0.000000   8.885766
  10  0.000000  6.283185  6.283185   8.885766
  11  6.283185  0.000000 -6.283185   8.885766
  12  0.000000  6.283185 -6.283185   8.885766
  13  6.283185  6.283185  0.000000   8.885766


   SHELL NUMBER    2 WITH   12 POINTS


  14  0.000000-12.566371  0.000000  12.566371
  15-12.566371  0.000000  0.000000  12.566371
  16  0.000000  0.000000 12.566371  12.566371
  17  0.000000  0.000000-12.566371  12.566371
  18 12.566371  0.000000  0.000000  12.566371
  19  0.000000 12.566371  0.000000  12.566371
```

```
SHELL NUMBER    3 WITH    6 POINTS


    20 -6.283185-12.566371  6.283185 15.390598
    21-12.566371 -6.283185  6.283185 15.390598

        <PRINTOUT SHORTENED>

    VOL      0.5000000  SWS      0.4923725  SOO      0.3125000


    MODE,NPX,NPY,NPZ    0    0   17    0

        <OUTPUT FROM POINT 1 AND 2 REMOVED>

    POINT NUMBER    3    PKX,PKY,PKZ    0.1250    0.1250    0.000

        QQ DIAGONAL BLOCK

-7.66703E+01
-1.87199E+01 -2.95949E+00
 4.24823E-14 -2.28786E-13  5.91899E+00
 1.87199E+01  3.71938E-12  2.28786E-13 -2.95949E+00
 6.96054E-14 -1.99035E-01 -4.84019E-16 -2.70105E-01  8.53155E-01
-1.07012E-13  9.69049E-14  5.62957E-01  6.84506E-16 -6.61847E-14 -3.412
 2.20588E+00  4.87535E-01 -1.67844E-13 -4.87535E-01 -2.02797E-13  1.621
 9.84539E-14  6.84506E-16 -5.62957E-01  9.69049E-14  2.41160E-14  3.311
-1.13177E-12  2.70105E-01 -4.84019E-16  1.99035E-01  4.56000E+00 -2.411



 7.17350E-14
-1.87199E+01 -4.50861E-13
-1.14215E-15  2.00914E-13  9.01722E-13
-1.87199E+01 -8.94808E+00 -2.00914E-13 -4.50861E-13
-2.72281E+00 -1.99035E-01  9.58357E-14  2.70105E-01  5.35883E-13
-7.79012E-14  0.           5.62957E-01 -1.35532E-13 -1.84518E-14 -2.143
 3.36052E-13 -4.87535E-01  0.          -4.87535E-01 -2.50353E-01  4.519
-8.64595E-14  1.36353E-13  5.62957E-01  0.           9.43786E-14  4.088
 2.72281E+00  2.70105E-01 -9.64163E-14 -1.99035E-01 -1.15236E-12 -9.437
```

9.3 Correction Structure Constant Programme COR

The COR is constructed to calculate those extra structure constants which
may be used to correct the ASA for approximate treatment of the region
between the sphere and the atomic polyhedron, and for the neglect of higher
ℓ components as described in Sect.6.9. The programme produces and stores on
disk or tape a set of correction-term matrices distributed on a suitable
grid in the irreducible wedge of the Brillouin zone. Whenever requested the
correction matrices may be retrieved by LMTO and used together with the
canonical structure constants to set up the corrected LMTO matrices.

Calculation of the correction structure constants (6.58,8.26) involves
reciprocal lattice sumations of spherical Bessel functions which are performed in a straightforward manner. In a typical application COR is executed

once for a given crystal structure, in analogy with STR, using input parameters consistent with the corresponding execution of STR. There is, however, one difference between STR and COR which stems from the Bessel functions in (6.59), and which is reflected as a difference in mode of operation.
The canonical structure constants (8.22) scale with $(S_t/S)^{\ell+1/2}$, and this
scaling can be applied after the structure constants have been calculated.
The correction constants do not have this convenient property, and they must
be calculated for the actual choice of the ratio S_t/S. Of course this applies
only to the cases with two or more inequivalent atoms in the cell.

The basis input for COR is the basis vectors giving the positions of the
atoms in the cell, and the reciprocal-space vectors generated by STR. The
basic output of COR is the correction-term structure constants used by LMTO
to perform the most accurate band calculations allowed by the present collection of computer programmes.

9.3.1 Listing of COR

```
C    *****************************************************************
C    *                                                               *
C    *    CALCULATION OF CORRECTION STRUCTURE CONSTANTS              *
C    *    VERSION DEC. 1982                                          *
C    *                                                               *
C    *    H.L.SKRIVER                                                *
C    *    RISOE NATIONAL LABORATORY                                  *
C    *    DK-4000 ROSKILDE                                           *
C    *    DENMARK                                                    *
C    *                                                               *
C    *****************************************************************
C    *                                                               *
C    *    FILES AND THEIR ATTRIBUTES                                 *
C    *                                                               *
C    *    FILE 2 :UNFORMATTED,TYPE=SEQUENTIAL,RECORDSIZE=VARIABLE,   *
C    *            NAME=VEC/XXX                                       *
C    *    FILE 5 :FORMATTED,TYPE=SEQUENTIAL,RECORDLENGTH=80,         *
C    *            NAME=DATA/COR/XXX                                  *
C    *    FILE 6 :PRINTER                                            *
C    *    FILE 9 :UNFORMATTED,TYPE=SEQUENTIAL,RECORDLENGTH=VARIABLE, *
C    *            NAME=COR/XXX/A                                     *
C    *                                                               *
C    *****************************************************************
C    *                                                               *
C    *    INPUT   (FILE 5)                                           *
C    *    -----                                                      *
C    *                                                               *
C    *    BOA     :B OVER A LATTICE PARAMETER                        *
C    *    COA     :C OVER A LATTICE PARAMETER                        *
C    *    GMAX    :RADIUS OF CONVERGENCE SPHERE IN RECIP. SPACE IN   *
C    *            UNITS OF 1/A                                       *
C    *    LAT     :CRYSTAL STRUCTURE, SEE <MESH>                     *
C    *    MODE    :=0:AUTOMATIC GENERATION OF K-POINTS, <MESH>       *
C    *            =1:K-POINTS FROM INPUT                             *
C    *    NL      :NUMBER OF L-QUANTUM NUMBERS, LMAX=NL-1            *
C    *    NOWRT   :=0:SHORT PRINTOUT                                 *
C    *            =1:PRINT CORRECTION MATRIX                         *
```

```
C    *    NPX                                                        *
C    *    NPY    :NUMBERS FOR K-MESH CONSTRUCTION                    *
C    *    NPZ                                                        *
C    *    NQ     :NUMBER OF ATOMS IN THE PRIMITIVE CELL              *
C    *    OKX                                                        *
C    *    OKY    :K-VECTOR IN UNITS OF PI/A, MODE=1. HEXAGONAL       *
C    *    OKZ     COORDINNATES IF LAT=4                              *
C    *    QX                                                         *
C    *    QY     :BASIS VECTORS IN REAL SPACE IN UNITS OF A          *
C    *    QZ                                                         *
C    *    SWP(I) :DETERMINES THE RATIO OF THE WIGNER-SEITZ RADII,    *
C    *            I.E. SW(I)/SW(1)=SWP(I)                            *
C    *    TXT    :60 CHARACTERS OF TEXT FOR IDENTIFICATION           *
C    *                                                              *
C    *    INPUT  (FILE 2)                                            *
C    *    -----                                                      *
C    *                                                              *
C    *            FILE 2 CONTAINS RECIPROCAL SPACE VECTORS TO BE USED *
C    *            IN LATTICE SUMS. IT IS GENERATED BY <STR> (FILE 1)  *
C    *            WHERE A SPECIFICATION OF THE CONTENT MAY BE FOUND   *
C    *                                                              *
C    ***********************************************************************
C    *                                                              *
C    *    OUTPUT (FILE 9)                                            *
C    *    ------                                                     *
C    *                                                              *
C    *    DDLR   :LOWER TRIANGLE OF CORRECTION TERM MATRIX,          *
C    *            REAL PART                                         *
C    *    DDLI   :LOWER TRIANGLE OF CORRECTION TERM MATRIX,          *
C    *            IMAGINARY PART                                     *
C    *    TXT    :60 CHARACTERS OF TEXT FOR IDENTIFICATION           *
C    *                                                              *
C    ***********************************************************************
C    *                                                              *
C    *    EXTERNAL ROUTINES                                          *
C    *                                                              *
C    *            THE PROGRAMME USES TWO ROUTINES, <KTRNSF> AND <MESH>, *
C    *            FROM <STR>                                          *
C    *                                                              *
C    ***********************************************************************
      IMPLICIT REAL*8 (A-H,O-Z)
      COMPLEX*16 CIM,DCMPLX
      COMMON/B1/CIM,CYLM(16),SWS,TWOPI,GMAX,NUMVG,NQ,NL,NLM,NOWRT,IDIM,
     1AKX(300),AKY(300),AKZ(300),QX(6),QY(6),QZ(6),SW(6),
     2DDLR(2916),DDLI(2916)
      COMMON/MSH/DKX,DKY,DKZ,DHX,BOA,COA,KX(2500),KY(2500),KZ(2500),
     1WW(2500)
      DIMENSION BKX(3),BKY(3),BKZ(3),SWP(6)
      INTEGER TXT(15),POINT
    1 FORMAT(15A4)
    2 FORMAT(3E15.7)
    3 FORMAT(F10.5,3I5)
    4 FORMAT(1H ,/,11X,'NUMVG,GMAX,NL,NQ,NLM,LMQ',//,11X,I5,F10.4,
     14I5,//)
    5 FORMAT(1H1,10X,15A4,/)
    6 FORMAT(1H ,//,11X,'SWS',7X,F10.6,//,11X,'SWS SCALED',6F10.6)
    7 FORMAT(4I5)
    8 FORMAT(3F10.6,A4)
    9 FORMAT(1H ,//,11X,'POINT NUMBER',I5,5X,'PKX,PKY,PKZ',3F10.4,5X,A4)
   10 FORMAT(6F10.6)
   11 FORMAT(1H ,11X,'(',F10.5,',',F10.5,',',F10.5,' )')
   12 FORMAT(1H ,10X,'BASIS VECTORS IN UNIT CELL',//)
   13 FORMAT(1H0,10X,'LAT,BOA,COA',I5,2F10.6,//,11X,'MODE,NPX,NPY,NPZ',
     14I5)
   14 FORMAT(1H0,10X,'**GMAX',F10.0,' .GT. DGMAX',F10.6,'**')
   15 FORMAT(I5,2E15.8)
C
160
```

```
          IDIM=54
          CIM=DCMPLX(0.D00,1.D00)
          PI=4.D00*DATAN(1.D00)
          TWOPI=2.D00*PI
          FOURPI=4.D00*PI
C
          CALL YLMCST
C
C         READ K-SPACE VECTORS GENERATED IN STR
C
          READ(2) BKX,BKY,BKZ,VOL,RMAX,DGMAX,NUMVR,NUMVG,NRO
          READ(2) (AKX(I),I=1,NUMVR)
          READ(2) (AKX(I),I=1,NUMVR)
          READ(2) (AKX(I),I=1,NUMVR)
          READ(2) (AKX(I),I=1,NUMVR)
          READ(2) (AKX(I),I=1,NUMVG)
          READ(2) (AKY(I),I=1,NUMVG)
          READ(2) (AKZ(I),I=1,NUMVG)
C
          READ(5,1) TXT
          READ(5,3) GMAX,NQ,NL,NOWRT
          NLM=NL**2
          LMQ=NQ*NLM
          READ(5,10) (SWP(LQ),LQ=1,NQ)
          READ(5,15) LAT,BOA,COA
          DO 21 I=1,NQ
       21 READ(5,2) QX(I),QY(I),QZ(I)
          READ(5,7) MODE,NPX,NPY,NPZ
C
          IF(GMAX.GT.DGMAX) GO TO 22
C
          WRITE(6,5) TXT
          WRITE(6,4) NUMVG,GMAX,NL,NQ,NLM,LMQ
          WRITE(6,12)
          DO 25 I=1,NQ
       25 WRITE(6,11) QX(I),QY(I),QZ(I)
          SWS=(3.D00*VOL/FOURPI/NQ)**(1.D00/3.D00)
C
          CALL SCASW(NQ,SWP,SW,SWS)
C
          WRITE(6,6) SWS,(SW(LQ),LQ=1,NQ)
          WRITE(6,13) LAT,BOA,COA,MODE,NPX,NPY,NPZ
          WRITE(9) TXT
          IF(MODE.EQ.1) GO TO 28
C
C         GENERATE MESH IN K-SPACE
C
          CALL MESH(LAT,NPX,NPY,NPZ,NPT)
C
          DO 38 NP=1,NPT
          KKX=KX(NP)
          KKY=KY(NP)
          KKZ=KZ(NP)
          CALL KTRNSF(LAT,KKX,KKY,KKZ,PKX,PKY,PKZ)
          IF(NOWRT.GT.0) WRITE(6,9) NP,PKX,PKY,PKZ,POINT
          PKXP=PKX*PI
          PKYP=PKY*PI
          PKZP=PKZ*PI
C
          CALL DDLLP(PKXP,PKYP,PKZP)
C
          J1=-IDIM
          DO 33 JJ=1,LMQ
          J1=J1+IDIM
          I1=J1+JJ
          I2=J1+LMQ
          WRITE(9) (DDLR(LIN),LIN=I1,I2)
```

```
   33 WRITE(9) (DDLI(LIN),LIN=I1,I2)
   38 CONTINUE
      CLOSE(9,DISP=CRUNCH)
      STOP
C
C     READ K-VECTOR IN UNITS OF PIA
C
   28 NPOINT=NPX
      FX=PI
      FY=PI/BOA
      FZ=PI/COA
      FHX=0.D0
      IF(LAT.NE.4) GO TO 23
      FX=2.D0/3.D0*PI
      FY=PI/DSQRT(3.D0)
      FHX=-0.5D0*FX
   23 DO 29 I=1,NPOINT
      READ(5,8) OKX,OKY,OKZ,POINT
      WRITE(6,9) I,OKX,OKY,OKZ,POINT
      PKXP=OKX*FX+OKY*FHX
      PKYP=OKY*FY
      PKZP=OKZ*FZ
C
      CALL DDLLP(PKXP,PKYP,PKZP)
C
      J1=-IDIM
      DO 30 J=1,LMQ
      J1=J1+IDIM
      I1=J1+J
      I2=J1+LMQ
      WRITE(9) (DDLR(LIN),LIN=I1,I2)
   30 WRITE(9) (DDLI(LIN),LIN=I1,I2)
   29 CONTINUE
      CLOSE (9,DISP=CRUNCH)
      STOP
   22 WRITE(6,14) GMAX,DGMAX
      STOP
      END
      BLOCK DATA
C     ******************************************************************
C     *                                                                *
C     *    INSERT LM-DATA                                              *
C     *                                                                *
C     ******************************************************************
      COMMON/LDATA/LL(25),MM(25)
      DATA LL/0,3*1,5*2,7*3,9*4/
      DATA MM/0,-1,0,1,-2,-1,0,1,2,-3,-2,-1,0,1,2,3,-4,-3,-2,-1,0,1,2,
     13,4/
      END
      SUBROUTINE DDLLP(PKX,PKY,PKZ)
C     ******************************************************************
C     *                                                                *
C     *    CALCULATE CORRECTION MATRIX TO BE USED IN LMTO              *
C     *                                                                *
C     ******************************************************************
      IMPLICIT REAL*8 (A-H,O-Z)
      COMPLEX*16 SPHRM,CLP,CL,SUM,CIM,SYLM,CDEXP,CQDK
      REAL*8 DREAL,DIMAG
      COMMON/B1/CIM,CYLM(16),SWS,TWOPI,GMAX,NUMVG,NQ,NL,NLM,NOWRT,IDIM,
     1AKX(300),AKY(300),AKZ(300),QX(6),QY(6),QZ(6),SW(6),
     2DDLR(2916),DDLI(2916)
      COMMON/LDATA/LL(25),MM(25)
      DIMENSION DLT(150,18),SYLM(150,16),CQDK(150,15)
    1 FORMAT(1H0,10X,'CORRECTION MATRIX',//)
    2 FORMAT(1H ,9(1PE13.5))
    3 FORMAT(1H ,10X,'**ID IN DDLLP IS DIMENSIONED',I5)
    4 FORMAT(1H0,10X,'QP,Q =',2I5,//)
    5 FORMAT(1H0)
```

```
C
      ID=150
      NQM=NQ-1
C

      IG=0
      DO 20 I=1,NUMVG
      X=AKX(I)+PKX
      Y=AKY(I)+PKY
      Z=AKZ(I)+PKZ
      DGI=DSQRT(X*X+Y*Y+Z*Z)
      IF(DGI.GT.GMAX) GO TO 20
      IG=IG+1
      IF(IG.GT.ID) GO TO 30
      UKX=X/DGI
      UKY=Y/DGI
      UKZ=Z/DGI
      MQ=-NL
      DO 25 IQ=1,NQ
      MQ=MQ+NL
      AKS=DGI*SW(IQ)
      AKSC=AKS**3
      CALL BESSJ(AKS,BONE,BTWO,BTHR,BFOU)
      DLT(IG,MQ+1)=3.D0/AKSC*BONE
      DLT(IG,MQ+2)=5.D0/AKSC*BTWO
      DLT(IG,MQ+3)=7.D0/AKSC*BTHR
   25 DLT(IG,MQ+4)=9.D0/AKSC*BFOU
      DO 23 ILM=1,NLM
   23 SYLM(IG,ILM)=SPHRM(ILM,UKX,UKY,UKZ)
      IF(NQ.EQ.1) GO TO 20
      LQ=0
      DO 24 JQ=1,NQM
      JQP=JQ+1
      QPX=QX(JQ)
      QPY=QY(JQ)
      QPZ=QZ(JQ)
      DO 24 IQ=JQP,NQ
      LQ=LQ+1
      QPPX=QX(IQ)-QPX
      QPPY=QY(IQ)-QPY
      QPPZ=QZ(IQ)-QPZ
      DOT=QPPX*X+QPPY*Y+QPPZ*Z
   24 CQDK(IG,LQ)=CDEXP(CIM*DOT)
   20 CONTINUE
C
C     QQP DIAGONAL BLOCKS
C
      IQQ=-(1+IDIM)*NLM
      LQ=-NL
      DO 21 IQ=1,NQ
      IQQ=IQQ+(1+IDIM)*NLM
      LQ=LQ+NL
      J1=-IDIM
      DO 21 JLM=1,NLM
      J1=J1+IDIM
      L=LL(JLM)
      TWOL=2*L+1
      M=MM(JLM)
      L1=L+LQ+1
      KLM=L*L+L+M+1
      KLM1=KLM-2*M
      DO 21 ILM=JLM,NLM
      LIN=J1+ILM
      LINQ=IQQ+LIN
      LP=LL(ILM)
      TWOLP=2*LP+1
      MP=MM(ILM)
      L2=LP+LQ+1
      KLMP=LP*LP+LP+MP+1
```

163

```
      SUM=DCMPLX(0.D0,0.D0)
      DO 22 I=1,IG
      DT=DLT(I,L1)*DLT(I,L2)
      CLP=SYLM(I,KLMP)
      CL=SYLM(I,KLM1)
      SUM=SUM+DT*CLP*CL
   22 CONTINUE
      CCTT=CYLM(KLM)*CYLM(KLMP)*TWOLP*TWOL
      SUM=SUM*CCTT
      DDLR(LINQ)=REAL(SUM)
   21 DDLI(LINQ)=AIMAG(SUM)
C
      IF(NOWRT.EQ.0) GO TO 40
      WRITE(6,1)
      IQQ=-(1+IDIM)*NLM
      DO 29 JQ=1,NQ
      WRITE(6,4)JQ,JQ
      IQQ=IQQ+(1+IDIM)*NLM
      I1=-IDIM
      DO 31 ILM=1,NLM
      I1=I1+IDIM
      ILMQ=ILM+IQQ
      I2=I1+ILMQ
   31 WRITE(6,2) (DDLR(LIN),LIN=ILMQ,I2,IDIM)
      WRITE(6,5)
      I1=-IDIM
      DO 32 ILM=1,NLM
      I1=I1+IDIM
      ILMQ=ILM+IQQ
      I2=I1+ILMQ
   32 WRITE(6,2) (DDLI(LIN),LIN=ILMQ,I2,IDIM)
   29 CONTINUE
   40 CONTINUE
      IF(NQ.EQ.1) RETURN
C
C     QQP OFF DIAGONAL BLOCKS
C
      JQQ=-NLM*IDIM
      LQ=0
      DO 26 JQ=1,NQM
      JLQ=(JQ-1)*NL
      JQP=JQ+1
      JQQ=JQQ+NLM*IDIM
      DO 26 IQ=JQP,NQ
      ILQ=(IQ-1)*NL
      IQQ=(IQ-1)*NLM+JQQ
      LQ=LQ+1
      J1=-IDIM
      DO 27 JLM=1,NLM
      J1=J1+IDIM
      L=LL(JLM)
      TWOL=2*L+1
      M=MM(JLM)
      L1=L+JLQ+1
      KLM=L*L+L+M+1
      KLM1=KLM-2*M
      DO 27 ILM=1,NLM
      LIN=J1+ILM
      LINQ=IQQ+LIN
      LP=LL(ILM)
      TWOLP=2*LP+1
      MP=MM(ILM)
      L2=LP+ILQ+1
      KLMP=LP*LP+LP+MP+1
      SUM=DCMPLX(0.D0,0.D0)
      DO 28 I=1,IG
      DT=DLT(I,L1)*DLT(I,L2)
      CLP=SYLM(I,KLMP)
```

```
      CL=SYLM(I,KLM1)
      SUM=SUM+DT*CLP*CL*CQDK(I,LQ)
   28 CONTINUE
      CCTT=CYLM(KLM)*CYLM(KLMP)*TWOLP*TWOL
      SUM=SUM*CCTT
      DDLR(LINQ)=REAL(SUM)
   27 DDLI(LINQ)=AIMAG(SUM)
C
      IF(NOWRT.EQ.0) GO TO 26
      WRITE(6,4) IQ,JQ
      DO 33 ILM=1,NLM
      J1=IQQ+ILM
      J2=J1+(NLM-1)*IDIM
   33 WRITE(6,2) (DDLR(LINQ),LINQ=J1,J2,IDIM)
      WRITE(6,5)
      DO 34 ILM=1,NLM
      J1=IQQ+ILM
      J2=J1+(NLM-1)*IDIM
   34 WRITE(6,2) (DDLI(LINQ),LINQ=J1,J2,IDIM)
   26 CONTINUE
      RETURN
   30 WRITE(6,3) ID
      STOP
      END
      SUBROUTINE YLMCST
C     ********************************************************************
C     *                                                                *
C     *    CALCULATE CONSTANTS FOR THE 16 FIRST YLM'S                  *
C     *                                                                *
C     *    CYLM(N), N=L*L+L+M+1                                        *
C     *                                                                *
C     ********************************************************************
      IMPLICIT REAL*8(A-H,O-Z)
      COMPLEX*16 CIM
      COMMON/B1/CIM,CYLM(16),SWS,TWOPI,GMAX,NUMVG,NQ,NL,NLM,NOWRT,IDIM,
     1AKX(300),AKY(300),AKZ(300),QX(6),QY(6),QZ(6),SW(6),
     2DDLR(2916),DDLI(2916)
C
      SQPI=DSQRT(TWOPI/2.D0)
      SQTPI=DSQRT(TWOPI)
      SQT=DSQRT(3.D0)
      SQF=DSQRT(5.D0)
      SQS=DSQRT(7.D0)
      SQFI=DSQRT(15.D0)
C
      CYLM(1)=0.5D0/SQPI
      CYLM(2)=SQT/2.D0/SQTPI
      CYLM(3)=SQT/2.D0/SQPI
      CYLM(4)=-CYLM(2)
      CYLM(5)=SQFI/4.D0/SQTPI
      CYLM(6)=SQFI/2.D0/SQTPI
      CYLM(7)=SQF/4.D0/SQPI
      CYLM(8)=-CYLM(6)
      CYLM(9)=CYLM(5)
      CYLM(10)=SQF*SQS/8.D0/SQPI
      CYLM(11)=SQS*SQFI/4.D0/SQTPI
      CYLM(12)=SQT*SQS/8.D0/SQPI
      CYLM(13)=SQS/4.D0/SQPI
      CYLM(14)=-CYLM(12)
      CYLM(15)=CYLM(11)
      CYLM(16)=-CYLM(10)
      RETURN
      END
      COMPLEX FUNCTION SPHRM*16(N,X,Y,Z)
```

```
C     ******************************************************************
C     *                                                                *
C     *      COMPLEX CONJUGATE OF SPHERICAL HARMONICS                  *
C     *                                                                *
C     *      (X,Y,Z) SHOULD BE A UNIT VECTOR AND                       *
C     *      N=L*L+L+M+1                                                *
C     *                                                                *
C     ******************************************************************
      IMPLICIT REAL*8 (A-H,O-Z)
      COMPLEX*16 DCMPLX
C
      GO TO (1,2,3,4,5,6,7,8,9,10,11,12,13,14,15,16),N
C
    1 SPHRM=DCMPLX(1.D00,0.D00)
      RETURN
    2 SPHRM=DCMPLX(X,Y)
      RETURN
    3 SPHRM=DCMPLX(Z,0.D00)
      RETURN
    4 SPHRM=DCMPLX(X,-Y)
      RETURN
    5 SPHRM=DCMPLX(X*X-Y*Y,2.D00*X*Y)
      RETURN
    6 SPHRM=DCMPLX(X*Z,Y*Z)
      RETURN
    7 SPHRM=DCMPLX(3.D00*Z*Z-1.D00,0.D00)
      RETURN
    8 SPHRM=DCMPLX(X*Z,-Y*Z)
      RETURN
    9 SPHRM=DCMPLX(X*X-Y*Y,-2.D00*X*Y)
      RETURN
   10 X2=X*X
      Y2=Y*Y
      SPHRM=DCMPLX(X*(X2-3.D00*Y2),Y*(3.D00*X2-Y2))
      RETURN
   11 SPHRM=DCMPLX(Z*(X*X-Y*Y),2.D00*X*Y*Z)
      RETURN
   12 Z2=5.D00*Z*Z-1.D00
      SPHRM=DCMPLX(Z2*X,Z2*Y)
      RETURN
   13 SPHRM=DCMPLX(Z*(5.D00*Z*Z-3.D00),0.D00)
      RETURN
   14 Z2=5.D00*Z*Z-1.D00
      SPHRM=DCMPLX(X*Z2,-Y*Z2)
      RETURN
   15 SPHRM=DCMPLX(Z*(X*X-Y*Y),-2.D00*X*Y*Z)
      RETURN
   16 X2=X*X
      Y2=Y*Y
      SPHRM=DCMPLX(X*(X2-3.D00*Y2),Y*(Y2-3.D00*X2))
      RETURN
      END
      SUBROUTINE BESSJ(AKS,B1,B2,B3,B4)
C     ******************************************************************
C     *                                                                *
C     *     CALCULATE THE 4 FIRST SPHERICAL BESSEL FUNCTIONS           *
C     *                                                                *
C     ******************************************************************
      IMPLICIT REAL*8 (A-H,O-Z)
C
      X=AKS
      IF(X.GT.0.5D0) GO TO 20
C
C     SERIES EXPANSION
C
      X1=X/3.D0
```

```
      X2=X1*X/5.D0
      X3=X2*X/7.D0
      X4=X3*X/9.D0
      Y=X*X*0.5D0
      Y2=Y*Y
      Y3=Y2*Y
      Y4=Y3*Y
      B1=X1*(-Y3/1890.D0+Y2/70.D0-Y/5.D0+1.D0)
      B2=X2*(-Y3/4158.D0+Y2/126.D0-Y/7.D0+1.D0)
      B3=X3*(-Y3/7722.D0+Y2/198.D0-Y/9.D0+1.D0)
      B4=X4*(-Y3/12870.D0+Y2/286.D0-Y/11.D0+1.D0)
      RETURN
C
C     RECURSION EXPRESSIONS
C
   20 B0=DSIN(X)/X
      B1=(B0-DCOS(X))/X
      B2=3.D0/X*B1-B0
      B3=5.D0*B2/X-B1
      B4=7.D0*B3/X-B2
      RETURN
      END
      SUBROUTINE SCASW(NQD,RMUF,SW,SWS)
C     **********************************************************************
C     *                                                                    *
C     *    SCALE WIGNER-SEITZ RADII ACCORDING TO THE NUMBERS RMUF(I)       *
C     *    CONSERVING THE VOLUME                                           *
C     *                                                                    *
C     **********************************************************************
      IMPLICIT REAL*8 (A-H,O-Z)
      DIMENSION F(10),RMUF(NQD),SW(NQD)
C
      NQ=NQD
      OM=NQ*SWS**3
      RMUF1=RMUF(1)
      SUM=0.D0
      DO 20 I=1,NQ
      F(I)=RMUF(I)/RMUF1
   20 SUM=SUM+F(I)**3
      R1=(OM/SUM)**(1.D0/3.D0)
      DO 21 I=1,NQ
   21 SW(I)=F(I)*R1
      RETURN
      END
      £
```

9.3.2 Execution of COR

The correction-term programme COR cannot be executed without a corresponding and previous execution of STR to generate the vectors needed in the re-
ciprocal-lattice summations. These vectors are stored in the file VEC/XXX.
An example of the additional input used is given in Table 9.3.

The first line of the input is a text meant to describe the correction
matrices being calculated, and which will be transmitted to and printed by
all later programmes.

The second line contains data for four variables. The number of recipro-
cal-lattice vectors included in the lattice sums (6.58, 8.26) at each k
point is determined by GMAX. It may be smaller than the corresponding GMAX

Table 9.3. Input for COR

Correction matrix for bcc structure			
17.8	1	3	1
1.0			
3	1.0		1.0
0.0	0.0		0.0
0	0	17	0

used in STR but never larger. Further, NQ = 1 and NL = 3 indicate one atom per cell and ℓ_{max} = 2. With NOWRT = 1 the execution results in a similar output to that shown in Sect.9.2.3. Normally one would use NOWRT = 0.

The third line contains NQ numbers used to scale the individual atomic radii such that the volume V of the cell is contained within the NQ atomic spheres belonging to the cell, i.e.

SW(I)/SW(1) = SWP(I)

$$\sum_{I=1}^{NQ} SW(I)^3 = 3 \ V/4\pi \ . \tag{9.10}$$

With SWP(I) = 1.0 all spheres have the same size, with radii equal to the average Wigner-Seitz radius (8.16).

In the fourth line, LAT selects the Brillouin zone to be used according to the table in subroutine MESH, and BOA and COA should be set equal to the actual lattice parameters b/a and c/a.

The next NQ lines contain the basis vectors giving the positions of the individual atoms in the cell. The last lines contain data related to the k vectors to be selected in the calculation as described in connection with the input data for STR in Sect.9.2.3.

As a final and obvious point we mention that the data for STR and COR for the same crystal structure should be consistent. If this is not the case LMTO will usually (fortunately) fail in the Cholesky decomposition of the overlap matrix.

9.3.3 Sample Output from COR

CORRECTION MATRIX FOR BCC STRUCTURE

NUMVG,GMAX,NL,NQ,NLM,LMQ

| 135 | 17.8000 | 3 | 1 | 9 | 9 |

BASIS VECTORS IN UNIT CELL

(0.00000, 0.00000, 0.00000)

SWS   0.492373

SWS SCALED 0.492373

LAT,BOA,COA   3 1.000000 1.000000

MODE,NPX,NPY,NPZ   0   0   17   0

<OUTPUT FROM POINTS 1 AND 2 REMOVED>

POINT NUMBER   3   PKX,PKY,PKZ   0.1250   0.1250   0.000

CORRECTION MATRIX

QP,Q =   1   1


```
 1.40221E+01
 3.32766E+00   1.58124E+00
 2.86921E-18   3.76882E-16   1.81223E-03
-3.32766E+00  -7.57090E-18  -4.11144E-16   1.58124E+00
 2.36282E-19   1.13500E-01   2.45552E-18   1.13656E-01   2.58189E-02
 1.98477E-17   4.16715E-16  -5.00398E-04  -3.47263E-18  -4.74314E-15   1.216
-3.92041E-01  -9.29612E-02   1.60394E-17   9.29612E-02   1.58503E-17  -3.804
-1.98477E-17  -3.47263E-18   5.00398E-04   4.16715E-16  -5.62964E-16   0.
 2.36282E-19  -1.13656E-01   2.45552E-18  -1.13500E-01  -1.94478E-02   5.111


 0.
 3.32766E+00   0.
 0.           -2.10761E-16   0.
 3.32766E+00   1.57950E+00   2.42387E-16   0.
 4.80167E-01   1.13500E-01  -1.12954E-16  -1.13656E-01  -4.17370E-16
 2.00840E-17   0.           -5.00398E-04   6.38963E-16  -2.64011E-15   0.
 0.            9.29612E-02   0.            9.29612E-02   1.36710E-02   2.443
 2.00840E-17  -6.38963E-16  -5.00398E-04   0.           -3.85339E-15  -3.531
-4.80167E-01  -1.13656E-01   1.12954E-16   1.13500E-01   2.11921E-15   5.403
```


9.4 Linear Muffin-Tin Orbital Programme LMTO

The LMTO programme is designed to establish the LMTO Hamiltonian and overlap matrices (5.46,47, 8.25) and to solve the LMTO eigenvalue problem (5.45).
The programme is constructed in agreement with the notation of Chap.5 valid
for a general κ^2, but in the actual programming the $\kappa^2 = 0$ limit has been
taken. Hence, we have inserted $D\{j\} - D\{n\} = 2\ell + 1$ in the three-centre
terms, and used (6.30,33) to construct the ℓ character. It is, however, a
simple matter to reintroduce $D\{j\} - D\{n\}$ for $2\ell + 1$ in the three-centre
terms and in the definition for the distortion parameter γ, Sect.4.6.

In a typical run, LMTO will store on disk eigenvalues and tℓ characters determined on a grid in the irreducible wedge of the Brillouin zone sufficiently densely to give a reasonable representation of the energy bands throughout k space. The k vectors which are actually included in this procedure are determined in a previous execution of STR. The energy bands may subsequently be retrieved for state-density determination DDNS, plotting purposes PLTBND, Fermi surface area calculations FSAR, etc., or they may simply be read and printed READB.

The basic input to LMTO is the structure-constant matrices generated by STR, the correction-term structure-constant matrices generated by COR, if the correction to the atomic-sphere approximation is to be included, and the potential parameters as given for instance in Table 4.1.

The basic output is the eigenvalues and tℓ characters evaluated at the k mesh established in STR. We mention in this context that if the correction to the ASA is included, the tℓ characters are renormalised as explained at the end of Sect.6.9.

9.4.1 LMTO and Cholesky Decomposition

The LMTO eigenvalue problem is a generalised eigenvalue problem of the form

$$(\underline{\underline{H}} - E\underline{\underline{O}})\mathbf{x} = \mathbf{0} \quad , \tag{9.11}$$

where $\underline{\underline{H}}$ and $\underline{\underline{O}}$ are the Hamiltonian and overlap matrices, respectively. To solve such a problem by standard matrix diagonalisation techniques one must transform (9.11) into a simple eigenvalue problem.

Since the overlap matrix is positive definite a matrix $\underline{\underline{L}}$ exists which is non-singular and lower triangular such that [9.3]

$$\underline{\underline{O}} = \underline{\underline{L}} \, \underline{\underline{L}}^h \quad , \tag{9.12}$$

where h indicates Hermitian conjugate. The LMTO eigenvalue problem is now

$$(\underline{\underline{L}}^{-1} \underline{\underline{H}} \, \underline{\underline{L}}^{-h} - E \, \underline{\underline{1}})\mathbf{y} = \mathbf{0}$$

$$\mathbf{y} = \underline{\underline{L}}^h \mathbf{x} \quad . \tag{9.13}$$

In practice, the elements l_{ij} of $\underline{\underline{L}}$ are found from

$$l_{ij} = \left(o_{ij} - \sum_{k=1}^{j-1} l_{ik} l_{jk}^* \right) / l_{jj}$$

$$l_{ii} = \left(o_{ii} - \sum_{k=1}^{i-1} l_{ik} l_{ik}^* \right)^{\frac{1}{2}} \tag{9.14}$$

and $\underline{\underline{L}}^{-1} \underline{\underline{H}} \underline{\underline{L}}^{-h}$ is calculated by defining the matrices

$$\underline{\underline{Z}} = \underline{\underline{H}} \underline{\underline{L}}^{-h}$$

$$\underline{\underline{W}} = \underline{\underline{L}}^{-1} \underline{\underline{Z}} \quad , \tag{9.15}$$

whose elements are found by solving

$$\underline{\underline{Z}} \underline{\underline{L}}^h = \underline{\underline{H}}$$

$$\underline{\underline{L}} \underline{\underline{W}} = \underline{\underline{Z}} \quad . \tag{9.16}$$

The explicit expressions for the elements of $\underline{\underline{Z}}$ are

$$z_{i1} = h_{i1}/l_{11}$$

$$z_{ij} = \left(h_{ij} - \sum_{k=1}^{j-1} z_{ik} l_{jk}^* \right) / l_{jj} \quad , \tag{9.17}$$

and for the elements of $\underline{\underline{W}}$

$$w_{i1} = \left(z_{i1} - \sum_{k=1}^{i-1} l_{ik} w_{k1} \right) / l_{ii}$$

$$w_{ii} = \left(z_{ii} - \sum_{k=1}^{i-1} l_{ik} w_{ik}^* \right) / l_{ii}$$

$$w_{ij} = \left(z_{ij} - \sum_{k-1}^{j-1} l_{ik} w_{jk}^* - \sum_{k=j}^{i-1} l_{ik} w_{kj} \right) / l_{ii} \quad . \tag{9.18}$$

The eigenvalue problem is now

$$(\underline{\underline{W}} - E \underline{\underline{1}})\mathbf{y} = 0 \quad . \tag{9.19}$$

which is a standard problem in matrix algebra.

If a decomposition in partial waves is wanted, the eigenvectors $\mathbf{x}$ of the original problem (9.11) must be evaluated from

$$\mathbf{x} = \underline{\underline{L}}^{-h} \mathbf{y} \quad . \tag{9.20}$$

To do this we invert $\underline{\underline{L}}$:

$$i_{ii} = 1/l_{ii}$$

$$i_{ij} = \left(- \sum_{k=j}^{i-1} l_{ik} i_{kj} \right) / l_{ii} \tag{9.21}$$

and the elements of $\mathbf{x}$ are then

$$x_i = \sum_{k=1}^{i} i_{ki}^* y_k \quad . \tag{9.22}$$

The technique described in this section is implemented in the LMTO programme.

9.4.2 Listing of LMTO

```
C     **********************************************************************
C     *                                                                    *
C     *    ENERGY BANDS BY THE LMTO METHOD                                  *
C     *    VERSION DEC. 1982                                                *
C     *                                                                    *
C     *    H.L.SKRIVER                                                      *
C     *    RISOE NATIONAL LABORATORY                                        *
C     *    DK-4000 ROSKILDE                                                 *
C     *    DENMARK                                                          *
C     *                                                                    *
C     **********************************************************************
C     *                                                                    *
C     *    FILES AND THEIR ATTRIBUTES                                       *
C     *                                                                    *
C     *    FILE 1 :UNFORMATTED,TYPE=SEQUENTIAL,RECORDLENGTH=VARIABLE,       *
C     *            NAME=STR/XXX/A                                           *
C     *    FILE 2 :FORMATTED,TYPE=SEQUENTIAL,RECORDLENGTH=80,               *
C     *            NAME=BND/YY/X                                            *
C     *    FILE 3 :UNFORMATTED,TYPE=SEQUENTIAL,RECORDLENGTH=VARIABLE,       *
C     *            NAME=COR/XXX/A                                           *
C     *    FILE 4 :UNFORMATTED,TYPE=RANDOM (DIRECT) ACCESS,                 *
C     *            RECORDLENGTH=MAX(NT*(NL+1),4),                           *
C     *            NAME=BND/YY/B                                            *
C     *    FILE 5 :FORMATTED,TYPE=SEQUENTIAL,RECORDLENGTH=80,               *
C     *            NAME=PTPRM/YY/B                                          *
C     *    FILE 6 :PRINTER                                                  *
C     *                                                                    *
C     **********************************************************************
C     *                                                                    *
C     *    INPUT  (FILE 5) <MAIN>                                           *
C     *    -----                                                           *
C     *                                                                    *
C     *    IPNCH  :=0:NO WRITE ON FILE 2                                    *
C     *           :=1:WRITE EIGENVALUES IN CARD FORMAT ON FILE 2            *
C     *    IPRNTO :=0:ONLY EIGENVALUES AT GAMMA ARE PRINTED                 *
C     *           :=1:ALL EIGENVALUES ARE PRINTED                          *
C     *    IPTP   :NOT USED                                                 *
C     *    IT(IQ) :TYPE LABEL FOR ATOM NUMBER IQ                            *
C     *    NOC    :=0:NO CORRECTION TERMS INCLUDED                          *
C     *           :=1:CORRECTION TERMS INCLUDED.READ ON FILE 3              *
C     *    NOEVC  :=0:EIGENVECTORS ARE NOT CALCULATED                       *
C     *           :=1:EIGENVECTORS AND T,L-CHARACTERS ARE CALCULATED        *
C     *    NOWRT  :=0:PRINT SELECTED BY IPRNTO ONLY                         *
C     *           :=1:PRINT HAMILTON AND OVERLAP MATRICES                   *
C     *           :=2:PRINT EIGENVECTOR MATRICES                            *
C     *    NPOINT :THE PROGRAMME CALCULATES ENERGY BANDS IN THE K-VECTOR*
C     *    NSTART  RANGE SPECIFIED BY NSTART-NPOINT. IF NSTART .GT. 1   *
C     *            AND NOEVC .NE. 0 THE PROGRAMME WILL WRITE INTO FILE 4*
C     *            FROM THE RECORD CORRESPONDING TO NSTART.             *
C     *    SWS    :AVERAGE ATOMIC WIGNER-SEITZ RADIUS, ATOMIC UNITS     *
C     *    SWSP   :ATOMIC RADIUS FOR ATOM NUMBER IQ, ATOMIC UNITS       *
C     *    TAIL   :=' T':INCLUDE LMAX ONLY IN THE INTERNAL L-SUMMATION  *
C     *                  IN THE THREE-CENTRE TERM,I.E. AS A TAIL.       *
C     *                  N.B. ONLY FOR L=3 AND NQ=1                     *
C     *           =' N':NO TAIL CORRECTION                              *
C     *    TTXT   :CHEMICAL LABEL FOR ATOM NUMBER IQ                    *
```

```
C   *    TXTP    :80 CHARACTERS OF TEXT IDENTIFYING THE POTENTIAL      *
C   *            PARAMETERS                                            *
C   *                                                                  *
C   *    ENY                                                           *
C   *    OMM                                                           *
C   *    SFIMSQ  :STANDARD POTENTIAL PARAMETERS                        *
C   *    FMOFP                                                         *
C   *    FIAV                                                          *
C   *                                                                  *
C   *    OMN                                                           *
C   *    OMJ                                                           *
C   *    FINQ                                                          *
C   *    FIJSQ   :HISTORICAL POTENTIAL PARAMETERS RELATED TO THE       *
C   *    LNN      LMTO MATRIX (5,15)                                   *
C   *    LNJ                                                           *
C   *    LJJ                                                           *
C   *                                                                  *
C   *    INPUT (FILE 1) <MAIN>                                         *
C   *    -----                                                         *
C   *                                                                  *
C   *            FILE 1 IS A STANDARD STRUCTURE CONSTANT FILE GENE-    *
C   *            RATED  BY <STR> WHERE A SPECIFICATION OF THE CONTENT  *
C   *            MAY BE FOUND                                          *
C   *                                                                  *
C   *    INPUT (FILE 3) <MAIN>                                         *
C   *    -----                                                         *
C   *                                                                  *
C   *            FILE 3 IS A STANDARD CORRECTION MATRIX FILE GENERATED *
C   *            BY <COR> WHERE A SPECIFICATION OF THE CONTENT MAY BE  *
C   *            FOUND                                                 *
C   *                                                                  *
C *******************************************************************
C   *                                                                  *
C   *    OUTPUT (FILE 2) <MAIN> AND <REIGNV>                           *
C   *    ------                                                        *
C   *            A SPECIFICATION OF OUTPUT NOT IN THE LISTS BELOW CAN  *
C   *            BE FOUND IN INPUT (FILE 5)                            *
C   *                                                                  *
C   *    IEV     :EIGENVALUES CONVERTED TO INTEGER NUMBERS             *
C   *    KX                                                            *
C   *    KY      :K-VECTOR                                             *
C   *    KZ                                                            *
C   *    W       :SYMMETRY WEIGHTS                                     *
C   *                                                                  *
C   *    OUTPUT (FILE 4) <MAIN> AND <REIGNV>                           *
C   *    ------                                                        *
C   *                                                                  *
C   *    AMSO    :T,S-CHARACTER                                        *
C   *    AMPO    :T,P-CHARACTER                                        *
C   *    AMDO    :T,D-CHARACTER                                        *
C   *    AMFO    :T,F-CHARACTER                                        *
C   *    BOA     :B OVER A LATTICE PARAMETER                           *
C   *    COA     :C OVER A LATTICE PARAMETER                           *
C   *    DUM1    :NOT USED                                             *
C   *    KX                                                            *
C   *    KY      :K-VECTOR                                             *
C   *    KZ                                                            *
C   *    LAT     :BRILLOUIN ZONE, SEE <MESH>                           *
C   *    NL      :NUMBER OF L-QUANTUM NUMBERS, LMAX=NL-1               *
C   *    NLM     :SUM OF 2*L+1, =NL**2                                 *
C   *    NLMT    :NUMBER OF EIGENVALUES PER K-POINT                    *
C   *    NPT     :NUMBER OF K-POINTS IN THE IRREDUCIBLE ZONE           *
C   *    NPX                                                           *
C   *    NPY     :NUMBERS FOR K-MESH CONSTRUCTION                      *
C   *    NPZ                                                           *
C   *    NQ      :NUMBER OF ATOMS IN THE CELL                          *
C   *    NT      :NUMBER OF INEQUIVALENT ATOMS IN THE CELL             *
C   *    TXTC    :TEXT IDENTIFYING THE CORRECTION MATRICES             *
```

```
C    *    TXTS    :TEXT IDENTIFYING THE STRUCTURE CONSTANTS              *
C    *    W       :SYMMETRY WEIGHTS                                      *
C    *                                                                  *
C    ********************************************************************
C    *                                                                  *
C    *    EXTERNAL ROUTINES                                             *
C    *                                                                  *
C    *         THE PROGRAMME USES FOUR ROUTINES, <IMTQL1>, <IMTQL2>,    *
C    *         <HTRIDI>, AND <HTRIBK> FROM THE SUBROUTINE PACKAGE       *
C    *         'EISPACK'. REF.:BOYLE,J.M,GARBOW,B.S.,KLEMA,V.C.,        *
C    *         AND SMITH,B.T.,'EISPACK EIGENSYSTEM PACKAGE---THE        *
C    *         USER AND HIS GUIDE',ARGONNE NATIONAL LABORATORY          *
C    *         REPORT (IN PREPARATION)                                  *
C    *                                                                  *
C    ********************************************************************
      IMPLICIT REAL*8 (A-H,O-Z)
      REAL*8 LNN,LNJ,LJJ
      INTEGER TXTS(15),TXTP(20,6),TXTC(15),TTXT(6),POINT,TAIL
      COMMON/B1/ SRL(2916),SIL(2916),ENY(18),OMN(18),OMJ(18),FINSQ(18) ,
     1FIJSQ(18),ONEH(18),TWOH(18),CRYH(18),ONEO(18),TWOO(18),CRYO(18),
     2NP,KX,KY,KZ,NOEVC,IDIM,DDLR(2916),DDLI(2916),HSWSCU(6),FCOR(6)
     3,SWSSQ(6),LNN(18),LNJ(18),LJJ(18),IT(6),IDUM,NT,SFIN(18),GAM(18) ,
     4BM(18),HR(54,54),HI(54,54),OR(54,54),OI(54,54)
      COMMON/LDAT/LLP(54),LL(54)
      DIMENSION SWSP(6),SWSU(6),SW(6)
      DATA TXTC/'NO C','ORRE','CTIO','N   ','TERM',10*'    '/
    1 FORMAT(F9.7,1X,I5,A2,3I1,4I5,6I5)
    2 FORMAT(1H ,10X,'**TAIL CORRECTION NOT OPERATIONAL FOR NQ .GT. 1',
     1/,11X,'TAIL ',A2,'NQ ',I4)
    3 FORMAT(1H ,5X,8F10.6)
    4 FORMAT(5F10.6)
    5 FORMAT(1H ,///,11X,20A4,//,9X,A4,3X,'SWS',F10.6,///11X,'ENY',11X ,
     1'OM(-)',9X,'S*FI**2'6X,'FI(-)/FI(+)',3X,'<FID**2>**1/2',//)
    6 FORMAT(20A4)
    7 FORMAT(1H0,10X,'** END OF DATA SET ',I5,'  K-POINTS READ **')
    8 FORMAT(1H ,10X,15A4,///,11X,'NL',I4,' NQ',I4,' NLM',I4,' NLMQ',I4,
     1' NLMT'I4,'  SWS',F10.6,'  ALAT',F10.6,//)
    9 FORMAT(1H ,10X,'**NLMQ LARGER THAN IDIM**',/,11X,'NLMQ,IDIM',2I5 )
   10 FORMAT(1H ,/,11X,'EIGENVALUES AT ',A4,5X,'(PKX,PKY,PKZ,) ',
     13F10.6,3X,'NO',I5,5X,'(KX,KY,KZ)',3I5,/)
   11 FORMAT(1H ,10X,'NPOINT',I5,' IPRNT',I4,' IPNCH',I4,' NOWRT',I4,
     1' NOC',I4,' NOEVC',I4,' NSTART',I4,//,11X,'IT',6I3)
   12 FORMAT(1H ,10X,15A4,//)
   13 FORMAT(F10.6,A4)
   14 FORMAT(1H1)
   15 FORMAT(1H,//,10X,'**EIGENVECTORS REQUESTED BUT NO IT SUPPLIED**' )
   18 FORMAT(1H ,5F15.6)
C
      IDIM=54
      READ(1) TXTS,NL,NQ,NLM,NLMQ,SWSOA
      READ(1) NPX,NPY,NPZ,LAT,NPT,BOA,COA,DUM1
C
      READ(5,1) SWS,NPOINT,TAIL,IPRNTO,IPNCH,NOWRT,NOC,NOEVC,
     1NSTART,IPTP,(IT(IQ),IQ=1,NQ)
C
      NT=IT(NQ)
      DO 34 IQ=1,NQ
      IF(IT(IQ).EQ.0.AND.NOEVC.EQ.1) GO TO 33
   34 CONTINUE
C
      CALL CRLMQ(NQ,NL)
C
      IF(NLMQ.GT.IDIM) GO TO 24
      NLQ=NQ*NL
      NLMQM=NLMQ-1
      NLMT=NLMQ
      IF(TAIL.EQ.' T') NLMT=NLMQ-2*NL+1
```

174

```
      ALAT=SWS/SWSOA
      FOURPI=16.D0*DATAN(1.D0)
      VOL=NQ*FOURPI/3.D0*SWS**3
      ACOR=DSQRT(VOL/2.D0)/FOURPI
C
      WRITE(6,14)
      IF(NOC.EQ.0) GO TO 29
      READ(3) TXTC
      WRITE(6,12) TXTC
   29 WRITE(6,8) TXTS,NL,NQ,NLM,NLMQ,NLMT,SWS,ALAT
      WRITE(6,11) NPOINT,IPRNTO,IPNCH,NOWRT,NOC,NOEVC,NSTART,
     1(IT(IQ),IQ=1,NQ)
C
C     READ, CONVERT, AND PRINT POTENTIAL PARAMETERS
C
      INL=-NL
      DO 20 IQ=1,NQ
      INL=INL+NL
      READ(5,6)(TXTP(I,IQ),I=1,20)
      READ(5,13) SWSP(IQ),TTXT(IQ)
      WRITE(6,5)(TXTP(I,IQ),I=1,20),TTXT(IQ),SWSP(IQ)
      DO 20 IL=1,NL
      I=INL+IL
      TWOL=2*IL-1
      READ(5,4) ENY(I),OMM,SFIMSQ,FMOFP,FIAV
      OMN(I)=OMM*SWSP(IQ)**2
      FINSQ(I)=SFIMSQ/SWSP(IQ)
      FIJSQ(I)=FINSQ(I)/FMOFP**2
      OMJ(I)=OMN(I)-TWOL*SWSP(IQ)**2*SFIMSQ/FMOFP
      BMA=FIAV**2/SWSP(IQ)**4
      LNN(I)=1.D0+BMA*OMN(I)**2
      LNJ(I)=1.D0+BMA*OMN(I)*OMJ(I)
      LJJ(I)=1.D0+BMA*OMJ(I)**2
      WRITE(6,18) ENY(I),OMM,SFIMSQ,FMOFP,FIAV
   20 CONTINUE
C
      CALL SCASW(NQ,SWSP,SW,SWS)
C
      IF(IPNCH.NE.0) WRITE(2,6) (TXTP(I,1),I=1,20)
C
      DO 31 IQ=1,NQ
      SWSI=SW(IQ)
      SWSU(IQ)=SWSI/SWS
      SWSSQ(IQ)=SWSI*SWSI
      FCOR(IQ)=SWSI**2.5D0/ACOR
   31 HSWSCU(IQ)=0.5D0*SWSI**3
C
      CALL POTTRM(NLQ,NL,NQ,NOC)
C
C     WRITE INFORMATION FOR STATE DENSITY CALCULATIONS ON FILE 4
C
C     THIS IS BYPASSED IF NOEVC=0 OR IF THE PROGRAMME IS RESTARTED
C
      IF(NOEVC.EQ.0) GO TO 39
      IF(NSTART.GT.1) GO TO 38
      WRITE(4) TXTS(1),TXTS(2),TXTS(3),TXTS(4)
      WRITE(4) TXTS(5),TXTS(6),TXTS(7),TXTS(8)
      WRITE(4) TXTS(9),TXTS(10),TXTS(11),TXTS(12)
      WRITE(4) TXTS(13),TXTS(14),TXTS(15),NPOINT
      WRITE(4) TXTC(1),TXTC(2),TXTC(3),TXTC(4)
      WRITE(4) TXTC(5),TXTC(6),TXTC(7),TXTC(8)
      WRITE(4) TXTC(9),TXTC(10),TXTC(11),TXTC(12)
      WRITE(4) TXTC(13),TXTC(14),TXTC(15),NL
      WRITE(4) NQ,NLM,NLMT,NT
      WRITE(4) NPX,NPY,NPZ,LAT
      WRITE(4) NPT,BOA,COA,DUM1
      LT1=0
```

175

```
      INL=-NL
      DO 35 IQ=1,NQ
      INL=INL+NL
      LT2=IT(IQ)
      IF(LT1.EQ.LT2) GO TO 35
      WRITE(4) TXTP(1,IQ),TXTP(2,IQ),TXTP(3,IQ),TXTP(4,IQ)
      WRITE(4) TXTP(5,IQ),TXTP(6,IQ),TXTP(7,IQ),TXTP(8,IQ)
      WRITE(4) TXTP(9,IQ),TXTP(10,IQ),TXTP(11,IQ),TXTP(12,IQ)
      WRITE(4) TXTP(13,IQ),TXTP(14,IQ),TXTP(15,IQ),TXTP(16,IQ)
      WRITE(4) TXTP(17,IQ),TXTP(18,IQ),TXTP(19,IQ),TXTP(20,IQ)
      WRITE(4) SWS,SW(IQ),TTXT(IQ)
      DO 36 IL=1,NL
      I=INL+IL
      WRITE(4) ENY(I),OMN(I),OMJ(I),FINSQ(I)
      WRITE(4) FIJSQ(I),LNN(I),LNJ(I),LJJ(I)
   36 CONTINUE
   35 LT1=LT2
C
      GO TO 39
C
C     RESTART
C
   38 IREC=11+NT*6+NT*NL*2
      IREC=IREC+1+(NSTART-1)*(NLMQ+1)
      FIND(4'IREC)
C
   39 WRITE(6,14)
      DO 21 LPOINT=1,NPOINT
      IPRNT=IPRNTO
      IF(LPOINT.EQ.1) IPRNT=1
C
C     READ STRUCTURE CONSTANTS
C
      READ(1,END=23) NP,PKX,PKY,PKZ,W,POINT,KX,KY,KZ
      J1=-IDIM
      DO 22 J=1,NLMQ
      J1=J1+IDIM
      I1=J1+J
      I2=J1+NLMQ
      READ(1) (SRL(LIN),LIN=I1,I2)
   22 READ(1) (SIL(LIN),LIN=I1,I2)
      IF(LPOINT.LT.NSTART) GO TO 32
C
C     SCALE TO SWS
C
      J1=-IDIM
      DO 30 J=1,NLMQ
      J1=J1+IDIM
      L=LL(J)
      JQ=(J-1)/NLM+1
      TWOL=2*L+1
      SWSJ=SWSU(JQ)**TWOL
      DO 30 I=J,NLMQ
      IQ=(I-1)/NLM+1
      LIN=J1+I
      LP=LL(I)
      TWOLP=2*LP+1
      SWSI=SWSU(IQ)**TWOLP
      FACS=DSQRT(SWSJ*SWSI)
      SRL(LIN)=SRL(LIN)*FACS
   30 SIL(LIN)=SIL(LIN)*FACS
C
C     FILL IN UPPER TRIANGLE
C
      J1=-IDIM
      DO 25 J=1,NLMQM
      J1=J1+IDIM
      JP=J+1
```

176

```
      DO 25 I=JP,NLMQ
      I1=(I-1)*IDIM
      LIN=J1+I
      LIH=I1+J
      SRL(LIH)=SRL(LIN)
   25 SIL(LIH)=-SIL(LIN)
   32 IF(NOC.EQ.0) GO TO 26
C
C     READ CORRECTION TERMS IF REQUESTED
C
      J1=-IDIM
      DO 27 J=1,NLMQ
      J1=J1+IDIM
      I1=J1+J
      I2=J1+NLMQ
      READ(3) (DDLR(LIN),LIN=I1,I2)
   27 READ(3) (DDLI(LIN),LIN=I1,I2)
C
C     FILL IN UPPER TRIANGLE
C
      J1=-IDIM
      DO 28 J=1,NLMQM
      J1=J1+IDIM
      JP=J+1
      DO 28 I=JP,NLMQ
      I1=(I-1)*IDIM
      LIN=J1+I
      LIH=I1+J
      DDLR(LIH)=DDLR(LIN)
   28 DDLI(LIH)=-DDLI(LIN)
   26 IF(LPOINT.LT.NSTART) GO TO 21
      IF(IPRNT.EQ.1) WRITE(6,10) POINT,PKX,PKY,PKZ,NP,KX,KY,KZ
C
      CALL HMTOLP(NLMQ,NLMT,NOWRT,NOC,NLM)
      CALL REIGNV(NLMT,TAIL,NLM,NL,NQ,NOWRT,IPNCH,IPRNT,NOC,W)
C
   21 CONTINUE
      STOP
   23 NPP=LPOINT-1
      WRITE(6,7) NPP
      STOP
   24 WRITE(6,9) NLMQ,IDIM
      STOP
   33 WRITE(6,15)
      STOP
   37 WRITE(6,2) TAIL,NQ
      STOP
      END
      SUBROUTINE CRLMQ(NQ,NL)
C     ****************************************************************
C     *                                                              *
C     *    INSERT DATA FOR INDEX GYMNASTICS                          *
C     *                                                              *
C     ****************************************************************
      COMMON/LDAT/LLP(54),LL(54)
C
      LA=0
      DO 20 LQ=1,NQ
      IQ=(LQ-1)*NL
      DO 20 LP=1,NL
      L=LP-1
      NM=2*L+1
      LLPQ=IQ+LP
      DO 20 MP=1,NM
      LA=LA+1
      LL(LA)=L
   20 LLP(LA)=LLPQ
```

177

```
      RETURN
      END
      SUBROUTINE POTTRM(NLQD,NLD,NQ,NOC)
C     ****************************************************************
C     *                                                              *
C     *    CALCULATES POTENTIAL-PARAMETER-DEPENDENT TERMS TO BE      *
C     *    INSERTED INTO THE ONE-,TWO-,AND THREE-CENTRE TERMS OF     *
C     *    THE HAMILTON AND OVERLAP MATRICES                         *
C     *                                                              *
C     ****************************************************************
      IMPLICIT REAL*8 (A-H,O-Z)
      REAL*8 LNN,LNJ,LJJ
      COMMON/B1/ SRL(2916),SIL(2916),ENY(18),OMN(18),OMJ(18),FINSQ(18) ,
     1FIJSQ(18),ONEH(18),TWOH(18),CRYH(18),ONEO(18),TWOO(18),CRYO(18),
     2NP,KX,KY,KZ,NOEVC,IDIM,DDLR(2916),DDLI(2916),HSWSCU(6),FCOR(6)
     3,SWSSQ(6),LNN(18),LNJ(18),LJJ(18),IT(6),IDUM,NT,SFIN(18),GAM(18) ,
     1BM(18),HR(54,54),HI(54,54),OR(54,54),OI(54,54)
      DIMENSION ONE(4),TWO(4),THRE(4)
      NLQ=NLQD
      NL=NLD
C
C     NORMALISATION TERMS
C
      L1=-NL
      DO 25 LQ=1,NQ
      L1=L1+NL
      DO 25 LP=1,NL
      LPQ=L1+LP
      SFINSQ=HSWSCU(LQ)*FINSQ(LPQ)
      SFIN(LPQ)=SFINSQ/SWSSQ(LQ)
      GAM(LPQ)=SFINSQ/(OMN(LPQ)-OMJ(LPQ))
      IF(DABS(OMJ(LPQ)).LT.1.D-03) GO TO 26
      BM(LPQ)=(LJJ(LPQ)-1.D0)/OMJ(LPQ)/OMJ(LPQ)
      GO TO 25
   26 BM(LPQ)=(LNN(LPQ)-1.D0)/OMN(LPQ)/OMN(LPQ)
   25 CONTINUE
C
C     ONE-CENTRE TERMS
C
      L1=-NL
      DO 20 LQ=1,NQ
      L1=L1+NL
      DO 20 LP=1,NL
      LPQ=L1+LP
      ANOM=HSWSCU(LQ)*FINSQ(LPQ)
      ALMO=LNN(LPQ)/ANOM*SWSSQ(LQ)
      ALMH=OMN(LPQ)/ANOM+ENY(LPQ)*ALMO
      ONEO(LPQ)=ALMO
   20 ONEH(LPQ)=ALMH
C
C     TWO-CENTRE TERMS
C
      L1=-NL
      DO 21 LQ=1,NQ
      L1=L1+NL
      DO 21 LP=1,NL
      LPQ=L1+LP
      ANOM=OMN(LPQ)-OMJ(LPQ)
      ALMO=LNJ(LPQ)/ANOM*SWSSQ(LQ)
      ALMH=OMJ(LPQ)/ANOM+ENY(LPQ)*ALMO
      TWOO(LPQ)=-ALMO
   21 TWOH(LPQ)=-ALMH
C
C     THREE-CENTRE TERMS
C
      L1=-NL
      DO 22 LQ=1,NQ
```

```
      L1=L1+NL
      DO 22 LP=1,NL
      LPQ=L1+LP
      L=LP-1
      ALJNS=(2*L+1)**2
      ANOM=4.*ALJNS*HSWSCU(LQ)*FIJSQ(LPQ)
      ALMO=LJJ(LPQ)/ANOM*SWSSQ(LQ)
      ALMH=OMJ(LPQ)/ANOM+ENY(LPQ)*ALMO
      CRYO(LPQ)=ALMO
   22 CRYH(LPQ)=ALMH
C
      IF(NOC.EQ.0) RETURN
C
C     CORRECTION TERMS
C
      DO 23 LP=1,NL
      L=LP-1
      TWOL=2*L
      TWOL1=TWOL+1.DO
      TWOL3=TWOL+3.DO
      TWOL5=TWOL+5.DO
      TWOL7=TWOL+7.DO
      ONE(LP)=8.DO*TWOL3/TWOL5/TWOL5+2.DO*TWOL1*TWOL1/TWOL7/TWOL5/TWOL 5
      TWO(LP)=2.DO/TWOL1/TWOL5
   23 THRE(LP)=0.5DO/TWOL1/TWOL1/TWOL3
      L1=-NL
      DO 24 LQ=1,NQ
      L1=L1+NL
      DO 24 LP=1,NL
      LPQ=L1+LP
      ONES=ONE(LP)*SWSSQ(LQ)
      TWOS=TWO(LP)*SWSSQ(LQ)
      TRES=THRE(LP)*SWSSQ(LQ)
      ONEO(LPQ)=ONEO(LPQ)-ONES
      TWOO(LPQ)=TWOO(LPQ)+TWOS
   24 CRYO(LPQ)=CRYO(LPQ)-TRES
      RETURN
      END
      SUBROUTINE REIGNV(LMTD,TAIL,NLMD,NLD,NQD,NOWRT,IPNCH,IPRNT,NOC,W )
C     **********************************************************************
C     *                                                                    *
C     *    CALCULATES EIGENVALUES,EIGENVECTORS AND T,L-CHARACTERS AND       *
C     *    STORES THE RESULTS FOR LATER USE                                 *
C     *                                                                    *
C     **********************************************************************
      IMPLICIT REAL*8 (A-H,O-Z)
      INTEGER TAIL
      REAL*8 LNN,LNJ,LJJ
      COMMON/B1/ SRL(2916),SIL(2916),ENY(18),OMN(18),OMJ(18),FINSQ(18) ,
     1FIJSQ(18),ONEH(18),TWOH(18),CRYH(18),ONEO(18),TWOO(18),CRYO(18),
     2NP,KX,KY,KZ,NOEVC,IDIM,DDLR(2916),DDLI(2916),HSWSCU(6),FCOR(6)
     3,SWSSQ(6),LNN(18),LNJ(18),LJJ(18),IT(6),IDUM,NT,SFIN(18),GAM(18) ,
     1BM(18),HR(54,54),HI(54,54),OR(54,54),OI(54,54)
      COMMON/LDAT/LLP(54),LL(54)
      DIMENSION D(54),E2(54),E(54),IEV(54),EVR(54,54),EVI(54,54),
     1EVRL(2916),EVIL(2916),PRL(2916),PIL(2916),DIAG(54),ORL(2916),
     2OIL(2916),TAU(2,54)
      DIMENSION AMSO(6),AMPO(6),AMDO(6),AMFO(6),BLJKS(54,6),BLJKP(54,6),
     1BLJKD(54,6),BLJKF(54,6)
      EQUIVALENCE (HR(1),PRL(1)),(HI(1),PIL(1)),(OR(1),ORL(1)),
     1(OI(1),OIL(1)),(EVR(1),EVRL(1)),(EVI(1),EVIL(1))
    1 FORMAT(3I3,1X,10I7)
    2 FORMAT(F10.4,10I7)
    3 FORMAT(5(10X,10I7))
    4 FORMAT(1H0,5X,F10.4,9F8.4,/,16X,9F8.4,/,16X,9F8.4,/,16X,9F8.4,/
     1,16X,9F8.4,/,16X,9F8.4)
    5 FORMAT(1H ,'***ERROR IN EIGENVALUE PACKAGE , IRET =',//,
     110X,6I5,/)
```

179

```
      6 FORMAT(1H0,9(1PD13.5),/,1X,9(1PD13.5))
      7 FORMAT(1H0,10X,'REAL PART OF EIGENVECTORS',//)
      8 FORMAT(1H0,10X,'IMAG PART OF EIGENVECTORS',//)
      9 FORMAT(1H ,//,6X,'E   ',9F10.5)
     10 FORMAT(1H0,5X,'S   ',9F10.5)
     11 FORMAT(1H ,5X,'P   ',9F10.5)
     12 FORMAT(1H ,5X,'D   ',9F10.5)
     13 FORMAT(1H ,5X,'F   ',9F10.5)
C
      LMQ=LMTD
      NLM=NLMD
      NL=NLD
      NQ=NQD
C
C     EVALUATE EIGENVALUES ONLY
C
      IF(NOEVC.NE.0) GO TO' 21
C
      CALL HTRIDI(IDIM,LMQ,HR,HI,D,E,E2,TAU)
      CALL IMTQL1(LMQ,D,E,IERR)
C
      IF(IPRNT.EQ.0) GO TO 45
      IF(IERR.EQ.0) GO TO 22
      WRITE(6,5) IERR
     22 DO 20 I=1,LMQ
      EVT=D(I)
     20 IEV(I)=EVT*1.D06+DSIGN(0.5D0,EVT)
      WRITE(6,4) W,(D(L),L=1,LMQ)
C
C     WRITE EIGENVALUES IN CARD FORMAT ON FILE 2
C
     45 IF(IPNCH.EQ.0) GO TO 36
      LO=MINO(LMQ,10)
      WRITE(2,1) KX,KY,KZ,(IEV(L),L=1,LO)
      IF(LMQ.LT.10) GO TO 36
      LO=LO+1
      L1=MINO(LMQ,20)
      WRITE(2,2) W,(IEV(L),L=LO,L1)
      IF(LMQ.LE.20) GO TO 36
      L1=MINO(LMQ,30)
      WRITE(2,3) (IEV(L),L=21,L1)
      IF(LMQ.LE.30) GO TO 36
      L1=MINO(LMQ,40)
      WRITE(2,3) (IEV(L),L=31,L1)
      IF(LMQ.LE.40) GO TO 36
      L1=MINO(LMQ,50)
      WRITE(2,3) (IEV(L),L=41,L1)
      IF(LMQ.LE.50) GO TO 36
      WRITE(2,3) (IEV(L),L=51,LMQ)
     36 RETURN
C
C     EVALUATE EIGENVALUES AND EIGENVECTORS
C
     21 CALL HTRIDI(IDIM,LMQ,HR,HI,D,E,E2,TAU)
C
      LMQM=LMQ-1
      I1=-IDIM
      DO 31 I=1,LMQ
      I1=I1+IDIM
      LIN=I1+I
     31 EVRL(LIN)=1.D0
      J1=0
      DO 33 J=2,LMQM
      JP=J+1
      JM=J-1
      J1=J1+IDIM
      DO 32 I=JP,LMQ
```

180

```
          LIN=J1+I
    32 EVRL(LIN)=0.D0
          DO 33 I=1,JM
          LIN=J1+I
    33 EVRL(LIN)=0.D0
          DO 34 I=2,LMQ
    34 EVRL(I)=0.D0
          I1=(LMQ-1)*IDIM
          DO 35 I=1,LMQM
          LIN=I1+I
    35 EVRL(LIN)=0.D0
C
          CALL IMTQL2(IDIM,LMQ,D,E,EVR,IERR)
          CALL HTRIBK(IDIM,LMQ,HR,HI,TAU,LMQ,EVR,EVI)
C
C     INVERT L BY SOLVING L*L(-1)=1
C             =             = =     =
C
          I1=-IDIM
          DO 24 I=1,LMQ
          I1=I1+IDIM
          LIN=I1+I
          DIT=1.D0/ORL(LIN)
          DIAG(I)=DIT
          PRL(LIN)=DIT
    24 PIL(LIN)=0.D0
          J1=-IDIM
          DO 25 J=1,LMQM
          JP=J+1
          J1=J1+IDIM
          DO 25 I=JP,LMQ
          IM=I-1
          LIN=J1+I
          PRIJ=0.D0
          PIIJ=0.D0
          DO 26 K=J,IM
          K1=(K-1)*IDIM
          LIK=K1+I
          LKJ=J1+K
          ORIK=ORL(LIK)
          OIIK=OIL(LIK)
          PRKJ=PRL(LKJ)
          PIKJ=PIL(LKJ)
          PRIJ=PRIJ-ORIK*PRKJ+OIIK*PIKJ
    26 PIIJ=PIIJ-ORIK*PIKJ-OIIK*PRKJ
          PRL(LIN)=PRIJ*DIAG(I)
    25 PIL(LIN)=PIIJ*DIAG(I)
C
C     CALCULATE THE EIGENVECTOR MATRIX APPROPRIATE FOR THE ORIGINAL
C     GENERALISED EIGENVALUE PROBLEM:
C
C     X=L(-H)*Y
C     = =      =
C
          J1=-IDIM
          DO 27 J=1,LMQ
          J1=J1+IDIM
          I1=-IDIM
          DO 27 I=1,LMQ
          I1=I1+IDIM
          LIN=J1+I
          ORIJ=0.D0
          OIIJ=0.D0
          DO 28 K=I,LMQ
          KI=I1+K
          KJ=J1+K
          PRKI=PRL(KI)
          PIKI=PIL(KI)
```

```
          EVRKJ=EVRL(KJ)
          EVIKJ=EVIL(KJ)
          ORIJ=ORIJ+PRKI*EVRKJ+PIKI*EVIKJ
       28 OIIJ=OIIJ+PRKI*EVIKJ-PIKI*EVRKJ
          ORL(LIN)=ORIJ
       27 OIL(LIN)=OIIJ
C
C         OR AND OI NOW CONTAIN THE TRANSFORMED EIGENVECTOR MATRIX
C
          IF(NOWRT.NE.2) GO TO 47
C
          IF(IERR.EQ.0) GO TO 23
          WRITE(6,5) IERR
       23 WRITE(6,4) W,(D(L),L=1,LMQ)
          WRITE(6,7)
          J1=-IDIM
          DO 30 J=1,LMQ
          J1=J1+IDIM
          I1=J1+1
          I2=J1+LMQ
       30 WRITE(6,6) (ORL(LIN),LIN=I1,I2)
          WRITE(6,8)
          J1=-IDIM
          DO 29 J=1,LMQ
          J1=J1+IDIM
          I1=J1+1
          I2=J1+LMQ
       29 WRITE(6,6) (OIL(LIN),LIN=I1,I2)
C
       47 I1=-IDIM
C
C         WRITE K-VECTOR INFORMATION ON FILE 4
C
          WRITE(4) W,KX,KY,KZ
C
          DO 37 J=1,LMQ
          I1=I1+IDIM
          DO 38 KT=1,NT
          AMSO(KT)=0.D0
          AMPO(KT)=0.D0
          AMDO(KT)=0.D0
       38 AMFO(KT)=0.D0
          IQQ=-NLM
          DO 39 IQ=1,NQ
          KT=IT(IQ)
          IQQ=IQQ+NLM
C
C         T,S CHARACTER
C
          IQP=IQQ+1
          LQP=LLP(IQP)
          SFIP=SFIN(LQP)
          I2=-IDIM
          CR=0.D0
          CI=0.D0
          DO 40 K=1,LMQ
          I2=I2+IDIM
          LINQ=IQP+I2
          LIN=I1+K
          CR=CR+ORL(LIN)*SRL(LINQ)-OIL(LIN)*SIL(LINQ)
       40 CI=CI+OIL(LIN)*SRL(LINQ)+ORL(LIN)*SIL(LINQ)
          CR=CR*GAM(LQP)
          CI=CI*GAM(LQP)
          LE=I1+IQP
          ARS=ORL(LE)-CR
          AIS=OIL(LE)-CI
          BRS=ORL(LE)*OMN(LQP)-OMJ(LQP)*CR
```

```
      BIS=OIL(LE)*OMN(LQP)-OMJ(LQP)*CI
      BSQ=BRS*BRS+BIS*BIS
      AMSO(KT)=AMSO(KT)+(ARS*ARS+AIS*AIS+BM(LQP)*BSQ)/SFIP
C
C     T,P CHARACTER
C
      DO 41 MP=2,4
      IQP=IQQ+MP
      LQP=LLP(IQP)
      SFIP=SFIN(LQP)
      I2=-IDIM
      CR=0.DO
      CI=0.DO
      DO 42 K=1,LMQ
      I2=I2+IDIM
      LINQ=IQP+I2
      LIN=I1+K
      CR=CR+ORL(LIN)*SRL(LINQ)-OIL(LIN)*SIL(LINQ)
   42 CI=CI+OIL(LIN)*SRL(LINQ)+ORL(LIN)*SIL(LINQ)
      CR=CR*GAM(LQP)
      CI=CI*GAM(LQP)
      LE=I1+IQP
      ARP=ORL(LE)-CR
      AIP=OIL(LE)-CI
      BRP=ORL(LE)*OMN(LQP)-OMJ(LQP)*CR
      BIP=OIL(LE)*OMN(LQP)-OMJ(LQP)*CI
      BSQ=BRP*BRP+BIP*BIP
   41 AMPO(KT)=AMPO(KT)+(ARP*ARP+AIP*AIP+BM(LQP)*BSQ)/SFIP
C
C     T,D CHARACTER
C
      DO 43 MP=5,9
      IQP=IQQ+MP
      LQP=LLP(IQP)
      SFIP=SFIN(LQP)
      I2=-IDIM
      CR=0.DO
      CI=0.DO
      DO 44 K=1,LMQ
      I2=I2+IDIM
      LINQ=IQP+I2
      LIN=I1+K
      CR=CR+ORL(LIN)*SRL(LINQ)-OIL(LIN)*SIL(LINQ)
   44 CI=CI+OIL(LIN)*SRL(LINQ)+ORL(LIN)*SIL(LINQ)
      CR=CR*GAM(LQP)
      CI=CI*GAM(LQP)
      LE=I1+IQP
      ARD=ORL(LE)-CR
      AID=OIL(LE)-CI
      BRD=ORL(LE)*OMN(LQP)-OMJ(LQP)*CR
      BID=OIL(LE)*OMN(LQP)-OMJ(LQP)*CI
      BSQ=BRD*BRD+BID*BID
   43 AMDO(KT)=AMDO(KT)+(ARD*ARD+AID*AID+BM(LQP)*BSQ)/SFIP
      IF(NL.LT.4) GO TO 39
C
C     T,F CHARACTER
C
      DO 53 MP=10,16
      IQP=IQQ+MP
      LQP=LLP(IQP)
      SFIP=SFIN(LQP)
      I2=-IDIM
      CR=0.DO
      CI=0.DO
      DO 54 K=1,LMQ
      I2=I2+IDIM
      LINQ=IQP+I2
      LIN=I1+K
      CR=CR+ORL(LIN)*SRL(LINQ)-OIL(LIN)*SIL(LINQ)
```

```
    54 CI=CI+OIL(LIN)*SRL(LINQ)+ORL(LIN)*SIL(LINQ)
       CR=CR*GAM(LQP)
       CI=CI*GAM(LQP)
       LE=I1+IQP
       IF(TAIL.NE.´ T´.OR.KT.NE.NT) GO TO 55
       ARF=CR
       AIF=CI
       BRF=OMJ(LQP)*CR
       BIF=OMJ(LQP)*CI
       BSQ=BRF*BRF+BIF*BIF
       GO TO 53
    55 ARF=ORL(LE)-CR
       AIF=OIL(LE)-CI
       BRF=ORL(LE)*OMN(LQP)-OMJ(LQP)*CR
       BIF=OIL(LE)*OMN(LQP)-OMJ(LQP)*CI
       BSQ=BRF*BRF+BIF*BIF
    53 AMFO(KT)=AMFO(KT)+(ARF*ARF+AIF*AIF+BM(LQP)*BSQ)/SFIP
    39 CONTINUE
C
C      RENORMALISE T,L-CHARACTERS IF CORRECTION TERMS ARE INCLUDED
C
       IF(NL.LT.4) GO TO 57
       IF(NOC.EQ.0) GO TO 59
       SUMA=0.D0
       DO 60 KT=1,NT
    60 SUMA=SUMA+AMSO(KT)+AMPO(KT)+AMDO(KT)+AMFO(KT)
       DO 61 KT=1,NT
       AMSO(KT)=AMSO(KT)/SUMA
       AMPO(KT)=AMPO(KT)/SUMA
       AMDO(KT)=AMDO(KT)/SUMA
    61 AMFO(KT)=AMFO(KT)/SUMA
C
C       WRITE EIGENVALUES AND T,L-CHARACTERS ON FILE 4
C
    59 WRITE(4) D(J),(AMSO(KT),AMPO(KT),AMDO(KT),AMFO(KT),KT=1,NT)
       GO TO 58
    57 IF(NOC.EQ.0) GO TO 64
       SUMA=0.D0
       DO 62 KT=1,NT
    62 SUMA=SUMA+AMSO(KT)+AMPO(KT)+AMDO(KT)
       DO 63 KT=1,NT
       AMSO(KT)=AMSO(KT)/SUMA
       AMPO(KT)=AMPO(KT)/SUMA
    63 AMDO(KT)=AMDO(KT)/SUMA
C
C       WRITE EIGENVALUES AND T,L-CHARACTERS ON FILE 4
C
    64 WRITE(4) D(J),(AMSO(KT),AMPO(KT),AMDO(KT),KT=1,NT)
    58 DO 46 KT=1,NT
       BLJKS(J,KT)=AMSO(KT)
       BLJKP(J,KT)=AMPO(KT)
       BLJKD(J,KT)=AMDO(KT)
       IF(NL.LT.4) GO TO 46
       BLJKF(J,KT)=AMFO(KT)
    46 CONTINUE
    37 CONTINUE
C
       IF(IPRNT.NE.1) GO TO 51
       NR=LMQ/9
       IQQ=-9
       DO 48 IQ=1,NR
       IQQ=IQQ+9
       I1=IQQ+1
       I2=IQQ+9
       WRITE(6,9) (D(J),J=I1,I2)
       DO 48 KT=1,NT
       WRITE(6,10) (BLJKS(J,KT),J=I1,I2)
```

```
      WRITE(6,11) (BLJKP(J,KT),J=I1,I2)
      WRITE(6,12) (BLJKD(J,KT),J=I1,I2)
      IF(NL.LT.4) GO TO 48
      WRITE(6,13) (BLJKF(J,KT),J=I1,I2)
   48 CONTINUE
      IDIF=LMQ-NR*9
      IF(IDIF.EQ.0) GO TO 51
      I1=I2+1
      I2=I2+IDIF
      WRITE(6,9) (D(J),J=I1,I2)
      DO 52 KT=1,NT
      WRITE(6,10) (BLJKS(J,KT),J=I1,I2)
      WRITE(6,11) (BLJKP(J,KT),J=I1,I2)
      WRITE(6,12) (BLJKD(J,KT),J=I1,I2)
      IF(NL.LT.4) GO TO 52
      WRITE(6,13) (BLJKF(J,KT),J=I1,I2)
   52 CONTINUE
C
C     WRITE EIGENVALUES IN CARD FORMAT ON FILE 2
C
   51 IF(IPNCH.EQ.0) GO TO 56
      DO 50 I=1,LMQ
      EVT=D(I)
   50 IEV(I)=EVT*1.D06+DSIGN(0.5D0,EVT)
      L0=MIN0(LMQ,10)
      WRITE(2,1) KX,KY,KZ,(IEV(L),L=1,L0)
      IF(LMQ.LT.10) GO TO 56
      L0=L0+1
      L1=MIN0(LMQ,20)
      WRITE(2,2) W,(IEV(L),L=L0,L1)
      IF(LMQ.LE.20) GO TO 56
      L1=MIN0(LMQ,30)
      WRITE(2,3) (IEV(L),L=21,L1)
      IF(LMQ.LE.30) GO TO 56
      L1=MIN0(LMQ,40)
      WRITE(2,3) (IEV(L),L=31,L1)
      IF(LMQ.LE.40) GO TO 56
      L1=MIN0(LMQ,50)
      WRITE(2,3) (IEV(L),L=41,L1)
      IF(LMQ.LE.50) GO TO 56
      WRITE(2,3) (IEV(L),L=51,LMQ)
   56 RETURN
      END
      SUBROUTINE HMTOLP(LQ,LT,NOWRT,NOC,NLMD)
C     ****************************************************************
C     *                                                              *
C     *   SET UP HAMILTON AND OVERLAP MATRICES, CHOLESKY DECOMPOSE    *
C     *   THE OVERLAP MATRIX, AND TRANSFORM THE HAMILTON MATRIX       *
C     *   ACCORDINGLY·                                                *
C     *                                                              *
C     ****************************************************************
C
      IMPLICIT REAL*8 (A-H,O-Z)
      REAL*8 LNN,LNJ,LJJ
      COMMON/B1/ SRL(2916),SIL(2916),ENY(18),OMN(18),OMJ(18),FINSQ(18) ,
     1FIJSQ(18),ONEH(18),TWOH(18),CRYH(18),ONEO(18),TWOO(18),CRYO(18),
     2NP,KX,KY,KZ,NOEVC,IDIM,DDLR(2916),DDLI(2916),HSWSCU(6),FCOR(6)
     3,SWSSQ(6),LNN(18),LNJ(18),LJJ(18),IT(6),IDUM,NT,SFIN(18),GAM(18) ,
     1BM(18),HR(54,54),HI(54,54),OR(54,54),OI(54,54)
      COMMON/LDAT/LLP(54),LL(54)
      DIMENSION ORD(54),PRL(2916),PIL(2916),HRL(2916),HIL(2916),
     1ORL(2916),OIL(2916)
      DATA PRL/2916*0.D0/,PIL/2916*0.D0/
      EQUIVALENCE (HR(1),HRL(1)),(HI(1),HIL(1)),(OR(1),ORL(1)),
     1(OI(1),OIL(1))
    1 FORMAT(1H ,9E12.4)
    2 FORMAT(1H ,//)
C
```

```
      LMQ=LT
      LMT=LQ
      LMQM=LMQ-1
      NLM=NLMD
C
      J1=-IDIM
      DO 20 JLQ=1,LMQM
C
C     DIAGONAL ELEMENTS
C
      J1=J1+IDIM
      LIND=J1+JLQ
      JLQP=JLQ+1
      LPQ=LLP(JLQ)
      TWOHP=TWOH(LPQ)
      TWOOP=TWOO(LPQ)
      SRJJ=SRL(LIND)
      HRT=(2.D0*TWOHP-1.D0)*SRJJ
      ORT=2.D0*TWOOP*SRJJ
C
C     THREE-CENTRE TERMS
C
      DO 23 ILP=1,LMT
      LIN=J1+ILP
      LPPP=LLP(ILP)
      SRII=SRL(LIN)
      SIII=SIL(LIN)
      SRIIS=SRII*SRII+SIII*SIII
      HRT=HRT+CRYH(LPPP)*SRIIS
   23 ORT=ORT+CRYO(LPPP)*SRIIS
      HRL(LIND)=HRT+ONEH(LPQ)
      HIL(LIND)=0.D0
      ORL(LIND)=ORT+ONEO(LPQ)
      OIL(LIND)=0.D0
C
C     OFF-DIAGONAL ELEMENTS
C
      DO 21 ILQ=JLQP,LMQ
      LIN=J1+ILQ
      LPPQ=LLP(ILQ)
      TWOHPP=TWOH(LPPQ)+TWOHP-1.D0
      TWOOPP=TWOO(LPPQ)+TWOOP
      SRIJ=SRL(LIN)
      SIIJ=SIL(LIN)
      HRT=TWOHPP*SRIJ
      HIT=TWOHPP*SIIJ
      ORT=TWOOPP*SRIJ
      OIT=TWOOPP*SIIJ
      K1=-IDIM
C
C     THREE-CENTRE TERMS
C
      DO 22 ILP=1,LMT
      K1=K1+IDIM
      L1=J1+ILP
      L2=K1+ILQ
      LPPP=LLP(ILP)
      SRIJ=SRL(L1)
      SIIJ=SIL(L1)
      SRQP=SRL(L2)
      SIQP=SIL(L2)
      CRYHP=CRYH(LPPP)
      CRYOP=CRYO(LPPP)
      DR=SRQP*SRIJ-SIQP*SIIJ
      DI=SRQP*SIIJ+SIQP*SRIJ
      HRT=HRT+CRYHP*DR
      HIT=HIT+CRYHP*DI
      ORT=ORT+CRYOP*DR
```

```
   22 OIT=OIT+CRYOP*DI
      HRL(LIN)=HRT
      HIL(LIN)=HIT
      ORL(LIN)=ORT
      OIL(LIN)=OIT
   21 CONTINUE
   20 CONTINUE
C
C     LAST DIAGONAL ELEMENT
C
      J1=J1+IDIM
      LIND=J1+LMQ
      LPQ=LLP(LMQ)
      TWOHP=TWOH(LPQ)
      TWOOP=TWOO(LPQ)
      SRJJ=SRL(LIND)
      HRT=(2.D0*TWOHP-1.D0)*SRJJ
      ORT=2.D0*TWOOP*SRJJ
C
C     THREE-CENTRE TERMS
C
      DO 24 ILP=1,LMT
      LIN=J1+ILP
      LPPP=LLP(ILP)
      SRII=SRL(LIN)
      SIII=SIL(LIN)
      SRIIS=SRII*SRII+SIII*SIII
      HRT=HRT+CRYH(LPPP)*SRIIS
   24 ORT=ORT+CRYO(LPPP)*SRIIS
      HRL(LIND)=HRT+ONEH(LPQ)
      HIL(LIND)=0.D0
      ORL(LIND)=ORT+ONEO(LPQ)
      OIL(LIND)=0.D0
C
C     CORRECTION TERMS
C
      IF(NOC.EQ.0) GO TO 27
      J1=-IDIM
      DO 28 JLQ=1,LMQ
      JQ=(JLQ-1)/NLM+1
      FQ=FCOR(JQ)
      J1=J1+IDIM
      DO 28 ILQ=JLQ,LMQ
      IQ=(ILQ-1)/NLM+1
      LIN=J1+ILQ
      FQQ=FQ*FCOR(IQ)
      DTR=FQQ*DDLR(LIN)
      DTI=FQQ*DDLI(LIN)
      ORL(LIN)=ORL(LIN)+DTR
   28 OIL(LIN)=OIL(LIN)+DTI
C
   27 IF(NOWRT.NE.1) GO TO 25
      WRITE(6,2)
      WRITE(6,1)((HR(JLM,ILM),ILM=1,JLM),JLM=1,LQ)
      WRITE(6,2)
      WRITE(6,1)((HI(JLM,ILM),ILM=1,JLM),JLM=1,LQ)
      WRITE(6,2)
      WRITE(6,1)((OR(JLM,ILM),ILM=1,JLM),JLM=1,LQ)
      WRITE(6,2)
      WRITE(6,1)((OI(JLM,ILM),ILM=1,JLM),JLM=1,LQ)
   25 CONTINUE
C
C     FIND THE MATRIX L WHICH FACTORISES THE OVERLAP MATRIX:
C                          =
C
C     O=L*L(H)
C     = = =
```

```
C
      OR11=DSQRT(ORL(1))
      ORD(1)=OR11
      ORL(1)=OR11
      ORL(2)=ORL(2)/OR11
      OIL(2)=OIL(2)/OR11
      J2=IDIM+2
      OR22=DSQRT(ORL(J2)-ORL(2)**2-OIL(2)**2)
      ORD(2)=OR22
      ORL(J2)=OR22
      DO 44 I=3,LMQ
      ORL(I)=ORL(I)/OR11
   44 OIL(I)=OIL(I)/OR11
      J1=0
      DO 40 J=2,LMQM
      JM=J-1
      JP=J+1
      J1=J1+IDIM
      LINP=J1+IDIM+JP
      ORJJ=ORD(J)
      DO 41 I=JP,LMQ
      LIN=J1+I
      ORIJ=ORL(LIN)
      OIIJ=OIL(LIN)
      K1=-IDIM
      DO 42 K=1,JM
      K1=K1+IDIM
      LI=K1+I
      LJ=K1+J
      ORIK=ORL(LI)
      ORJK=ORL(LJ)
      OIIK=OIL(LI)
      OIJK=OIL(LJ)
      ORIJ=ORIJ-ORIK*ORJK-OIIK*OIJK
   42 OIIJ=OIIJ+ORIK*OIJK-OIIK*ORJK
      ORL(LIN)=ORIJ/ORJJ
   41 OIL(LIN)=OIIJ/ORJJ
      ORII=ORL(LINP)
      K1=-IDIM
      DO 43 K=1,J
      K1=K1+IDIM
      JKP=K1+JP
   43 ORII=ORII-ORL(JKP)**2-OIL(JKP)**2
      ORII=DSQRT(ORII)
      ORD(JP)=ORII
      ORL(LINP)=ORII
   40 CONTINUE
C
C     OR AND OI NOW CONTAIN L
C                            =
C
C     SOLVE Z*L(H)=H
C           = =    =
C
      PRL(1)=HRL(1)/OR11
      DO 45 I=2,LMQ
      PRL(I)=HRL(I)/OR11
   45 PIL(I)=HIL(I)/OR11
      J1=0
      DO 46 J=2,LMQ
      JM=J-1
      J1=J1+IDIM
      DO 46 I=J,LMQ
      LIN=J1+I
      PRIJ=HRL(LIN)
      PIIJ=HIL(LIN)
      K1=-IDIM
```

188

```
      DO 47 K=1,JM
      K1=K1+IDIM
      LI=K1+I
      LJ=K1+J
      PRIK=PRL(LI)
      PIIK=PIL(LI)
      ORJK=ORL(LJ)
      OIJK=OIL(LJ)
      PRIJ=PRIJ-PRIK*ORJK-PIIK*OIJK
   47 PIIJ=PIIJ+PRIK*OIJK-PIIK*ORJK
      PRL(LIN)=PRIJ/ORD(J)
   46 PIL(LIN)=PIIJ/ORD(J)
C
C     SOLVE L*W=Z
C           = = =
C
      HRL(1)=PRL(1)/OR11
      DO 48 I=2,LMQ
      IM=I-1
      HRIJ=PRL(I)
      HIIJ=PIL(I)
      K1=-IDIM
      DO 49 K=1,IM
      K1=K1+IDIM
      LI=K1+I
      ORIK=ORL(LI)
      OIIK=OIL(LI)
      HRKJ=HRL(K)
      HIKJ=HIL(K)
      HRIJ=HRIJ-ORIK*HRKJ+OIIK*HIKJ
   49 HIIJ=HIIJ-ORIK*HIKJ-OIIK*HRKJ
      HRL(I)=HRIJ/ORD(I)
   48 HIL(I)=HIIJ/ORD(I)
      J1=0
      DO 50 J=2,LMQM
      J1=J1+IDIM
      JM=J-1
      JP=J+1
      LIN=J1+J
      HRJJ=PRL(LIN)
      HIJJ=PIL(LIN)
      K1=-IDIM
      DO 54 K=1,JM
      K1=K1+IDIM
      LJ=K1+J
      ORJK=ORL(LJ)
      OIJK=OIL(LJ)
      HRJK=HRL(LJ)
      HIJK=HIL(LJ)
      HRJJ=HRJJ-ORJK*HRJK-OIJK*HIJK
   54 HIJJ=HIJJ+ORJK*HIJK-OIJK*HRJK
      HRL(LIN)=HRJJ/ORD(J)
      HIL(LIN)=HIJJ/ORD(J)
      DO 50 I=JP,LMQ
      IM=I-1
      LIN=J1+I
      HRIJ=PRL(LIN)
      HIIJ=PIL(LIN)
      K1=-IDIM
      DO 51 K=1,JM
      K1=K1+IDIM
      LI=K1+I
      LJ=K1+J
      ORIK=ORL(LI)
      OIIK=OIL(LI)
      HRJK=HRL(LJ)
      HIJK=HIL(LJ)
      HRIJ=HRIJ-ORIK*HRJK-OIIK*HIJK
```

```
    51 HIIJ=HIIJ+ORIK*HIJK-OIIK*HRJK
       DO 52 K=J,IM
       K1=(K-1)*IDIM
       LI=K1+I
       LK=J1+K
       ORIK=ORL(LI)
       OIIK=OIL(LI)
       HRKJ=HRL(LK)
       HIKJ=HIL(LK)
       HRIJ=HRIJ-ORIK*HRKJ+OIIK*HIKJ
    52 HIIJ=HIIJ-ORIK*HIKJ-OIIK*HRKJ
       HRL(LIN)=HRIJ/ORD(I)
    50 HIL(LIN)=HIIJ/ORD(I)
       J1=J1+IDIM
       LIN=J1+LMQ
       HRJJ=PRL(LIN)
       HIJJ=PIL(LIN)
       K1=-IDIM
       DO 55 K=1,LMQM
       K1=K1+IDIM
       LJ=K1+LMQ
       ORJK=ORL(LJ)
       OIJK=OIL(LJ)
       HRJK=HRL(LJ)
       HIJK=HIL(LJ)
       HRJJ=HRJJ-ORJK*HRJK-OIJK*HIJK
    55 HIJJ=HIJJ+ORJK*HIJK-OIJK*HRJK
       HRL(LIN)=HRJJ/ORD(LMQ)
       HIL(LIN)=HIJJ/ORD(LMQ)
C
C
C     HR AND HI NOW CONTAIN THE TRANSFORMED HAMILTON MATRIX APPROPRIATE
C     FOR A STANDARD EIGENVALUE PROBLEM:
C
C     (L(-1)*H*L(-H) - E)*Y=0
C      =       = =       -
C
       IF(NOWRT.NE.1) GO TO 26
       WRITE(6,2)
       WRITE(6,1)((HR(JLM,ILM),ILM=1,JLM),JLM=1,LQ)
       WRITE(6,2)
       WRITE(6,1)((HI(JLM,ILM),ILM=1,JLM),JLM=1,LQ)
    26 CONTINUE
       RETURN
       END
       SUBROUTINE SCASW(NQD,RMUF,SW,SWS)
C     ******************************************************************
C     *                                                                *
C     *    SCALE WIGNER-SEITZ RADII ACCORDING TO THE NUMBERS RMUF(I)   *
C     *    CONSERVING THE VOLUME                                       *
C     *                                                                *
C     ******************************************************************
       IMPLICIT REAL*8 (A-H,O-Z)
       DIMENSION F(10),RMUF(NQD),SW(NQD)
C
       NQ=NQD
       OM=NQ*SWS**3
       RMUF1=RMUF(1)
       SUM=0.D0
       DO 20 I=1,NQ
       F(I)=RMUF(I)/RMUF1
    20 SUM=SUM+F(I)**3
       R1=(OM/SUM)**(1.D0/3.D0)
       DO 21 I=1,NQ
    21 SW(I)=F(I)*R1
       RETURN
       END
```

190

9.4.3 Execution of LMTO

The execution of LMTO requires a data set with the appropriate structure constant matrices, and, if requested, a data set containing correction-term
structure constant matrices. These matrices must be generated in previous
runs of STR and COR and stored in STR/XXX/A and COR/XXX/B, respectively.
They form the bulk of the input data for LMTO. The additional input needed
is essentially the appropriate potential parameters, exemplified in Table
9.4.

Table 9.4. Input for LMTO

2.684	1000 N100	1	1	01	0	1	

POTENTIAL PARAMETERS FOR NON-MAGNETIC CR   100381

2.684000 CR				
0.461561	0.208272	0.396336	0.868125	0.179964
0.620718	1.080268	0.361979	0.710534	0.153411
0.639058	0.151915	0.040556	0.063157	0.994676

The first line of this additional input contains data for control variables which govern the mode of operation. SWS is the average Wigner-Seitz
radius (8.16) in atomic units. NPOINT is the number of the last k vector to
be read from STR/XXX/A (and COR/XXX/A) and included in the eigenvalue procedure. It may be smaller, equal to, or larger than the actual number of k
points specified in STR. In the latter case, LMTO will stop execution when
End Of File is encountered by the READ(1,END=23) statement. TAIL = ' T' indicates that the $\ell = 3$ terms are included only in the internal ℓ'' summation
in the three-centre terms of (5.46,47), whereas max $(\ell,\ell') = 2$. Hence the
tails of non-occupied f orbitals may be included without increasing the
size of the eigenvalue problem. This option works only for monoatomic solids.
IPRNTO, IPNCH and NOWRT are print and punch options. IPRNTO = 0 gives the
shortest printout. IPNCH = 1 may be used to store eigenvalues in card format
and is typically used when eigenvectors and $t\ell$ characters are not calculated,
see NOEVC below. NOWRT = 1,2 may be used to display the contents of the
Hamiltonian, overlap and eigenvector matrices for testing purposes. Normally
NOWRT = 0. Inclusion of the correction to the ASA is governed by NOC. When
NOC = 1 the correction-term matrix data set COR/XXX/A must exist. NOEVC = 1
indicates that eigenvectors and $t\ell$ characters will be calculated. In this

case, the eigenvalues and the $t\ell$ characters will be stored together with some extra information in the standard energy-band data set BND/YY/B, which may be used by DDNS to generate projected state densities. If only eigenvalues are needed, the execution time for LMTO can be reduced by a factor of 2-3. This requires NOEVC = 0, and the eigenvalues are stored in card format by IPNCH = 1. Usually one would also take IPRNTO = 1 in this case. NSTART is the number of the first k vector included in the eigenvalue procedure. Hence, eigenvalues, etc. are generated in the k-vector range specified by NSTART, NPOINT. The type number of each atom in the primitive cell is given by IT(IQ), i.e. for an alloy like $AuCu_3$ it contains the four numbers 1,2,2,2, and is used to generate the $t\ell$ characters.

The second line of the input contains 80 characters of text describing the potential parameters to follow. It is often useful to include here the date at which these parameters were generated. The text will be included in the energy-band file BND/YY/B and printed by programmes which access this file.

The third line contains the atomic Wigner-Seitz radius S_t in atomic units for the atom of type t here labelled by the type number IT(IQ). Instead of the actual radius one may give an arbitrary number SWP(I), which will be used by the subroutine SCASW to generate the radius (radii) according to (9.10). TTXT(IQ) is the usual chemical symbol for the atom with type number IT(IQ).

The next NL lines contain E_ν and the four standard potential parameters defined by (4.1). To be able to specify $<\dot{\phi}^2> = 0$ the fourth parameter must be given as $<\dot{\phi}^2>^{\frac{1}{2}}$. If the primitive cell holds more than one atom, each extra atom should be described by data analogous to the above NL + 2 lines. In this case, there should be agreement between the sequence of potential-parameter data and IT(IQ) given in the first line.

9.4.4 Sample Output from LMTO

CORRECTION MATRIX FOR BCC STRUCTURE

STRUCTURE CONSTANTS FOR BCC LATTICE

NL   3 NQ   1 NLM   9 NLMQ   9 NLMT   9 SWS 2.684000 ALAT 5.45115

NPOINT 1000 IPRNT   1 IPNCH   0 NOWRT   0 NOC   1 NOEVC   1 NSTART

IT 1

POTENTIAL PARAMETERS FOR NON-MAGNETIC CR   100381

CR   SWS   2.684000

ENY	OM(-)	S*FI**2	FI(-)/FI(+)	<FID**2>**1/2
0.461561	0.208272	0.396336	0.868125	0.179964
0.620718	1.080268	0.361979	0.710534	0.153411
0.639058	0.151915	0.040556	0.063157	0.994676

EIGENVALUES AT   (PKX,PKY,PKZ,)   0.000000   0.000000   0.010

E	0.21382	0.73793	0.73793	0.73795	0.86993	0.86997	3.23
S	0.99994	0.00000	0.00000	0.00000	0.00000	0.00000	0.00
P	0.00006	0.00001	0.00001	0.00000	0.00000	0.00000	0.99
D	0.00000	0.99999	0.99999	1.00000	1.00000	1.00000	0.00

EIGENVALUES AT   (PKX,PKY,PKZ,)   0.125000   0.000000   0.000

E	0.21997	0.73791	0.73791	0.74003	0.86632	0.87160	3.20
S	0.99063	0.00000	0.00000	0.00000	0.00000	0.00003	0.01
P	0.00932	0.00104	0.00104	0.00000	0.00000	0.00005	0.98
D	0.00005	0.99896	0.99896	1.00000	1.00000	0.99992	0.00

EIGENVALUES AT   (PKX,PKY,PKZ,)   0.125000   0.125000   0.000

E	0.22610	0.73365	0.73744	0.74636	0.86552	0.87064	3.15
S	0.98129	0.00000	0.00004	0.00000	0.00003	0.00000	0.02
P	0.01844	0.00000	0.00225	0.00201	0.00002	0.00007	0.94
D	0.00026	1.00000	0.99771	0.99799	0.99995	0.99993	0.03

9.4.5 Energy-Band File BND/YY/B

The file BND/YY/B, which after execution of LMTO (if NOEVC = 1) contains
the eigenvalues and tℓ characters, is a random (direct) access file. This
file organisation very often causes problems when LMTOPACK is implemented
on machines other than Burroughs B7800. The problem is that the Burroughs,
for which the programmes were written, allows sequential write and subsequent random read in a file with fixed record length, while most other
machines do not.

   The reason for the special file organisation is that LMTO generates the
eigenvalues k point by k point while the state density programme needs the
data band by band. Let me illustrate this with the example of one atom per
cell and s, p, and d partial waves. The record length is then 4, enough to

193

accommodate eigenvalues, s, p, and d characters, and the file holds

23 records of text and potential paramters

plus 10 records per k point of the form

	1	2	...	10
1	W, KX, KY, KZ	E_1, S, P, D	...	E_9, S, P, D
2	W, KX, KY, KZ	E', S, P, D	...	E_9, S, P, D
⋮	⋮	⋮		⋮
NPT	W, KX, KY, KZ	E_1, S, P, D	...	E_9, S, P, D

containing information about the k point and the 9 eigenvalues with corresponding ℓ characters. However, LMTO writes this information row-wise
while DDNS reads column-wise, Column 2 when Band 1 is treated and column 10
when finally Band 9 is needed. On most machines except Burroughs this requires in the write and read statements a record pointer which is updated
at each read and write. Hence, instead of the WRITE (4) statements one must
substitute

IREC = IREC + 1

WRITE(4'IREC)   (9.23)

everywhere in LMTO. Similar substitutions should be made in DDNS for the
READ(1) statements.

9.5 Projected State–Density Programme DDNS

The DDNS programme is designed to evaluate tℓ-projected state densities
and corresponding number of state functions by means of the tetrahedron
technique [9.4,5]. The basic input is the eigenvalues and tℓ characters
generated by LMTO. The calculated functions are stored on disk, and may
be retrieved later for use in the ASA self-consistency procedure, or simply
for plotting purposes.[1] The programme is based on the paper by *Jepsen* and
Andersen [9.4], the unpublished thesis by *Jepsen* [9.6] and some private notes
by *O.K. Andersen*.

9.5.1 The Tetrahedron Method

The density N(E) of electronic states is defined such that N(E)dE is the
number of one-electron states per unit cell and per spin, with energy between
E and E + dE. It may be calculated from

---

[1] Note that for historical reasons the programme evaluates 2N(E) where N(E)
is defined in (9.25), cf. also (1.13).

$$N(E) = \frac{\Omega}{(2\pi)^3} \sum_j \int_{E_j(\mathbf{k})=E} \frac{dA}{|\nabla_\mathbf{k} E_j(\mathbf{k})|} \quad , \tag{9.24}$$

where the integral is taken over the surface of constant energy E, and j is the band index.

In the tetrahedron method, one divides the irreducible wedge of the Brillouin zone into tetrahedra, and the integral (9.24) is then evaluated as the sum

$$N(E) = \frac{\Omega}{(2\pi)^3} \sum_{j,i} N_{j,i}(E) \tag{9.25}$$

over these tetrahedra of partial state densities. The factor $\Omega/(2\pi)^3$ is included explicitly so that

$$\sum_i n_{j,i}(E) \equiv \sum_i \int_{-\infty}^{\infty} N_{j,i}(E)dE$$

$$= \Omega_{BZ}$$

$$= (2\pi)^3/\Omega \quad , \tag{9.26}$$

whereby one band will contain one electron per unit cell and per spin. Furthermore, the partial number of states $n_{j,i}(E)$ is the volume of that part of the i'th tetrahedron in which $E_j(\mathbf{k})$ is less than E.

When the eigenvalues in the j'th band at the four corners of the i'th tetrahedron are ordered such that

$$E_1 \leq E_2 \leq E_3 \leq E_4 \quad , \tag{9.27}$$

the partial state density and number of states at the energy E from this single tetrahedron are given by the simple analytic expressions in Table 9.5. The particular combinations of the energies D, M, and D_n also listed in the table are introduced to ensure that the analytic expressions do not diverge if two or more of the corner energies are equal. The projected number of states and density of states is obtained by multiplying n and N in Table 9.5 by the average of the tℓ characters (6.33, 8.28) at the four corners of the i'th tetrahedron.

9.5.2 Mesh and Tetrahedra

It now remains to construct a mesh in the irreducible wedge of the Brillouin zone and define the corresponding tetrahedra. This is done in the subroutine TGEN, and we shall briefly illustrate the procedure with the example of a simple cubic lattice.

Table 9.5. The number of states n and the density of states N in the tetra-hedron scheme as functions of the four corner energies E_1, E_2, E_3, E_4, the volume of the i'th tetrahedron V_i and the energy E [9.6]

	$N(E_1,E_2,E_3,E_4,V_i,E)$	$n(E_1,E_2,E_3,E_4,V_i,E)$
$E \leq E_1$	0	0
$E_1 \leq E \leq E_2$	$\dfrac{3V_i}{D} \dfrac{(E-E_1)^2}{(E_1-E_2)D_1}$	$\dfrac{V_i}{D} \dfrac{(E-E_1)^3}{(E_1-E_2)D_1}$
$E_2 \leq E \leq E_3$	$\dfrac{3V_i}{D} \dfrac{D_1D_2-(E-M)^2}{D_1D_2}$	$\dfrac{3V_i}{D}(E-M) - \dfrac{V_i}{D}\left[D_1+D_2 + \dfrac{(E-M)^3}{D_1D_2}\right]$
$E_3 \leq E \leq E_4$	$\dfrac{3V_i}{D} \dfrac{(E-E_4)^2}{(E_3-E_4)D_4}$	$\dfrac{V_i}{D}\left[D + \dfrac{(E-E_4)^3}{(E_4-E_3)D_4}\right]$
$E_4 \leq E$	0	V_i

$$D = E_4 + E_3 - E_2 - E_1$$
$$M = (E_4E_3 - E_2E_1)/D$$
$$D_n = E_n - M \; ; \; n = 1,2,4$$

The irreducible wedge of the Brillouin zone of the simple cubic structure may be defined by

$$k_z \leq k_y \leq k_x \leq \pi/a \qquad (9.28)$$

and is shown in Fig.9.2. Inside this wedge we define a cubic mesh, and num-ber the mesh points consecutively. Generally, there are seven other points associated with a given mesh point, namely those seven defining the additi-onal corners of the cube in front of and above the given point, Fig.9.3d. Hence, a general mesh point defines six tetrahedra. A point on Λ defines one tetrahedron, Fig.9.3a, a point in the ΓMR plane defines three tetrahedra, Fig.9.3b, and a point in the ΓXR plane defines three tetrahedra, Fig.9.3c. Finally, points in the XMR plane do not define any tetrahedra.

The subroutine TGEN constructs the mesh and establishes the mesh-point numbers N1,...,N7 for the seven neighbours of a given point. Some of these numbers are zero in special cases, e.g. Fig.9.3b where N6 = N7 = 0. It is

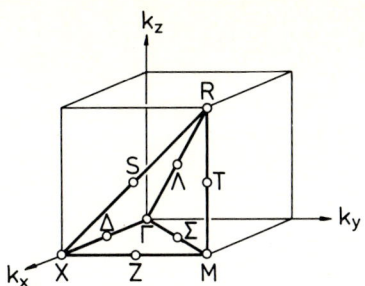

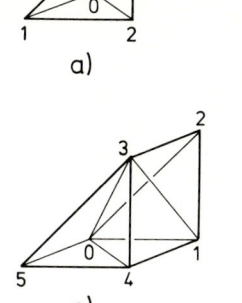

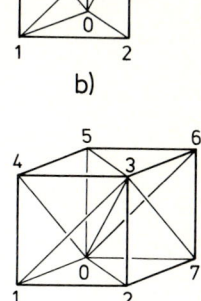

Fig.9.2. The irreducible wedge of the simple cubic Brillouin zone

Fig.9.3. Tetrahedra associated with a given lattice point, here designated 0, in the simple cubic structure. The summation over tetrahedra in DENS is based on the following numbering scheme: one tetrahedron: 0123; b) three tetrahedra: 0123, 0134, 0345; c) three tetrahedra: 0123, 0134, 0345; d) six tetrahedra: 0123, 0134, 0345, 0356, 0367, 0237

this process and the actual summation in DENS which must be reconsidered if Brillouin zones other than those already included in the programmes are needed.

9.5.3 Listing of DDNS

```
C    ************************************************************
C    *                                                          *
C    *    PROJECTED (DECOMPOSED) STATE-DENSITY PROGRAMME        *
C    *    VERSION DEC. 1982                                     *
C    *                                                          *
C    *    H.L.SKRIVER                                           *
C    *    RISOE NATIONAL LABORATORY                             *
C    *    DK-4000 ROSKILDE                                      *
C    *    DENMARK                                               *
C    *                                                          *
C    ************************************************************
C    *                                                          *
C    *    FILES AND THEIR ATTRIBUTES                            *
C    *                                                          *
C    *    FILE 1 :UNFORMATTED,TYPE=RANDOM (DIRECT) ACCESS,      *
C    *            RECORDLENGTH=MAX(NT*(NL+1),4),                *
C    *            NAME=BND/YY/B, BND/CAN/XXX, OR BND/UNH/YY     *
C    *    FILE 2 :UNFORMATTED,TYPE=SEQUENTIAL,RECORDLENGTH=VARIABLE, *
C    *            NAME=DOS/YY/B                                 *
C    *    FILE 5 :FORMATTED,TYPE=SEQUENTIAL,RECORDLENGTH=80,    *
C    *            NAME=DATA/DDNS                                *
C    *    FILE 6 :PRINTER                                       *
C    *                                                          *
C    ************************************************************
C    *                                                          *
C    *    INPUT  (FILE 5)                                       *
C    *    -----                                                 *
```

197

```
C    *                                                                      *
C    *      EB          THE STATE DENSITY IS CALCULATED IN THE              *
C    *      EF          :ENERGY RANGE EB-EF WITH STEPLENGTH                 *
C    *      NE          (EF-EB)/(NE-1)                                      *
C    *      NB1         FROM BAND NUMBER NB1 TO BAND                        *
C    *      NB2         NUMBER NB2                                          *
C    *      NOPRT       :=0:STATE DENSITIES NOT PRINTED                     *
C    *                  =1:STATE DENSITIES PRINTED                          *
C    *      NOWRT       :=0:ORIGINAL EIGENVALUES NOT PRINTED                *
C    *                  =1:EIGENVALUES PRINTED                              *
C    *      MODE        :=0:EB IS MODIFIED TO BE THE BOTTOM OF BAND NUMBER  *
C    *                    NB1. EF IS (EF-EB) ABOVE THE MODIFIED EB          *
C    *                  =1:EB AND EF NOT MODIFIED                           *
C    *                                                                      *
C    ************************************************************************
C    *                                                                      *
C    *      INPUT   (FILE 1)                                                 *
C    *      -----                                                           *
C    *                                                                      *
C    *                                                                      *
C    *              FILE 1 IS A STANDARD ENERGY BAND FILE GENERATED BY      *
C    *              <LMTO> (FILE 4) OR <CANON> (FILE 9). A SPECIFICATION    *
C    *              OF THE CONTENT IS GIVEN BELOW                           *
C    *                                                                      *
C    *      BOA     :B OVER A LATTICE PARAMETER                             *
C    *      COA     :C OVER A LATTICE PARAMETER                             *
C    *      DUM1    :NOT USED                                               *
C    *      E       :EIGENVALUES                                            *
C    *      FTS     :T,S-CHARACTER                                          *
C    *      FTP     :T,P-CHARACTER                                          *
C    *      FTD     :T,D-CHARACTER                                          *
C    *      FTF     :T,F-CHARACTER                                          *
C    *      IDUM1   :NOT USED                                               *
C    *      LAT     :BRILLOUIN ZONE                                         *
C    *      NL      :NUMBER OF L-QUANTUM NUMBERS                            *
C    *      NLM     :SUM OF 2*L+1, =NL**2                                   *
C    *      NLMQ    :NUMBER OF EIGENVALUES PER K-POINT                      *
C    *      NPX                                                             *
C    *      NPY     :NUMBERS FOR K-MESH CONSTRUCTION                        *
C    *      NPZ                                                             *
C    *      NQ      :NUMBER OF ATOMS IN THE CELL                            *
C    *      NT      :NUMBER OF INEQUIVALENT ATOMS IN THE CELL               *
C    *      SWSO    :AVERAGE WIGNER-SEITZ RADIUS, ATOMIC UNITS              *
C    *      SWS     :WIGNER-SEITZ RADIUS FOR ATOM KT, ATOMIC UNITS          *
C    *      TXT     :CHEMICAL LABEL FOR ATOM KT                             *
C    *      TXTC    :TEXT IDENTIFYING THE CORRECTION MATRICES               *
C    *      TXTP    :TEXT IDENTIFYING THE POTENTIAL PARAMETERS              *
C    *      TXTS    :TEXT IDENTIFYING THE STRUCTURE CONSTANTS               *
C    *                                                                      *
C    *      ENY                                                             *
C    *      OMN                                                             *
C    *      OMJ                                                             *
C    *      FINSQ   :HISTORICAL POTENTIAL PARAMETERS                        *
C    *      FIJSQ                                                           *
C    *      LNN                                                             *
C    *      LNJ                                                             *
C    *      LJJ                                                             *
C    *                                                                      *
C    *                                                                      *
C    ************************************************************************
C    *                                                                      *
C    *      OUTPUT (FILE 2)                                                  *
C    *      ------                                                          *
C    *                                                                      *
C    *              A SPECIFICATION OF OUTPUT NOT ON THE LIST BELOW CAN     *
C    *              BE FOUND IN INPUT (FILE 5) OR INPUT (FILE 1)            *
C    *                                                                      *
C    *      CNT     :TOTAL NUMBER OF STATES                                 *
C    *      CTS     :NUMBER OF T,S-STATES                                   *
```

```
C    *    CTP    :NUMBER OF T,P-STATES                              *
C    *    CTD    :NUMBER OF T,D-STATES                              *
C    *    CTF    :NUMBER OF T,F-STATES                              *
C    *    DE     :ENERGY STEP, =(EF-EB)/(NE-1)                      *
C    *    SNT    :TOTAL STATE DENSITY                               *
C    *    STS    :T,S-STATE DENSITY                                 *
C    *    STP    :T,P-STATE DENSITY                                 *
C    *    STD    :T,D-STATE DENSITY                                 *
C    *    STF    :T,F-STATE DENSITY                                 *
C    *                                                              *
C    ****************************************************************

      IMPLICIT REAL*8 (A-H,O-Z)
      REAL*8 LNN,LNJ,LJJ
      INTEGER*4 TLAT(8)/´   SC´,´  FCC´,´  BCC´,´  HCP´,´   ST´,´  BCT´,´AL-U´
     1,´TRIG´/
      INTEGER TXTS(15),TXTP(20,6),TXTC(15),TXT(6),POINT
      DATA TXT/6*´         ´/
      COMMON/B1/ E(1000),EM(1001),CNT(1001),SNT(1001),DE,V,BOA,COA,NE,
     1NPT,NL,NEMAX,NT,STS(1001,4),CTS(1001,4),STP(1001,4),CTP(1001,4),
     2STD(1001,4),CTD(1001,4),STF(1001,4),CTF(1001,4),FTS(1000,4),
     3FTP(1000,4),FTD(1000,4),FTF(1000,4),
     3N1(1000),N2(1000),N3(1000),N4(1000),N5(1000),N6(1000),N7(1000)
      DIMENSION ENY(18),OMN(18),OMJ(18),FINSQ(18),FIJSQ(18),SWS(6),
     ,SWSSQ(6),LNN(18),LNJ(18),LJJ(18)
    1 FORMAT(F10.6,I5)
    2 FORMAT(2F10.6,3I5,2X,3I1)
    3 FORMAT(11F7.6)
    4 FORMAT(1H ,3X,F8.4,2F10.5,3X,2F10.5,3X,2F10.5,3X,2F10.5,
     13X,2F10.5)
    6 FORMAT(1H1,10X,´DENSITY OF STATES PROGRAMME FOR ´,A4,´  LATTICE´,/
     1/11X,´BOA,COA,V,EB,EF,NE,NPT,NPX,NPY,NPZ,NB1,NB2´,//,11X,5F10.6,
     12I5,5I3,//)
    7 FORMAT(1H ,10X,I4,5X,2F10.6,2I5)
    8 FORMAT(1H ,15F7.4)
    9 FORMAT(1H ,10X,´ENERGY RANGE GENERATED AUTOMATICALLY , EB =´,
     1F10.6,´  EF =´,F10.6,´  DE=´,F10.6,´  NE=´,I5,//)
   10 FORMAT(1H1,10X,´DENSITY OF STATES FOR ´,A4,´ IN´,6A4,///,8X,´E´,
     17X,´NOS´,7X,´DOS´,10X,´S-NOS´,5X,´S-DOS´,8X,´P-NOS´,5X,´P-DOS´,
     28X,´D-NOS´,5X,´D-DOS´,8X,´F-NOS´,5X´F-DOS´,//)
   11 FORMAT(1H1)
   12 FORMAT(1H ,///,11X,20A4,//,11X,´SWS´,F10.6,///,11X,´ENY´,7X,´OMN´,
     17X,´OMJ´,6X,´FINSQ´,5X,´FIJSQ´,6X,´LNN´,7X,´LNJ´,7X,´LJJ´,//)
   13 FORMAT(1H ,5X,8F10.6)
   14 FORMAT(1H ,10X,15A4,/)
   15 FORMAT(1H ,10X,´BAND NO´,6X,´EMIN´,6X,´EMAX´,5X,´N1´,3X,´N2´,//)
C
C     READ TEXTS IDENTIFYING STRUCTURE CONSTANTS AND CORRECTION
C     MATRICES USED IN BAND CALCULATION
C
      READ(1) TXTS(1),TXTS(2),TXTS(3),TXTS(4)
      READ(1) TXTS(5),TXTS(6),TXTS(7),TXTS(8)
      READ(1) TXTS(9),TXTS(10),TXTS(11),TXTS(12)
      READ(1) TXTS(13),TXTS(14),TXTS(15),NPOINT
      READ(1) TXTC(1),TXTC(2),TXTC(3),TXTC(4)
      READ(1) TXTC(5),TXTC(6),TXTC(7),TXTC(8)
      READ(1) TXTC(9),TXTC(10),TXTC(11),TXTC(12)
      READ(1) TXTC(13),TXTC(14),TXTC(15),NL
      READ(1) NQ,NLM,NLMQ,NT
      READ(1) NPX,NPY,NPZ,LAT
      READ(1) NPT1,BOA,COA,DUM1
C
      READ(5,2) EB,EF,NE,NB1,NB2,NOPRT,NOWRT,MODE
      DE=(EF-EB)/(NE-1)
      CALL TGEN(LAT,NPX,NPY,NPZ)
      WRITE(6,6) TLAT(LAT),BOA,COA,V,EB,EF,NE,NPT,NPX,NPY,NPZ,NB1,NB2
C
```

```
C      READ POTENTIAL PARAMETERS USED IN BAND CALCULATION
C
       INL=-NL
       DO 35 KT=1,NT
       INL=INL+NL
       READ(1) TXTP(1,KT),TXTP(2,KT),TXTP(3,KT),TXTP(4,KT)
       READ(1) TXTP(5,KT),TXTP(6,KT),TXTP(7,KT),TXTP(8,KT)
       READ(1) TXTP(9,KT),TXTP(10,KT),TXTP(11,KT),TXTP(12,KT)
       READ(1) TXTP(13,KT),TXTP(14,KT),TXTP(15,KT),TXTP(16,KT)
       READ(1) TXTP(17,KT),TXTP(18,KT),TXTP(19,KT),TXTP(20,KT)
       READ(1) SWO,SWS(KT),TXT(KT)
       DO 35 IL=1,NL
       I=INL+IL
       READ(1) ENY(I),OMN(I),OMJ(I),FINSQ(I)
       READ(1) FIJSQ(I),LNN(I),LNJ(I),LJJ(I)
   35 CONTINUE
C
       WRITE(6,14) TXTS
       WRITE(6,14) TXTC
C
C      WRITE INFORMATION FOR STATE DENSITY DATA SET
C
       WRITE(2) TXTS,TXTC,SWO,NL,NQ,NLM,NLMQ,NT
       WRITE(2) NPX,NPY,NPZ,LAT,NPT,BOA,COA,DUM1
       INL=-NL
       DO 26 KT=1,NT
       INL=INL+NL
       WRITE(6,12)(TXTP(I,KT),I=1,20),SWS(KT)
       WRITE(2)(TXTP(I,KT),I=1,20),SWS(KT),TXT(KT)
       DO 26 IL=1,NL
       I=INL+IL
       WRITE(6,13) ENY(I),OMN(I),OMJ(I),FINSQ(I),FIJSQ(I),LNN(I),LNJ(I),
      1LJJ(I)
       WRITE(2) ENY(I),OMN(I),OMJ(I),FINSQ(I),FIJSQ(I),LNN(I),LNJ(I),
      1LJJ(I)
   26 CONTINUE
C
       WRITE(6,11)
       NLMQP=NLMQ+1
       IREC1=11+NT*6+NT*NL*2
C
C      CALCULATE STATE DENSITY BAND BY BAND
C
       DO 21 LB=NB1,NB2
       IR=IREC1+LB-NLMQP
       IF(NL.EQ.4) GO TO 30
       DO 25 IP=1,NPT
       IR=IR+NLMQP
       READ(1'IR) W,KX,KY,KZ
   25 READ(1) E(IP),(FTS(IP,KT),FTP(IP,KT),FTD(IP,KT),KT=1,NT)
       GO TO 31
   30 DO 32 IP=1,NPT
       IR=IR+NLMQP
       READ(1'IR) W,KX,KY,KZ
   32 READ(1) E(IP),(FTS(IP,KT),FTP(IP,KT),FTD(IP,KT),FTF(IP,KT),
      1KT=1,NT)
C
C      FIND MAX AND MIN ENERGY OF BAND NUMBER LB
C
   31 EMAX=-100.DO
       EMIN=100.DO
       DO 22 LP=1,NPT
       ET=E(LP)
       IF(ET.GE.EMIN) GO TO 23
       EMIN=ET
   23 IF(ET.LE.EMAX) GO TO 22
       EMAX=ET
   22 CONTINUE
```

```
C
C       MODIFY ENERGY RANGE IF REQUESTED
C
        IF(LB.NE.NB1) GO TO 33
        IF(MODE.EQ.1) GO TO 41
        IEB=EMIN*10.D0
        IF(EMIN.LT.0.D0) IEB=IEB-1
        EBB=IEB/10.D0
        EF=(EF-EB)+EBB
        EB=EBB
        WRITE(6,9) EB,EF,DE,NE
C
   41   WRITE(2) EB,EF,DE,NE,NB1,NB2
C
C       CONSTRUCT ENERGY MESH
C
        DO 20 LE=1,NE
   20   EM(LE)=EB+(LE-1)*DE
        WRITE(6,15)
   33   IF(EMAX.LE.EB) GO TO 38
        IF(EMIN.GE.EF) GO TO 42
        EMAX=DMIN1(EMAX,EM(NE-1))
        NEMAX=(EMAX-EB)/DE
        NEMAX=NEMAX+2
        IF(NEMAX.GT.NE) NEMAX=NE
        NEMIN=(EMIN-EB)/DE
        IF(EMIN.LT.EB) NEMIN=-1
        NEMIN=NEMIN+2
        GO TO 39
   38   NEMIN=1
        NEMAX=1
C
   39   WRITE(6,7) LB,EMIN,EMAX,NEMIN,NEMAX
        IF(NOWRT.EQ.0) GO TO 40
        WRITE(6,8) (E(I),I=1,NPT)
        WRITE(6,11)
   40   CONTINUE
C
C       PERFORM K-SPACE SUMMATIONS
C
        IF(LAT.EQ.2) GO TO 34
        IF(LAT.EQ.4) GO TO 36
        IF(LAT.EQ.8) GO TO 36
        CALL DENS
        GO TO 37
   34   CALL FDENS
        GO TO 37
   36   CALL HDENS
C
   37   CONTINUE
        IF(NEMAX.GE.NE) GO TO 21
        NEP=NEMAX+1
        DO 28 LE=NEP,NE
        SNT(LE)=SNT(NEMAX)
        DO 28 KT=1,NT
        STS(LE,KT)=STS(NEMAX,KT)
        STP(LE,KT)=STP(NEMAX,KT)
        STD(LE,KT)=STD(NEMAX,KT)
        IF(NL.LT.4) GO TO 28
        STF(LE,KT)=STF(NEMAX,KT)
   28   CONTINUE
   21   CONTINUE
C
C       WRITE DENSITIES ON SYSOUT AND FILE 2
C
   42   IF(NOPRT.EQ.0) GO TO 43
        DO 27 KT=1,NT
        WRITE(6,10) TXT(KT),TXT
```

```
   27 WRITE(6,4) (EM(LE),SNT(LE),CNT(LE),STS(LE,KT),CTS(LE,KT),
     1STP(LE,KT),CTP(LE,KT),STD(LE,KT),CTD(LE,KT),STF(LE,KT),CTF(LE,KT),
     2LE=1,NE)
   43 WRITE(2) (SNT(LE),LE=1,NE)
      WRITE(2) (CNT(LE),LE=1,NE)
      DO 29 KT=1,NT
      WRITE(2) (STS(LE,KT),LE=1,NE)
      WRITE(2) (CTS(LE,KT),LE=1,NE)
      WRITE(2) (STP(LE,KT),LE=1,NE)
      WRITE(2) (CTP(LE,KT),LE=1,NE)
      WRITE(2) (STD(LE,KT),LE=1,NE)
      WRITE(2) (CTD(LE,KT),LE=1,NE)
      IF(NL.LT.4) GO TO 29
      WRITE(2) (STF(LE,KT),LE=1,NE)
      WRITE(2) (CTF(LE,KT),LE=1,NE)
   29 CONTINUE
      STOP
      END
      BLOCK DATA
C     ****************************************************************
C     *                                                              *
C     *    SET ARRAYS TO ZERO                                        *
C     *                                                              *
C     ****************************************************************
      COMMON/B1/ E(1000),EM(1001),CNT(1001),SNT(1001),DE,V,BOA,COA,NE,
     1NPT,NL,NEMAX,NT,STS(1001,4),CTS(1001,4),STP(1001,4),CTP(1001,4),
     2STD(1001,4),CTD(1001,4),STF(1001,4),CTF(1001,4),FTS(1000,4),
     3FTP(1000,4),FTD(1000,4),FTF(1000,4),
     3N1(1000),N2(1000),N3(1000),N4(1000),N5(1000),N6(1000),N7(1000)
      DATA STS/4004*0.D0/,CTS/4004*0.D0/,STP/4004*0.D0/,CTP/4004*0.D0/
      DATA STD/4004*0.D0/,CTD/4004*0.D0/,SNT/1001*0.D0/,CNT/1001*0.D0/
      DATA STF/4004*0.D0/,CTF/4004*0.D0/
      END
      SUBROUTINE TGEN(LAT,NPXD,NPYD,NPZD)
C     ****************************************************************
C     *                                                              *
C     *    DEFINE TETRAHEDRA ASSOCIATED WITH EACH K-POINT            *
C     *                                                              *
C     *    THE K-SPACE MESH IMPLIED SHOULD BE CONSISTENT WITH THAT   *
C     *    GENERATED BY SUBROUTINE <MESH> IN <STR> AND <COR>         *
C     *                                                              *
C     *    LAT=1  SIMPLE CUBIC                                       *
C     *                                                              *
C     *    LAT=2  FACE CENTRED CUBIC                                 *
C     *                                                              *
C     *    LAT=3  BODY CENTRED CUBIC                                 *
C     *                                                              *
C     *    LAT=4  HEXAGONAL CLOSE PACKED                             *
C     *                                                              *
C     *    LAT=5  SIMPLE TETRAGONAL                                  *
C     *                                                              *
C     *    LAT=6  BODY CENTRED TETRAGONAL                           *
C     *                                                              *
C     *    LAT=7  ALPHA URANIUM                                      *
C     *                                                              *
C     *    LAT=8  TRIGONAL ZONE FOR ANTIFERROMAGNETIC OXIDES         *
C     *                                                              *
C     ****************************************************************
      IMPLICIT REAL*8 (A-H,O-Z)
      INTEGER*4 X,Y,Z
      COMMON/B1/ E(1000),EM(1001),CNT(1001),SNT(1001),DE,V,BOA,COA,NE,
     1NPT,NL,NEMAX,NT,STS(1001,4),CTS(1001,4),STP(1001,4),CTP(1001,4),
     2STD(1001,4),CTD(1001,4),STF(1001,4),CTF(1001,4),FTS(1000,4),
     3FTP(1000,4),FTD(1000,4),FTF(1000,4),
     3N1(1000),N2(1000),N3(1000),N4(1000),N5(1000),N6(1000),N7(1000)
C
```

```
        NPX=NPXD
        NPY=NPYD
        NPZ=NPZD
        NPXM=NPX-1
        NPYM=NPY-1
        NPZM=NPZ-1
        GO TO (20,21,22,23,24,25,26,27),LAT
    20 NP=0
C
C       SIMPLE CUBIC IRREDUCIBLE ZONE DEFINED BY
C       0 .LE. KZ .LE. KY .LE. KX .LE. PI/A
C
        DKX=1.D0/NPYM
        DKY=DKX
        DKZ=DKX
        DHX=0.D0
        V=2.D0*DKX**3
        DO 30 I=1,NPY
        NI=((I+1)*I)/2
        DO 30 J=1,I
        DO 30 K=1,J
        NP=NP+1
        IF(I.EQ.NPY) GO TO 42
        IF(I.EQ.J.AND.I.EQ.K.AND.K.EQ.J) GO TO 40
        IF(J.EQ.K) GO TO 41
        IF(I.EQ.J) GO TO 43
C
C       GENERAL POINT
C
        N1(NP)=NP+NI
        N2(NP)=N1(NP)+J
        N3(NP)=N2(NP)+1
        N4(NP)=N1(NP)+1
        N5(NP)=NP+1
        N6(NP)=N5(NP)+J
        N7(NP)=NP+J
        GO TO 30
C
C       THREE TETRAHEDRA
C
    41 N1(NP)=NP+J
        N2(NP)=NP+J+1
        N3(NP)=N2(NP)+NI
        N4(NP)=N1(NP)+NI
        N5(NP)=NP+NI
        N6(NP)=0
        N7(NP)=0
        GO TO 30
C
C       THREE TETRAHEDRA
C
    43 N1(NP)=NP+NI
        N2(NP)=N1(NP)+J
        N3(NP)=N2(NP)+1
        N4(NP)=N1(NP)+1
        N5(NP)=NP+1
        N6(NP)=0
        N7(NP)=0
        GO TO 30
C
C       ONE TETRAHEDRON
C
    40 N1(NP)=NP+NI
        N2(NP)=N1(NP)+J
        N3(NP)=N2(NP)+1
        N4(NP)=0
        N5(NP)=0
        N7(NP)=0
```

```
          GO TO 30
    42 N1(NP)=0
       N4(NP)=0
       N5(NP)=0
       N7(NP)=0
    30 CONTINUE
       NPT=NP
       RETURN
    21 NP=0
C
C      FCC IRREDUCIBLE ZONE DEFINED BY
C      0 .LE. KZ .LE. KY .LE. KX .LE. 2PI/A
C      KX + KY + KZ .LE. 3PI/A
C
       DKX=2.D0/NPYM
       DKY=DKX
       DKZ=DKX
       DHX=0.D0
       V=.5D0*DKX**3
       NPX=NPY
       NTF=3*(NPYM/4)+1
       NTH=(NPYM/2)*3+3
       NPH=NPYM/2+1
       DO 31 I=1,NPX
       M1=NPX-I+NPH
       NPY=MINO(I,M1)
       M2=(M1+1)/2
       M3=M1/2
       M4=M1-I
       IP=I+1
       M1P=M1-1
       M2P=M3
       M3P=M1P/2
       M4P=M1P-IP
       M4S=0
       IF(I.LE.NPH) GO TO 44
       MI=(M2*(M2+1))/2+(M3*(M3+1))/2
       IF(I.GE.NTF) GO TO 45
       M4S=(M4*(M4+1))/2
       MI=MI-M4S
       GO TO 45
    44 MI=(I*(I+1))/2
    45 DO 31 J=1,NPY
       M5=M1-J+1
       NPZ=MINO(J,M5)
       M5P=M1P-J+1
       NPZP=MINO(J,M5P)
       MJ1=(M5*(M5+1))/2-M4S
       MJ2=(M2P*(M2P+1))/2+(M3P*(M3P+1))/2-(M5P*(M5P+1))/2
       IF(J.LE.M2) GO TO 46
       NI=MJ1+MJ2
       NJA=NPZ
       NJB=NPZP
       GO TO 47
    46 NI=MI
       NJA=J
       NJB=J
       IF(J.GT.M2P) NJB=NPZP
    47 DO 31 K=1,NPZ
       NP=NP+1
       IJK=I+J+K
       IF(I.EQ.NPX) GO TO 48
       IF(IJK.EQ.NTH) GO TO 48
       IF(IJK+1.EQ.NTH) GO TO 49
       IF(IJK+2.EQ.NTH) GO TO 50
       IF(I.EQ.J.AND.J.EQ.K) GO TO 53
       IF(K.EQ.J) GO TO 54
       IF(I.EQ.J) GO TO 55
```

204

```
C
C      GENERAL POINT
C
       N1(NP)=NP+NI
       N2(NP)=N1(NP)+NJB
       N3(NP)=N2(NP)+1
       N4(NP)=N1(NP)+1
       N5(NP)=NP+1
       N6(NP)=N5(NP)+NJA
       N7(NP)=NP+NJA
       GO TO 31
C
C      THREE TETAHEDRA
C
   55  N1(NP)=NP+NI
       N2(NP)=N1(NP)+NJB
       N3(NP)=N2(NP)+1
       N4(NP)=N1(NP)+1
       N5(NP)=NP+1
       N6(NP)=0
       N7(NP)=0
       GO TO 31
C
C      THREE TETRAHEDRA
C
   54  N1(NP)=NP+NJA
       N2(NP)=N1(NP)+1
       N3(NP)=NP+NI+NJB+1
       N4(NP)=NP+NI+NJB
       N5(NP)=NP+NI
       N6(NP)=0
       N7(NP)=0
       GO TO 31
C
C      ONE TETRAHEDRON
C
   53  N1(NP)=NP+NI
       N2(NP)=N1(NP)+NJB
       N3(NP)=N2(NP)+1
       N4(NP)=0
       N5(NP)=0
       N6(NP)=0
       N7(NP)=0
       GO TO 31
   50  IF(K.EQ.NPZ) GO TO 56
       IF(J.EQ.NPY) GO TO 57
C
C      FIVE TETRAHEDRA
C
       N1(NP)=NP+1
       N2(NP)=N1(NP)+NJA
       N3(NP)=NP+NJA
       N4(NP)=NP+NI+NJB
       N5(NP)=NP+NI
       N6(NP)=N5(NP)+1
       N7(NP)=0
       GO TO 31
C
C      TWO TETRAHEDRA
C
   57  N1(NP)=NP+NI+NJB
       N2(NP)=NP+NI
       N3(NP)=N2(NP)+1
       N4(NP)=NP+1
       N5(NP)=0
       N6(NP)=0
       N7(NP)=0
       GO TO 31
```

205

```
C
C     TWO TETRAHEDRA
C
   56 N1(NP)=NP+NJA+1
      N2(NP)=NP+NJA
      N3(NP)=NP+NI+NJB
      N4(NP)=NP+NI
      N5(NP)=0
      N6(NP)=0
      N7(NP)=0
      GO TO 31
   49 IF(K.EQ.NPZ) GO TO 58
      IF(J.EQ.NPY) GO TO 59
C
C     ONE TETRAHEDRON
C
      N1(NP)=NP+NI
      N2(NP)=NP+1
      N3(NP)=NP+NJA
      N4(NP)=0
      N5(NP)=0
      N6(NP)=0
      N7(NP)=0
      GO TO 31
C
C     ONE TETRAHEDRON
   59 N1(NP)=NP+NI
      N2(NP)=N1(NP)+NJB-1
      N3(NP)=NP+1
      N4(NP)=0
      N5(NP)=0
      N6(NP)=0
      N7(NP)=0
      GO TO 31
C
C     ONE TETRAHEDRON
C
   58 N1(NP)=NP+NI
      N2(NP)=NP+NJA
      N3(NP)=M6
      N4(NP)=0
      N5(NP)=0
      N6(NP)=0
      N7(NP)=0
      GO TO 31
   48 N1(NP)=0
      N2(NP)=0
      N3(NP)=0
      N4(NP)=0
      N5(NP)=0
      N6(NP)=0
      N7(NP)=0
      IF(J.EQ.M2) M6=NP
   31 CONTINUE
      NPT=NP
      RETURN
   22 NP=0
C
C     BCC IRREDUCIBLE ZONE DEFINED BY
C     O .LE. KZ .LE. KY .LE. KX .LE. 2PI/A
C     KX + KY .LE. 2PI/A
C
      DKX=2.D0/NPYM
      DKY=DKX
      DKZ=DKX
      DHX=0.D0
      V=DKX**3
```

206

```
      NPH=NPY/2+1
      NPYH=NPYM/2
      DO 32 I=1,NPY
      X=I-1
      JM=MINO(NPY-I+1,I)
      NI=((JM+1)*JM)/2
      DO 32 J=1,JM
      Y=J-1
      DO 32 K=1,J
      Z=K-1
      NP=NP+1
      IF(X+Y.GE.NPYM-1) GO TO 64
      IF(X.EQ.Y.AND.X.EQ.Z) GO TO 65
      IF(Y.EQ.Z) GO TO 66
      IF(X.EQ.Y) GO TO 67
C
C     GENERAL POINT
C
      N1(NP)=NP+NI
      N2(NP)=N1(NP)+J
      N3(NP)=N2(NP)+1
      N4(NP)=N1(NP)+1
      N5(NP)=NP+1
      N6(NP)=N5(NP)+J
      N7(NP)=NP+J
      GO TO 32
C
C     THREE TETRAHEDRA
C
   66 N1(NP)=NP+J
      N2(NP)=NP+J+1
      N3(NP)=N2(NP)+NI
      N4(NP)=N1(NP)+NI
      N5(NP)=NP+NI
      N6(NP)=0
      N7(NP)=0
      GO TO 32
C
C     THREE TETRAHEDRA
C
   67 N1(NP)=NP+NI
      N2(NP)=N1(NP)+J
      N3(NP)=N2(NP)+1
      N4(NP)=N1(NP)+1
      N5(NP)=NP+1
      N6(NP)=0
      N7(NP)=0
      GO TO 32
C
C     ONE TETRAHEDRON
C
   65 N1(NP)=NP+NI
      N2(NP)=N1(NP)+J
      N3(NP)=N2(NP)+1
      N4(NP)=0
      N5(NP)=0
      N7(NP)=0
      GO TO 32
   64 IF(X+Y.EQ.NPYM) GO TO 63
      IF(Y.EQ.Z) GO TO 68
C
C     THREE TETRAHEDRA
C
      N1(NP)=NP+NI+1
      N2(NP)=NP+1
      N3(NP)=NP+J+1
      N4(NP)=NP+NI
```

```
            N5(NP)=NP+J
            N6(NP)=0
            N7(NP)=0
            GO TO 32
C
C     ONE TETRAHEDRON
C
     68 N1(NP)=NP+NI
            N2(NP)=NP+J
            N3(NP)=N2(NP)+1
            N4(NP)=0
            N5(NP)=0
            N6(NP)=0
            N7(NP)=0
            GO TO 32
     63 N1(NP)=0
            N4(NP)=0
            N5(NP)=0
            N7(NP)=0
     32 CONTINUE
            NPT=NP
            RETURN
     23 NP=0
C     HEXAGONAL CLOSE PACKED IRREDUCIBLE ZONE DEFINED BY
C
C
C     0 .LE. 2KY .LE. KX .LE. 4/3 PI/A
C     0 .LE. KZ .LE. A/C PI/A
C
            NPH=NPY/2+1
            NZ=NPH*NPH
            AOC=1.D0/COA
            DKX=4.D0/3.D0/NPYM
            DKY=DSQRT(3.D0)/2.D0*DKX
            DKZ=AOC/NPZM
            V=.75D0*DKX*DKX*DKZ*COA
            DHX=-0.5D0*DKX
            DO 33 K=1,NPZ
            DO 33 J=1,NPH
            Y=J-1
            NX=NPY-2*Y
            IM=2*Y+1
            DO 33 I=IM,NPY
            NP=NP+1
            IF(I.EQ.NPY) GO TO 60
            IF(K.EQ.NPZ) GO TO 60
            IF(I.EQ.NPYM) GO TO 61
            IF(I.EQ.IM) GO TO 62
C
C     GENERAL POINT
C
            N1(NP)=NP+1
            N2(NP)=NP+NX
            N3(NP)=N1(NP)+NZ
            N4(NP)=N2(NP)+NZ
            N5(NP)=NP+NZ
            N6(NP)=N4(NP)-1
            N7(NP)=N2(NP)-1
            GO TO 33
C
C     THREE TETRAHEDRA
C
     61 N1(NP)=NP+1
            N2(NP)=NP+NX-1
            N3(NP)=N1(NP)+NZ
            N4(NP)=N2(NP)+NZ
            N5(NP)=NP+NZ
            N6(NP)=0
```

208

```
      N7(NP)=0
      GO TO 33
C
C     THREE TETRAHEDRA
C
   62 N1(NP)=NP+1
      N2(NP)=NP+NX
      N3(NP)=N1(NP)+NZ
      N4(NP)=N2(NP)+NZ
      N5(NP)=NP+NZ
      N6(NP)=0
      N7(NP)=0
      GO TO 33
   60 N1(NP)=0
      N2(NP)=0
      N3(NP)=0
      N4(NP)=0
      N5(NP)=0
      N6(NP)=0
      N7(NP)=0
   33 CONTINUE
      NPT=NP
      RETURN
   24 NP=0
C
C     SIMPLE TETRAGONAL
C
      AOC=1.D0/COA
      DKX=1.D0/NPYM
      DKY=DKX
      DKZ=AOC/NPZM
      DHX=0.D0
      V=2.D0/3.D0*DKX*DKY*DKZ*COA
      DO 34 I=1,NPY
      NI=I*NPZ
      DO 34 J=1,I
      DO 34 K=1,NPZ
      NP=NP+1
      IF(I.EQ.NPY.OR.K.EQ.NPZ)GO TO 52
      IF(I.EQ.J)GO TO 51
C
C     GENERAL POINT
C
      N1(NP)=NP+NI
      N2(NP)=N1(NP)+NPZ
      N3(NP)=N2(NP)+1
      N4(NP)=N1(NP)+1
      N5(NP)=NP+1
      N6(NP)=N5(NP)+NPZ
      N7(NP)=NP+NPZ
      GO TO 34
C
C     THREE TETRAHEDRA
C
   51 N1(NP)=NP+NI
      N2(NP)=N1(NP)+NPZ
      N3(NP)=N2(NP)+1
      N4(NP)=N1(NP)+1
      N5(NP)=NP+1
      N6(NP)=0
      N7(NP)=0
      GO TO 34
   52 N1(NP)=0
      N4(NP)=0
      N5(NP)=0
      N7(NP)=0
```

```
   34 CONTINUE
      NPT=NP
      RETURN
   25 NP=0
C
C     BODY CENTRED TETRAGONAL
C
      AOC=1.D0/COA
      DKX=1.D0/NPYM
      DKY=DKX
      DKZ=2.D0*AOC/NPZM
      DHX=0.D0
      V=1.D0/3.D0*DKX*DKY*DKZ*COA
      DO 35 I=1,NPY
      NI=I*NPZ
      DO 35 J=1,I
      DO 35 K=1,NPZ
      NP=NP+1
      IF(I.EQ.NPY.OR.K.EQ.NPZ)GO TO 74
      IF(I.EQ.J)GO TO 73
C
C     GENERAL POINT
C
      N1(NP)=NP+NI
      N2(NP)=N1(NP)+NPZ
      N3(NP)=N2(NP)+1
      N4(NP)=N1(NP)+1
      N5(NP)=NP+1
      N6(NP)=N5(NP)+NPZ
      N7(NP)=NP+NPZ
      GO TO 35
C
C     THREE TETRAHEDRA
C
   73 N1(NP)=NP+NI
      N2(NP)=N1(NP)+NPZ
      N3(NP)=N2(NP)+1
      N4(NP)=N1(NP)+1
      N5(NP)=NP+1
      N6(NP)=0
      N7(NP)=0
      GO TO 35
   74 N1(NP)=0
      N4(NP)=0
      N5(NP)=0
      N7(NP)=0
   35 CONTINUE
      NPT=NP
      RETURN
   26 NP=0
C
C     ALPHA URANIUM
C
      DKX=1.D0/NPXM
      DKY=2.D0/NPYM/BOA
      DKZ=1.D0/NPZM/COA
      DHX=0.D0
      V=1.D0/6.D0*DKX*DKY*DKZ*BOA*COA
      NI=NPY*NPZ
      DO 36 I=1,NPX
      DO 36 J=1,NPY
      DO 36 K=1,NPZ
      NP=NP+1
      IF(I.EQ.NPX.OR.J.EQ.NPY.OR.K.DQ.NPZ)  GO TO 69
C
C     GENERAL POINT
```

210

```
C
      N1(NP)=NP+NI
      N2(NP)=N1(NP)+NPZ
      N3(NP)=N2(NP)+1
      N4(NP)=N1(NP)+1
      N5(NP)=NP+1
      N6(NP)=N5(NP)+NPZ
      N7(NP)=NP+NPZ
      GO TO 36
   69 N1(NP)=0
      N2(NP)=0
      N3(NP)=0
      N4(NP)=0
      N5(NP)=0
      N6(NP)=0
      N7(NP)=0
   36 CONTINUE
      NPT=NP
      RETURN
   27 NP=0
C
C     TRIGONAL ZONE FOR ANTIFERROMAGNETIC OXIDES
C
C     0. < KX < 2.*SQRT(2.)/3.
C     0. < KY < SQRT(2./3.)/2.
C     -0.75/SQRT(3.) < KZ < 0.75/SQRT(3.)
      NPH=NPY/2+1
      NZ=NPH*NPH
      DKX=2.D0*DSQRT(2.D0)/3.D0/NPYM
      DKY=DSQRT(3.D0)/2.D0*DKX
      DKZ=3.D0/2.D0/DSQRT(3.D0)/NPZM
      V=2.0D0*DKX*DKY*DKZ
      DHX=-0.5D0*DKX
      DO 37 K=1,NPZ
      DO 37 J=1,NPH
      Y=J-1
      NX=NPY-2*Y
      IM=2*Y+1
      DO 37 I=IM,NPY
      NP=NP+1
      IF(I.EQ.NPY) GO TO 70
      IF(K.EQ.NPZ) GO TO 70
      IF(I.EQ.NPYM) GO TO 71
      IF(I.EQ.IM) GO TO 72
C
C     GENERAL POINT
C
      N1(NP)=NP+1
      N2(NP)=NP+NX
      N3(NP)=N1(NP)+NZ
      N4(NP)=N2(NP)+NZ
      N5(NP)=NP+NZ
      N6(NP)=N4(NP)-1
      N7(NP)=N2(NP)-1
      GO TO 37
C
C     THREE TETRAHEDRA
C
   71 N1(NP)=NP+1
      N2(NP)=NP+NX-1
      N3(NP)=N1(NP)+NZ
      N4(NP)=N2(NP)+NZ
      N5(NP)=NP+NZ
      N6(NP)=0
      N7(NP)=0
      GO TO 37
C
```

```
C      THREE TETRAHEDRA
C
   72 N1(NP)=NP+1
      N2(NP)=NP+NX
      N3(NP)=N1(NP)+NZ
      N4(NP)=N2(NP)+NZ
      N5(NP)=NP+NZ
      N6(NP)=0
      N7(NP)=0
      GO TO 37
   70 N1(NP)=0
      N2(NP)=0
      N3(NP)=0
      N4(NP)=0
      N5(NP)=0
      N6(NP)=0
      N7(NP)=0
   37 CONTINUE
      NPT=NP
      RETURN
      END
      SUBROUTINE PDNS(I1,I2,I3,I4)
C     ***************************************************************
C     *                                                             *
C     *    CALCULATE STATE DENSITY AND NUMBER OF STATES FOR A SINGLE *
C     *    TETRAHEDRON                                              *
C     *                                                             *
C     *    CN : STATE DENSITY                                       *
C     *    SN : NUMBER OF STATES                                    *
C     *                                                             *
C     ***************************************************************
      IMPLICIT REAL*8 (A-H,O-Z)
      COMMON/B1/ E(1000),EM(1001),CNT(1001),SNT(1001),DE,V,BOA,COA,NE,
     1NPT,NL,NEMAX,NT,STS(1001,4),CTS(1001,4),STP(1001,4),CTP(1001,4),
     2STD(1001,4),CTD(1001,4),STF(1001,4),CTF(1001,4),FTS(1000,4),
     3FTP(1000,4),FTD(1000,4),FTF(1000,4),
     3N1(1000),N2(1000),N3(1000),N4(1000),N5(1000),N6(1000),N7(1000)
      DIMENSION S(4)
C
      S(1)=E(I1)
      S(2)=E(I2)
      S(3)=E(I3)
      S(4)=E(I4)
C
      CALL SORT(S)
C
      S1=S(1)
      IF(S1.GE.EM(NE)) RETURN
      EB=EM(1)
      NE1=(S1-EB)/DE
      IF(S1.LT.EB) NE1=-1
      NE1=NE1+2
      NE2=NEMAX
      DO 25 LE=NE1,NE2
      EE=EM(LE)
      IF(EE.LE.S(1)) GO TO 25
      IF(EE.GE.S(4)) GO TO 20
      IF(EE.LE.S(2)) GO TO 21
      IF(EE.GE.S(3)) GO TO 22
      DELTA=S(4)+S(3)-S(2)-S(1)
      AM=(S(4)*S(3)-S(2)*S(1))/DELTA
      D1=S(1)-AM
      D2=S(2)-AM
      V1=V/DELTA
      DEE=EE-AM
      FRAC=DEE*DEE/D2/D1
      CN=3.D0*V1*(1.D0-FRAC)
```

212

```
         SN=3.D0*V1*DEE-V1*(D1+D2+DEE*FRAC)
         GO TO 23
   22 DELTA=S(4)+S(3)-S(2)-S(1)
         AM=(S(4)*S(3)-S(2)*S(1))/DELTA
         D4=S(4)-AM
         V1=V/DELTA
         DEE=EE-S(4)
         FRAC=DEE*DEE/D4/(S(4)-S(3))
         CN=3.D0*V1*FRAC
         SN=V1*(DELTA+DEE*FRAC)
         GO TO 23
   21 DELTA=S(4)+S(3)-S(2)-S(1)
         AM=(S(4)*S(3)-S(2)*S(1))/DELTA
         D1=S(1)-AM
         V1=V/DELTA
         DEE=EE-S(1)
         FRAC=DEE*DEE/D1/(S(1)-S(2))
         CN=3.D0*V1*FRAC
         SN=V1*FRAC*DEE
         GO TO 23
   20 SN=V
         CN=0.D0
C
C
C        CALCULATE AND SUM PROJECTED DENSITY AND NUMBER OF STATES
C
   23 DO 24 KT=1,NT
         GS=(FTS(I1,KT)+FTS(I2,KT)+FTS(I3,KT)+FTS(I4,KT))/4.D0
         GP=(FTP(I1,KT)+FTP(I2,KT)+FTP(I3,KT)+FTP(I4,KT))/4.D0
         GD=(FTD(I1,KT)+FTD(I2,KT)+FTD(I3,KT)+FTD(I4,KT))/4.D0
         STS(LE,KT)=STS(LE,KT)+SN*GS
         STP(LE,KT)=STP(LE,KT)+SN*GP
         STD(LE,KT)=STD(LE,KT)+SN*GD
         CTS(LE,KT)=CTS(LE,KT)+CN*GS
         CTP(LE,KT)=CTP(LE,KT)+CN*GP
         CTD(LE,KT)=CTD(LE,KT)+CN*GD
         IF(NL.LT.4) GO TO 24
         GF=(FTF(I1,KT)+FTF(I2,KT)+FTF(I3,KT)+FTF(I4,KT))/4.D0
         STF(LE,KT)=STF(LE,KT)+SN*GF
         CTF(LE,KT)=CTF(LE,KT)+CN*GF
   24 CONTINUE
         SNT(LE)=SNT(LE)+SN
         CNT(LE)=CNT(LE)+CN
   25 CONTINUE
         RETURN
         END
         SUBROUTINE DENS
C     **********************************************************************
C     *                                                                    *
C     *     SUM OVER ALL TETRAHEDRA                                        *
C     *                                                                    *
C     *     LAT=1,3,5,6,7                                                  *
C     *                                                                    *
C     **********************************************************************
         IMPLICIT REAL*8 (A-H,O-Z)
         COMMON/B1/ E(1000),EM(1001),CNT(1001),SNT(1001),DE,V,BOA,COA,NE,
        1NPT,NL,NEMAX,NT,STS(1001,4),CTS(1001,4),STP(1001,4),CTP(1001,4),
        2STD(1001,4),CTD(1001,4),STF(1001,4),CTF(1001,4),FTS(1000,4),
        3FTP(1000,4),FTD(1000,4),FTF(1000,4),
        3N1(1000),N2(1000),N3(1000),N4(1000),N5(1000),N6(1000),N7(1000)
C
         DO 21 LP=1,NPT
         IF(N7(LP).NE.0) GO TO 23
         IF(N5(LP).NE.0) GO TO 22
         IF(N1(LP).EQ.0) GO TO 21
         I1=N1(LP)
         I2=N2(LP)
         I3=N3(LP)
```

```
          CALL PDNS(LP,I1,I2,I3)
          GO TO 21
       22 I1=N1(LP)
          I2=N2(LP)
          I3=N3(LP)
          CALL PDNS(LP,I1,I2,I3)
          I2=N4(LP)
          CALL PDNS(LP,I1,I2,I3)
          I1=N5(LP)
          CALL PDNS(LP,I1,I2,I3)
          GO TO 21
       23 I1=N1(LP)
          I2=N2(LP)
          I3=N3(LP)
          CALL PDNS(LP,I1,I2,I3)
          I2=N4(LP)
          CALL PDNS(LP,I1,I2,I3)
          I1=N5(LP)
          CALL PDNS(LP,I1,I2,I3)
          I2=N6(LP)
          CALL PDNS(LP,I1,I2,I3)
          I1=N7(LP)
          CALL PDNS(LP,I1,I2,I3)
          I2=N2(LP)
          CALL PDNS(LP,I1,I2,I3)
       21 CONTINUE
C
C
          RETURN
          END
          SUBROUTINE FDENS
C
C    ****************************************************************
C    *                                                              *
C    *     SUM OVER ALL TETRAHEDRA                                  *
C    *                                                              *
C    *     LAT=2                                                    *
C    *                                                              *
C    ****************************************************************
          IMPLICIT REAL*8 (A-H,O-Z)
          COMMON/B1/ E(1000),EM(1001),CNT(1001),SNT(1001),DE,V,BOA,COA,NE,
         1NPT,NL,NEMAX,NT,STS(1001,4),CTS(1001,4),STP(1001,4),CTP(1001,4),
         2STD(1001,4),CTD(1001,4),STF(1001,4),CTF(1001,4),FTS(1000,4),
         3FTP(1000,4),FTD(1000,4),FTF(1000,4),
         3N1(1000),N2(1000),N3(1000),N4(1000),N5(1000),N6(1000),N7(1000)
C
          DO 21 LP=1,NPT
          IF(N7(LP).NE.0) GO TO 23
          IF(N6(LP).NE.0) GO TO 24
          IF(N5(LP).NE.0) GO TO 22
          IF(N4(LP).NE.0) GO TO 25
          IF(N1(LP).EQ.0) GO TO 21
          I1=N1(LP)
          I2=N2(LP)
          I3=N3(LP)
          CALL PDNS(LP,I1,I2,I3)
          GO TO 21
       25 I1=N1(LP)
          I2=N2(LP)
          I3=N3(LP)
          CALL PDNS(LP,I1,I2,I3)
          I2=N4(LP)
          CALL PDNS(LP,I1,I2,I3)
          GO TO 21
       22 I1=N1(LP)
          I2=N2(LP)
          I3=N3(LP)
          CALL PDNS(LP,I1,I2,I3)
```

214

```
      I2=N4(LP)
      CALL PDNS(LP,I1,I2,I3)
      I1=N5(LP)
      CALL PDNS(LP,I1,I2,I3)
      GO TO 21
   24 I1=N1(LP)
      I2=N2(LP)
      I3=N6(LP)
      CALL PDNS(LP,I1,I2,I3)
      I1=N5(LP)
      CALL PDNS(LP,I1,I2,I3)
      I3=N3(LP)
      CALL PDNS(LP,I1,I2,I3)
      I4=N4(LP)
      CALL PDNS(I4,I1,I2,I3)
      I3=N6(LP)
      CALL PDNS(I4,I1,I2,I3)
      GO TO 21
   23 I1=N1(LP)
      I2=N2(LP)
      I3=N3(LP)
      CALL PDNS(LP,I1,I2,I3)
      I2=N4(LP)
      CALL PDNS(LP,I1,I2,I3)
      I1=N5(LP)
      CALL PDNS(LP,I1,I2,I3)
      I2=N6(LP)
      CALL PDNS(LP,I1,I2,I3)
      I1=N7(LP)
      CALL PDNS(LP,I1,I2,I3)
      I2=N2(LP)
      CALL PDNS(LP,I1,I2,I3)
   21 CONTINUE
      RETURN
      END
      SUBROUTINE HDENS
C     ****************************************************************
C     *                                                              *
C     *    SUM OVER ALL TETRAHEDRA                                   *
C     *    LAT=4,8                                                   *
C     *                                                              *
C     ****************************************************************
      IMPLICIT REAL*8 (A-H,O-Z)
      COMMON/B1/ E(1000),EM(1001),CNT(1001),SNT(1001),DE,V,BOA,COA,NE,
     1NPT,NL,NEMAX,NT,STS(1001,4),CTS(1001,4),STP(1001,4),CTP(1001,4),
     2STD(1001,4),CTD(1001,4),STF(1001,4),CTF(1001,4),FTS(1000,4),
     3FTP(1000,4),FTD(1000,4),FTF(1000,4),
     3N1(1000),N2(1000),N3(1000),N4(1000),N5(1000),N6(1000),N7(1000)
C
      DO 21 LP=1,NPT
      IF(N7(LP).NE.0) GO TO 23
      IF(N1(LP).EQ.0) GO TO 21
      I1=N1(LP)
      I2=N2(LP)
      I3=N3(LP)
      CALL PDNS(LP,I1,I2,I3)
      I1=N4(LP)
      CALL PDNS(LP,I1,I2,I3)
      I2=N5(LP)
      CALL PDNS(LP,I1,I2,I3)
      GO TO 21
   23 I1=N1(LP)
      I2=N2(LP)
      I3=N3(LP)
      CALL PDNS(LP,I1,I2,I3)
      I1=N4(LP)
      CALL PDNS(LP,I1,I2,I3)
```

```
      I2=N5(LP)
      CALL PDNS(LP,I1,I2,I3)
      I3=N6(LP)
      CALL PDNS(LP,I1,I2,I3)
      I2=N2(LP)
      CALL PDNS(LP,I1,I2,I3)
      I1=N7(LP)
      CALL PDNS(LP,I1,I2,I3)
   21 CONTINUE
      RETURN
      END
      SUBROUTINE SORT(S)
C     ************************************************************
C     *                                                          *
C     *    SORT ELEMENTS OF S IN INCREASING ORDER                *
C     *                                                          *
C     ************************************************************
      IMPLICIT REAL*8 (A-H,O-Z)
      DIMENSION S(4)
      DO 20 J=1,3
      NP=4-J
      DO 20 I=1,NP
      IF(S(I+1).GE.S(I)) GO TO 20
      ST=S(I)
      S(I)=S(I+1)
      S(I+1)=ST
   20 CONTINUE
      RETURN
      END
```

9.5.4 Execution of DDNS

The execution of the projected state density programme DDNS requires a data
set with energy bands and $t\ell$ characters in the format described in Sect.9.4.5.
The additional input shown in Table 9.6 consists of a few control parameters.

Table 9.6. Input for DDNS

0.	.6	601	1	1	100

With the sample input the state density and the number of states will be
calculated at NE = 601 points distributed over an energy range EF-EB = 0.6 Ry,
but shifted, since MODE = 0, such that the modified EB is below the bottom
of Band 1, i.e. NB1 = 1. The state densities and the number-of-states functions will be printed, NOPRT = 1, but the original eigenvalues will not,
NOWRT = 0.

The NB1, NB2 parameters may be used to select certain bands from the calculated band structure. In the example, whose output is listed in Sect.
9.5.5, only the first band is included, NB2 = 1. Usually one would take
NB1 = 1 and NB2 = $(\ell_{max} + 1)^2$, i.e. include band numbers 1 through 9 in the
case of chromium.

9.5.5 Sample Output from DDNS

```
DENSITY OF STATES PROGRAMME FOR  BCC  LATTICE

BOA,COA,V,EB,EF,NE,NPT,NPX,NPY,NPZ,NB1,NB2

   1.000000   1.000000   0.001953   0.000000   0.500000   11   285   0 17


STRUCTURE CONSTANTS FOR BCC LATTICE

CORRECTION MATRIX FOR BCC STRUCTURE



POTENTIAL PARAMETERS FOR NON-MAGNETIC CR              100381

SWS   2.684000


ENY        OMN        OMJ        FINSQ      FIJSQ      LNN        LNJ


0.461561   1.500361 -1.788505   0.147666   0.195937   1.001405   0.998325
0.620718   7.782095 -3.227841   0.134865   0.267135   1.027465   0.988608
0.639058   1.094374-22.035254   0.015110   3.788171   1.022833   0.540255 1

ENERGY RANGE GENERATED AUTOMATICALLY , EB =   0.200000  EF =   0.7000


BAND NO      EMIN      EMAX      N1    N2


   1       0.213825   0.632737    2    10
DENSITY OF STATES FOR   CR IN   CR


E        NOS        DOS          S-NOS      S-DOS        P-NOS      P-DOS


0.2000   0.00000    0.00000      0.00000    0.00000      0.00000    0.00000
0.2500   0.01323    0.59812      0.01271    0.56152      0.00050    0.03457
0.3000   0.05534    1.06560      0.05047    0.91583      0.00444    0.13134
0.3500   0.12019    1.54256      0.10322    1.19088      0.01447    0.27800
0.4000   0.21245    2.20062      0.16943    1.47017      0.03340    0.49047
0.4500   0.35143    3.58738      0.25248    1.90600      0.06571    0.83054
0.5000   0.73741   11.81305      0.38779    3.04980      0.12427    1.47624
0.5500   1.52688   14.65123      0.52589    1.87132      0.23289    2.05190
0.6000   1.95886    3.40070      0.57189    0.21011      0.28886    0.44451
0.6500   2.00000    0.00000      0.57364    0.00000      0.29510    0.00000
0.7000   2.00000    0.00000      0.57364    0.00000      0.29510    0.00000
```

9.6 The Self-Consistency Programme SCFC

The SCFC programme is designed to solve the energy-band problem (7.10,11,18, 27,28) self-consistently by means of the canonical scaling principle including hybridisation as outlined in Sect.2.6. The programme treats only the conduction states, i.e. the outermost s, p, d, and f electrons, while the charge density of the remaining electrons is kept fixed. In this frozen-

core approximation the core charge density is obtained from atomic calculations renormalised to the relevant atomic volume. Obviously, such a procedure breaks down at very high pressures when the outermost core states
broaden into energy bands.

The basic input includes the projected state densities and number-of-
states functions generated by DDNS, and the atomic charge densities calculated by RHFS. The main output is the self-consistent potential parameters and the electronic pressure.

The first step in the programme is to construct a trial charge density.
This is done simply by renormalising the atomic charge density to the atomic
volume specified by the current Wigner-Seitz radius.

The second step is to solve the Poisson equation (1.3) with the trial
charge density which gives the electrostatic Hartree potential (7.27, 8.38).
To this is added $-2Z/r$ from the nuclei, the Madelung term (8.39) if more
inequivalent atoms are included, and an exchange-correlation potential of
some kind, e.g. that in [9.7], forming the effective one-electron trial potential (7.28, 8.39).

The third step is to generate radial wave functions and the corresponding potential parameters. To this end, the programme solves the Dirac
equation without the spin-orbit interaction (Sect.9.6.1) using the trial
potential. Hence, the programme includes the important relativistic mass-
velocity and Darwin shifts. The potential parameters are calculated from
(3.33-35) and then converted to standard parameters by the formulae in Sect.
4.6. The energy derivatives are calculated from the solutions of the Dirac
equation at two energies, $E_\nu + \varepsilon$ and $E_\nu - \varepsilon$, where ε is some small fraction
of the relevant bandwidth.

The fourth step is to construct a new trial charge density. To this end
the programme evaluates the moments of the projected state densities and
inserts these together with the partial waves and their energy derivatives
into (6.41, 8.30). The moment calculation is based on a set of projected
state densities obtained in previous runs of LMTO and DDNS where the potential parameters used in LMTO could be those given in step three, i.e. constructed from a renormalised atomic potential, in a previous separate run
of SCFC. The moments are evaluated by

$$\int^{n(E_F^*)} (E^* - E_\nu)^q \, dn(E^*) \quad , \quad q = 0,1,2,\ldots$$

$$n(E^*) = n(E^*(D(E))) \quad , \tag{9.29}$$

where n is the number-of-states function, E the energy scale of the original
band calculation, D(E) the logarithmic-derivative function obtained from

218

the Laurent expansion (3.30,32) with the original potential parameters, and $E^*(D)$ is the current energy scale given by the variational estimate (3.50) with the current potential parameters.

At this stage the loop may be closed and steps two through four iterated to self-consistency. The self-consistency criterion is that the first-order moments, i.e. $q = 1$, vanish. In that case, $E_{\nu\ell}$ is the centre of gravity of the occupied part of the ℓ band, and hence that range of the band structure which is important for self-consistency is described with reasonable accuracy.

Upon completion, the programme will print the four standard potential parameters per ℓ, some other potential parameters, e.g. masses and band edges, zeroth-, first-, and second-order moments of the state density, and partial and total pressures. If the potential parameters in the original run of LMTO have been suitably chosen, one will have a reasonably converged result after just one execution of SCFC. If this is not the case, one must use the output potential parameters to perform new band-structure and state-density calculations, and then repeat SCFC with the new state densities. Hence, we have in effect two iteration loops, a band loop consisting of a consecutive execution of LMTO, DDNS, and SCFC, and a scaling loop formed by the scaling procedure inside SCFC, see Fig.9.1. For many applications the latter will suffice.

9.6.1 The Dirac Equation Without Spin-Orbit Coupling

For atoms with high atomic numbers it is important that relativistic effects are included in the band-structure calculations. In SCFC we therefore solve the Dirac rather than the Schrödinger equation, but leave out the spin-orbit interaction. In doing so we obtain an effective one-electron equation which is essentially the Schrödinger equation with the important mass-velocity and Darwin corrections included. The present technique is based on unpublished work by O.K. Andersen and U.K. Poulsen. *Koelling* and *Harmon* [9.8] have taken a related approach.

The equations for the two radial wave functions g_κ and f_κ in the Dirac central field problem are [9.9,10]

$$c \frac{df_\kappa}{dr} = \frac{\kappa - 1}{r} cf_\kappa - (E - v)g$$

$$\frac{dg_\kappa}{dr} = \left(\frac{E - v}{c^2} + 1\right)cf - \frac{\kappa + 1}{r} g \quad , \tag{9.30}$$

where c is the velocity of light, κ is now the usual non-zero combination of angular momentum and spin 1/2 quantum numbers, and v the spherically symmetric potential. By straightforward manipulations the coupled equations may be cast in the form of a single second-order differential equation for g_κ, i.e.

$$\frac{d^2}{dr^2}(rg_\kappa) = \left[\frac{\ell(\ell + 1)}{r} + (v - E) - \frac{(v - E)^2}{c^2}\right.$$

$$\left. - B\left(\frac{\kappa + 1}{r} + \frac{d}{dr} - \frac{1}{r}\right)\right]rg_\kappa$$

$$B = \frac{1}{c^2}\frac{dv}{dr}\left(1 - \frac{v - E}{c^2}\right)^{-1} \quad . \tag{9.31}$$

We note that the usual Pauli equation is obtained from (9.31) in the approximation where $B = c^{-2}$ dv/dr. Inspection shows that (9.31) is independent of κ except for the spin-orbic term $-B(\kappa + 1)/r$ which we therefore neglect.

We now wish to have once more two coupled first-order differential equations similar to (9.30). To this end we define the functions

$$P_\kappa = rg_\kappa \tag{9.32}$$

$$Q_\kappa = [P'_\kappa + (\kappa/r)P_\kappa]/v \tag{9.33}$$

$$v = 1 - \frac{v - E}{c^2} \tag{9.34}$$

and find

$$P'_\kappa = vQ_\kappa - \frac{\kappa}{r}P_\kappa$$

$$Q'_\kappa = \frac{\kappa}{r}Q_\kappa + \left(v - E - \frac{\kappa + 1}{r}\frac{v'}{v^2}\right)P_\kappa \quad . \tag{9.35}$$

Here we differentiated (9.33) and used (9.31) without the spin-orbit term.
If we define

$$P_\ell = r\phi_\kappa$$

$$Q_\ell = P'_\ell - (\ell + 1)P_\ell/r \tag{9.36}$$

and use the radial Schrödinger equation (1.17), we find

$$P'_\ell = Q_\ell + \frac{\ell + 1}{r}P_\ell$$

$$Q'_\ell = -\frac{\ell + 1}{r}Q_\ell + (v - E)P_\ell \quad . \tag{9.37}$$

Since $v \to 1$ and $v' \to 0$ in the non-relativistic case, $c \to \infty$, we realise that the Dirac equations (9.35) in that limit are identical to the usual radial

Schrödinger equation provided $\kappa = -\ell - 1$. We may therefore identify $P_{\kappa=-\ell-1}/r$ as the partial wave ϕ_ℓ, and find that the logarithmic derivative $S\phi_\ell'/\phi_\ell$ is obtained from

$$D_\ell(E) = \ell + S\left[1 - \frac{v(S) - E}{c^2}\right] \frac{Q(S)}{P(S)} \quad . \tag{9.38}$$

The radial Dirac equations without spin-orbit interaction (9.35) are in a form which, except for the term $-(\kappa + 1)v'/(rv^2)P$, is identical to the equations (4-90,4-91) solved by *Loucks* [9.10]. In the subroutine WAVEFC a technique similar to that used by *Loucks* is therefore applied to solve the case without spin-orbit interaction.

9.6.2 Listing of SCFC

```
$SET $
$RESET FREE
$SET DBLTOSNGL
FILE   1(KIND=DISK,TITLE='DOS/CR/DUMMY',FILETYPE=7)
FILE   2(KIND=DISK,TITLE='DOS/CR/DUMMY',FILETYPE=7)
FILE   4(KIND=DISK,TITLE='PTPRM/CR/DUMMY',FILETYPE=7)
FILE   5(KIND=DISK,TITLE='DATA/SCFC/CR',FILETYPE=7)
FILE   6(KIND=PRINTER)
C    **********************************************************************
C    *                                                                    *
C    *    SELF-CONSISTENT SOLUTION OF THE LOCAL DENSITY ONE-ELECTRON       *
C    *    SCHROEDINGER EQUATION BY MEANS OF THE CANONICAL SCALING          *
C    *    PRINCIPLE. EVALUATION OF THE CORRESPONDING POTENTIAL             *
C    *    PARAMETERS, ENERGY-BAND PROPERTIES AND PRESSURE                  *
C    *    VERSION DEC. 1982                                                *
C    *                                                                    *
C    *    H.L.SKRIVER                                                      *
C    *    RISOE NATIONAL LABORATORY                                        *
C    *    DK-4000 ROSKILDE                                                 *
C    *    DENMARK                                                          *
C    *                                                                    *
C    **********************************************************************
C    *                                                                    *
C    *    FILES AND THEIR ATTRIBUTES                                       *
C    *                                                                    *
C    *    FILE 1 :UNFORMATTED,TYPE=SEQUENTIAL,RECORDLENGTH=VARIABLE,       *
C    *            NAME=DOS/YY/B OR DOS/YY/B-DOWN                           *
C    *    FILE 2 :UNFORMATTED,TYPE=SEQUENTIAL,RECORDLENGTH=VARIABLE,       *
C    *            NAME=DOS/YY/B OR DOS/YY/B-UP                             *
C    *    FILE 4 :FORMATTED,TYPE=SEQUENTIAL,RECORDLENGTH=80,               *
C    *            NAME=PTPRM/YY/B                                          *
C    *    FILE 5 :FORMATTED,TYPE=SEQUENTIAL,RECORDLENGTH=80,               *
C    *            NAME=DATA/SCFC/YY                                        *
C    *    FILE 6 :PRINTER                                                  *
C    *                                                                    *
C    **********************************************************************
C    *                                                                    *
C    *    INPUT   (FILE 5) <READT>                                         *
C    *    -----                                                           *
C    *                                                                    *
C    *            SEE LIST IN <READT>                                      *
C    *                                                                    *
C    **********************************************************************
C    *                                                                    *
```

```
C     *     INPUT   (FILE 1) <READT> AND <RPOT>                        *
C     *     -----                                                      *
C     *                                                                *
C     *             FILE 1 IS A STANDARD STATE-DENSITY FILE GENERATED BY *
C     *             <DDNS> (FILE 2) WHERE A SPECIFICATION OF THE CONTENT *
C     *             MAY BE FOUND. IN THE SPIN-POLARISED CASE IT SHOULD  *
C     *             CONTAIN THE SPIN-DOWN STATE DENSITY                 *
C     *                                                                *
C     ****************************************************************************
C     *                                                                *
C     *     INPUT   (FILE 2) <READT> AND <RPOT>                        *
C     *     -----                                                      *
C     *                                                                *
C     *             FILE 1 IS A STANDARD STATE-DENSITY FILE GENERATED BY *
C     *             <DDNS> (FILE 2) WHERE A SPECIFICATION OF THE CONTENT *
C     *             MAY BE FOUND. IN THE SPIN-POLARISED CASE IT SHOULD  *
C     *             CONTAIN THE SPIN-UP STATE DENSITY                   *
C     *                                                                *
C     ****************************************************************************
C     *                                                                *
C     *     INPUT   (FILE 4) <PRMW>                                    *
C     *     -----                                                      *
C     *                                                                *
C     *             FILE 4 IS A STANDARD POTENTIAL PARAMETER FILE USED  *
C     *             BY <LMTO> (FILE 5) WHERE A SPECIFICATION OF THE     *
C     *             CONTENT MAY BE FOUND                               *
C     *                                                                *
C     ****************************************************************************
      COMMON/CHARA/GAR(2,250),GARS(250,2,2),QTI(2),CHAR(2,250),
     1CHARS(250,2,2),WRKTLS(2,4,2),AMAG(2,4)
      COMMON/POT/COR(2,250),SPLIT(2),NZ(2),NEL(2)
      COMMON/POTENT/BGX(250,2,2),POTS(2,2),WS(2),DEO(4),ITER,ICTRM
      COMMON/MOM/EMOM(4,2,4,2),PINDEX(2,4,2)
      COMMON/PARAMT/OMML1T(4,4,2),SPHI2T(4,4,2),PHIRT(4,4,2),
     1FAVT(4,4,2),DNYT(4,4,2),ENYT(4,4,2),SWST,SWT(4),EMINT(2),DET(2),
     2NE(2),SPDNOS(1001,4,4,2),SPDDOS(1001,4,2)
      COMMON/NET/RI(2,250),SWP(2),SWS,R1(2),QTR(2),AMDL(2),XWS(2),BLX,
     1CESUM,VOL,EXCHF,CRIT,ICRIT,IXCH,JRIS(2),NQ,NL,NT,NS,NW,NTLFX(6,4),
     2ITP,MT(2),ITXT(4)
      COMMON/ATM/AMTC(2,250),CORE(2,250)
      DIMENSION SLPCT(8),SGX(2,250),ITXCH(5,4),QTRO(2),AMDLO(2)
      DATA ITXCH/'BART','H-HE','DIN ','EXCH','ANGE','X-AL','PHA ',
     1'EXCH','ANGE','    ','BART','H-HE','DIN-','JANA','K XC','VOSK',
     2'O,WI','LK,N','USAI','R XC'/
    1 FORMAT(1H ,'MAIN:  AMDL=',F10.6,'  SW=',F10.6,'  XWS=',F10.6,
     1'  JRI=',I5,'  VOL=',F10.4)
    2 FORMAT(1H ,2X,A4,6F16.6)
    5 FORMAT('0  ITER=',I3,'  N=',I3,'  BLX=',F13.6)
    7 FORMAT(1H1,20X,'ORIGINAL POTENTIAL PARAMETERS',//,16X,'ENY',14X,
     1'DNY-L',11X,'OM(-)',10X,'S*FI**2',6X,'FI(-)/FI(+)',3X,
     2'1/<FID**2>**1/2',//)
    8 FORMAT(1H ,//,36X,'**START OF CASE NO :',I3,' **',//,38X,5A4,//,
     139X,'S( 0.0 %)',2X,'S(',F5.1,' %)',//,30X,'AVERAGE:',F10.6,2X,
     1F10.6)
    9 FORMAT(1H ,30X,A4,'  :',F10.6,2X,F10.6)
   10 FORMAT(1H1)
C
C     READ ALL DATA
C
      CALL READT(SLPCT,QTRO,AMDLO,ITERX,ICD,IER)
      IF(IER.EQ.-1) STOP
C
C     WS - LOOP
C
      DO 22 II=1,NW
      IEXIT=0
      SWS=SWST*(1.+SLPCT(II)*0.01)
```

222

```
      CALL SWSC(NT,NQ,MT,SWP,WS,SWS)
      VOL=0.
      DO 21 IT=1,NT
      VOL=VOL+4.1887902*MT(IT)*WS(IT)**3
      XWS(IT)=1.+20.*ALOG(WS(IT)/R1(IT))
      JRI=XWS(IT)
      JRIS(IT)=JRI+1
      AMDL(IT)=AMDLO(IT)/SWS
      QTR(IT)=QTRO(IT)
   21 WRITE(6,1) AMDL(IT),WS(IT),XWS(IT),JRIS(IT),VOL
C
C     CALCULATE RENORMALISED ATOM POTENTIAL
C
      CALL SXCPOT
      CALL RENORM
      CALL POISON(GAR,NZ,SGX)
C
      IF(NS.EQ.1) GO TO 28
C
C     CONSTRUCT STARTING POTENTIAL AND CHARGE DENSITY FOR
C     SPIN-POLARISED CASE
C
      DO 24 IT=1,NT
      FAC1=SPLIT(IT)/NEL(IT)/2.
      FAC2=.5-FAC1
      JRI=JRIS(IT)
      IF(GAR(IT,JRI).LT.1.E-10) GAR(IT,JRI)=GAR(IT,JRI-1)*1.5+
     1GAR(IT,JRI-2)*0.5
      DO 25 IR=1,JRI
      R=RI(IT,IR)
      RCE=R*R
      RHO=GAR(IT,IR)/RCE
      RHO1=0.5*RHO
      RHO2=RHO1
      CALL XCPOT(RHO1,RHO2,RHO,CORX,V2,EXC)
      SGX(IT,IR)=RCE*(SGX(IT,IR)+CORX)
C
C     GARS=4*PI*RADIUS**2*CHARGE DENSITY
C     BGX=RADIUS**2*POTENTIAL
C
      BGX(IR,IT,1)=SGX(IT,IR)
      BGX(IR,IT,2)=SGX(IT,IR)
      GARS(IR,IT,1)=FAC1*COR(IT,IR)+FAC2*GAR(IT,IR)
   25 GARS(IR,IT,2)=GAR(IT,IR)-GARS(IR,IT,1)
   24 CONTINUE
      GO TO 29
C
C     CONSTRUCT STARTING POTENTIAL AND CHARGE DENSITY FOR
C     NON-POLARISED CASE
C
   28 DO 26 IT=1,NT
      JRI=JRIS(IT)
      IF(GAR(IT,JRI).LT.1.E-10) GAR(IT,JRI)=GAR(IT,JRI-1)*1.5+
     1GAR(IT,JRI-2)*0.5
      DO 27 IR=1,JRI
      R=RI(IT,IR)
      RCE=R*R
      RHO=GAR(IT,IR)/RCE
      RHO1=0.5*RHO
      RHO2=RHO1
      CALL XCPOT(RHO1,RHO2,RHO,CORX,V2,EXC)
      SGX(IT,IR)=RCE*(SGX(IT,IR)+CORX)
C
C     GARS=4*PI*RADIUS**2*CHARGE DENSITY
C     BGX=RADIUS**2*POTENTIAL
C
```

```
      BGX(IR,IT,1)=SGX(IT,IR)
   27 GARS(IR,IT,1)=.5*GAR(IT,IR)
   26 CONTINUE
C
C     WRITE ORIGINAL POTENTIAL PARAMETERS
C
   29 WRITE(6,7)
      DO 23 IS=1,NS
      DO 23 IT=1,NT
      DO 23 IL=1,NL
      VOAT=-1.
      IF(FAVT(IT,IL,IS).GT.1.E-06) VOAT=1./FAVT(IT,IL,IS)
   23 WRITE(6,2) ITXT(IT),ENYT(IT,IL,IS),DNYT(IT,IL,IS),
     1OMML1T(IT,IL,IS),SPHI2T(IT,IL,IS),PHIRT(IT,IL,IS),VOAT
      WRITE(6,8) II,(ITXCH(I,IXCH+1),I=1,5),SLPCT(II),SWST,SWS
      DO 30 IT=1,NT
   30 WRITE(6,9) ITXT(IT),SWT(IT),WS(IT)
      N=1
      DMAGNE=0.
C
C     ITERATION LOOP FOR GIVEN RADIUS
C
      DO 31 ITER=1,ITERX
      CALL CBLX(IEXIT,N,DMAGNE,DMAGOL)
      WRITE(6,5) ITER,N,BLX
C
      CALL WAVEFC
      CALL NWCHAR(IEXIT,ITERX)
      IF(IEXIT.EQ.1) GO TO 35
      CALL FORWRD
C
      DMAGOL=DMAGNE
      DMAGNE=0.
      DO 32 IT=1,NT
      DO 32 IS=1,NS
   32 DMAGNE=DMAGNE+EMOM(1,IT,3,IS)
      IF(ABS(100.*(DMAGNE-DMAGOL)).LE.2.) N=N+1
   31 CONTINUE
C
C     CONVERGENCE OR NUMBER OF ITERATIONS EXHAUSTED
C
   35 CALL WORK
      CALL CBLX(IEXIT,0,0.,0.)
C
C     UPDATE POTENTIAL PARAMETER DATA SET
C
      CALL PRMW
C
      DO 36 IT=1,NT
      SPLIT(IT)=0.
      DO 36 IL=1,NL
   36 SPLIT(IT)=SPLIT(IT)+AMAG(IT,IL)
C
      IF(IEXIT.EQ.1.AND.ICD.NE.0) CALL PRTCD(ICD)
      WRITE(6,10)
      IF(NW.EQ.1) GO TO 22
C
C     PREPARE INPUT FOR NEXT ITERATION
C
      DO 37 IT=1,NT
      JRI=JRIS(IT)
      I1=JRI+1
      I2=JRI+5
      I2=MINO(I2,250)
      DO 34 IR=1,JRI
   34 AMTC(IT,IR)=GAR(IT,IR)
      GARST=GAR(IT,JRI)
```

```
      SLOPE=(GAR(IT,JRI)-GAR(IT,JRI-1))/(RI(IT,JRI)-RI(IT,JRI-1))
      DO 37 IR=I1,I2
   37 AMTC(IT,IR)=GARST+SLOPE*(RI(IT,IR)-RI(IT,JRI))
   22 CONTINUE
      STOP
      END
      SUBROUTINE READT(SLPCT,QTRO,AMDLO,ITRX,ICD,IER)
C     ******************************************************************
C     *                                                                *
C     *                     ***READ INPUT DATA***                      *
C     *                                                                *
C     *     AMDLO  :MADELUNG POTENTIAL AT SITE T = AMDLO/SWS           *
C     *     AMTC   :4*PI*R**2*TOTAL ATOMIC ELECTRON DENSITY            *
C     *     BLXA   :THE CHARGE DENSITY FOR ITERATION NO. I+1 IS OBTAINED *
C     *     BLXB    AS BLX*C(I-1)+(1.-BLX)*C(I) WHERE BLX=BLXB IF IBLX=0.*
C     *             IF IBLX=1 BLX IS DECREASED FROM BLXA TO BLXB IN STEPS*
C     *             OF DBLX WHEN THE MOMENTS SETTLE                     *
C     *     BNDBLX :WHEN THE SELF-CONSISTENCY LOOP IS SATISFIED THE     *
C     *             POTENTIAL PARAMETERS OBTAINED ARE MIXED ACCORDING TO *
C     *             BNDBLX WITH THE ORIGINAL PARAMETERS FOR USE IN      *
C     *             SUBSEQUENT CALCULATIONS                            *
C     *     CORE   :4*PI*R**2*ELECTRON DENSITY IN CORE                 *
C     *     CRIT   :CONVERGENCE IS OBTAINED WHEN THE SUM OF THE FIRST  *
C     *             MOMENTS, NUMERICALLY, IS SMALLER THAN CRIT         *
C     *     DBLX   :SEE BLXA                                           *
C     *     DEO    :ENERGY STEP USED FOR DIFFERENTIATION IN FIRST      *
C     *             TWO ITERATIONS                                     *
C     *     DET    :ENERGY STEP IN STATE-DENSITY CALCULATION           *
C     *     DUMMY  :NOT USED                                           *
C     *     EF     :GUESS AT THE FERMI LEVEL                           *
C     *     EMAX   :STATE DENSITIES CALCULATED IN THE ENERGY RANGE     *
C     *     EMINT  (EMINT - EMAX)                                      *
C     *     ENY    :GUESS AT ENY                                       *
C     *     ESHIFT :OM(-)=OM(-)+ESHIFT(IL), USED TO CALCULATE THE      *
C     *             SUSCEPTIBILITY                                     *
C     *     EXCHF  :SLATER ALPHA. ONLY USED WHEN IXCH = 1              *
C     *     IBLX   :SEE BLXA                                           *
C     *     ICRIT  :PERFORM AT LEAST ICRIT ITERATIONS BEFORE EXIT      *
C     *     IDENT  :IDENTIFIER USED TO TEST DATA SET                   *
C     *     IDUM   :NOT USED                                           *
C     *     IFORM  :SELECTS FORMAT FOR ATOMIC CHARGE DENSITIES         *
C     *     INO    :=0:READ STANDARD NOS & DOS                         *
C     *             =2:INTERCHANGE <IT> FOR SPIN-UP DOS IN             *
C     *                 ANTIFERROMAGNETIC CALCULATIONS ( CR )          *
C     *             =3:INTERCHANGE <IT> FOR SPIN-UP DOS IN             *
C     *                 ANTIFERROMAGNETIC CALCULATIONS ( 3D METAL OXIDES )*
C     *     ION(IT):IONICITY DEFINED AS ATOMIC NUMBER MINUS NUMBER     *
C     *             OF ELECTRONS                                       *
C     *     ITRX   :NUMBER OF ITERATIONS TO BE PERFORMED               *
C     *     ICD    :CONTROLS PRINT AND PUNCH OF CHARGE DENSITIES , ETC. *
C     *     IXCH   :=0:BARTH-HEDIN EXCHANGE                            *
C     *             =1:SLATER EXCHANGE                                 *
C     *             =2:BARTH-HEDIN EXCHANGE WITH JANAK PARAMETERS      *
C     *             =3:VOSKO,WILK,NUSAIR IMPROVED CORRELATION          *
C     *     MT(IT) :NUMBER OF TYPE T ATOMS                             *
C     *     NAME   :IDENTIFIER USED TO TEST CHARGE DENSITIES           *
C     *     NAMEA  :SAME AS NAME                                       *
C     *     NB1    :STATE DENSITY CALCULATED USING BAND NUMBER NB1     *
C     *     NB2     THROUGH NB2                                        *
C     *     NE     :NUMBER OF POINTS ON ENERGY MESH FOR NOS & DOS      *
C     *     NEL    :NUMBER OF CONDUCTION ELECTRONS                     *
C     *     NL     :NUMBER OF L-QUANTUM NUMBERS                        *
C     *     NQ     :NUMBER OF ATOMS IN UNIT CELL                       *
C     *     NS     :=1:NON-SPIN-POLARISED CALCULATIONS                 *
C     *             =2:SPIN-POLARISED CALCULATIONS                     *
C     *     NT     :NUMBER OF INEQUIVALENT ATOMS IN UNIT CELL          *
C     *     NTLFX  :IF NTLFX(IT,IL) IS ONE THEN ENY(IT,IL,IS)          *
C     *             IS FIXED AT THE START VALUE                        *
```

225

```
C   *   NW      :THE SELF-CONSISTENCY LOOP IS PERFORMED FOR <NW>        *
C   *           DIFFERENT WIGNER-SEITZ RADII                            *
C   *   NZ      :ATOMIC NUMBER                                          *
C   *   QTRO    :GUESS AT CHARGE TRANSFER PER ATOM                      *
C   *   R1      :STARTING POINT ON THE RADIAL MESH                      *
C   *   RKEY    :SEE SLPCT                                              *
C   *   SHIFTE  :ENERGY SHIFT USED IN SUSCEPTIBILITY CALCULATION        *
C   *   SLPCT   :WIGNER-SEITZ RADIUS IN ATOMIC UNITS IF RKEY='   R',    *
C   *           OTHERWISE THE RADIUS IS (1.+SLPCT/100.)*SWST            *
C   *   SPDDOS  :DENSITY OF STATES FUNCTIONS                            *
C   *   SPDNOS  :NUMBER OF STATES FUNCTIONS                             *
C   *   SPLIT   :GUESS AT MAGNETIC MOMENT                               *
C   *   SWP     :RATIO OF INDIVIDUAL WIGNER-SEITZ RADII                 *
C   *   SWST    :AVERAGE ATOMIC WIGNER-SEITZ RADIUS USED IN BAND CALC.  *
C   *                                                                   *
C   *********************************************************************
      REAL MY
      COMMON/FERMI/EF,SHIFTE,NELTOT
      COMMON/PARAM/OMMLM1(2,4,2),SPHISQ(2,4,2),PHIRAT(2,4,2),CC(2,4,2),
     1FAV(2,4,2),MY(2,4,2),DNY(2,4,2),ENY(2,4,2),VL(2,4,2),TL(2,4,2),
     2BOT(2,4,2),TOP(2,4,2),D1(2,4,2),D2(2,4,2),D3(2,4,2),XMR(2,4,2),
     1AR(2,4,2),BR(2,4,2)
      COMMON/PARAMT/OMML1T(4,4,2),SPHI2T(4,4,2),PHIRT(4,4,2),
     1FAVT(4,4,2),DNYT(4,4,2),ENYT(4,4,2),SWST,SWT(4),EMINT(2),DET(2),
     2NE(2),SPDNOS(1001,4,4,2),SPDDOS(1001,4,4,2)
      COMMON/NET/RI(2,250),SWP(2),SWS,R1(2),QTR(2),AMDL(2),XWS(2),BLX,
     1CESUM,VOL,EXCHF,CRIT,ICRIT,IXCH,JRIS(2),NQ,NL,NT,NS,NW,NTLFX(6,4),
     2ITP,MT,ITXT(4)
      COMMON/POT/COR(2,250),SPLIT(2),NZ(2),NEL(2)
      COMMON/POTENT/BGX(250,2,2),POTS(2,2),WS(2),DEO(4),ITER,ICTRM
      COMMON/ATM/AMTC(2,250),CORE(2,250)
      COMMON/ABLX/BLXA,BLXB,DBLX,IBLX,BNDBLX
      DIMENSION SLPCT(8),NAME(2,10),NAMEA(10),QTRO(2),AMDLO(2),ION(2),
     1ITIS(4,2,3)
      DATA ITIS/1,2,3,4,1,2,3,4,1,2,3,4,2,1,3,4,1,2,3,4,3,4,1,2/
C
    1 FORMAT(4I5,I4,I1,6(1X,4I1))
    2 FORMAT(1H ,'READT:',4X,'SWP=',2F10.6)
    3 FORMAT(1H0,10X,A4,': ',10A4,/,11X,'NEL=',I5,' R1=',1PE15.6,/)
    4 FORMAT(10A4,A4,I2,F10.7,E15.8,I1)
    5 FORMAT(5E16.9)
    6 FORMAT(A3)
    7 FORMAT(5E15.8)
    8 FORMAT(8F10.6)
    9 FORMAT(1H ,'READT:',4X,'EXCHF=',F10.6)
   10 FORMAT(1H1,10X,'INPUT DATA FOR SELF-CONSISTENCY PROGRAMME',//)
   11 FORMAT(1H0,10X,A4,': ',10A4,/,11X,'NZ=',I5,'ION=',I5,' CESUM='
     11PE18.9,/)
   12 FORMAT(1H0,'READT:    EB=',F10.6,' EF=',F10.6,' DE=',F10.6,' NE=',
     1I5,' NB1=',I5,' NB2=',I5,//)
   13 FORMAT(1H ,'READT:',4X,'QTR=',2F10.6)
   14 FORMAT(1H ,'READT:',4X,'MT=',2I5)
   15 FORMAT(8F10.6)
   16 FORMAT(1H ,'READT:',4X,'SPLIT=',2F10.6)
   17 FORMAT(1H ,'READT:',4X,'BNDBLX='F10.4,'  AMDL=',2F10.6)
   18 FORMAT(5F10.6,A4)
   19 FORMAT(3F10.6,2I5,2F10.6,I5)
  100 FORMAT(1H ,'READT:',4X,'NTLFX=',6(1X,4I1))
  101 FORMAT(1H ,'READT:',4X,'NW=',I3,' NS=',I2,' ITRX=',I3,' IXCH=',
     1I2,' ICD=',I2,' INO=',I2,' IFRM=',I3)
  102 FORMAT(1H ,'READT:',4X,'CRIT=',F10.6,'  ICRIT=',I4)
  103 FORMAT(5D14.8)
  104 FORMAT(10A4,A4,I2,I4,E20.8,I2)
  105 FORMAT(6I5)
  106 FORMAT(1H ,'READT:',4X,'EF=',F10.6,' SHIFTE=',F10.6)
 2001 FORMAT(' INPUT ERROR 2000: IDENT=',A4)
 3001 FORMAT(' INPUT ERROR 3000: INCONSISTENT CORE AND ATOMIC DENSITIES'
     %/',IT=',I1,' NAMES=',2(2X,10A4))
```

```
 5001 FORMAT(´ INPUT ERROR 5000: INCORRECT NUMBER OF CARDS READ.
      %IDENT=´,A4)
C
      WRITE(6,10)
      IER=-1
C
      READ(5,1) NW,NS,ITRX,IXCH,ICD,INO,((NTLFX(IT,IL),IL=1,4),IT=1,6)
C
C     READ STATE-DENSITY DATA SETS, AND STORE NUMBER OF STATES AND
C     STATE DENSITIES IN SPDNOS> AND <SPDDO> ACCORDING TO
C     <ITIS(IT,IS,INO)>
C
      IF(INO.EQ.0) INO=1
      DO 31 IS=1,NS
      CALL RPOT(IS,INO)
      READ(IS) EMINT(IS),EMAX,DET(IS),NE(IS),NB1,NB2
      WRITE(6,12) EMINT(IS),EMAX,DET(IS),NE(IS),NB1,NB2
      NES=NE(IS)
C
C     NEXT TWO LINES: DUMMY READ
C
      READ(IS) (SPDNOS(IE,1,1,IS),IE=1,NES)
      READ(IS) (SPDDOS(IE,1,1,IS),IE=1,NES)
      DO 24 ITT=1,NT
      IT=ITIS(ITT,IS,INO)
      DO 24 IL=1,NL
      READ(IS) (SPDNOS(IE,IT,IL,IS),IE=1,NES)
   24 READ(IS) (SPDDOS(IE,IT,IL,IS),IE=1,NES)
      LOCK IS
   31 CONTINUE
C
      IF(INO.GE.2) NT=NT/2
      IF(INO.GE.2) NQ=NQ/2
C
   33 IF(NS.EQ.1) GO TO 30
C
C     STATE DENSITY AND NUMBER OF STATES PER SPIN IN THE
C     SPIN-POLARISED CASE
C
      DO 25 IS=1,NS
      NES=NE(IS)
      DO 25 IT=1,NT
      DO 25 IL=1,NL
      DO 25 IE=1,NES
      SPDNOS(IE,IT,IL,IS)=SPDNOS(IE,IT,IL,IS)*0.5
   25 SPDDOS(IE,IT,IL,IS)=SPDDOS(IE,IT,IL,IS)*0.5
C
   30 READ(5,18) EXCHF,DEO,RKEY
      READ(5,19) BLXA,BLXB,DBLX,IBLX,IDUM,BNDBLX,CRIT,ICRIT
      READ(5,8) (SLPCT(IW),IW=1,NW)
      READ(5,105) (MT(IT),IT=1,NT)
      READ(5,8) (AMDLO(IT),IT=1,NT)
      READ(5,8) (QTRO(IT),IT=1,NT)
      READ(5,8) (SPLIT(IT),IT=1,NT)
      READ(5,8) (SWP(IT),IT=1,NT)
      READ(5,8) EF,SHIFTE
      DO 26 IT=1,NT
   26 READ(5,15) ((ENY(IT,IL,IS),IL=1,NL),IS=1,NS)
C
      IF(ICRIT.EQ.0) ICRIT=8
      IF(CRIT.EQ.0.) CRIT=0.001
C
      IF(RKEY.NE.´    R´) GO TO 36
      DO 37 IW=1,NW
   37 SLPCT(IW)=(SLPCT(IW)-SWST)/SWST*100.
C
```

```
   36 WRITE(6,101) NW,NS,ITRX,IXCH,ICD,INO,IFRM
      WRITE(6,17) BNDBLX,(AMDLO(IT),IT=1,NT)
      WRITE(6,102) CRIT,ICRIT
      WRITE(6,2) (SWP(IT),IT=1,NT)
      WRITE(6,16) (SPLIT(IT),IT=1,NT)
      WRITE(6,13) (QTRO(IT),IT=1,NT)
      WRITE(6,14) (MT(IT),IT=1,NT)
      IF(IXCH.EQ.1) WRITE(6,9) EXCHF
      WRITE(6,106) EF,SHIFTE
      WRITE(6,100) ((NTLFX(IT,IL),IL=1,4),IT=1,NT)
C
C     READ CORE AMD ATOMIC CHARGE DENSITIES
C
      DO 21 IT=1,NT
      READ(5,4) (NAME(IT,I),I=1,10),IDENT,NEL(IT),DUMMY,R1(IT),IFORM
      IF(IDENT.NE.'CORE') GOTO 2000
      IF(IFORM.EQ.0) READ(5,7) (CORE(IT,IR),IR=1,250)
      IF(IFORM.EQ.1) READ(5,5) (CORE(IT,IR),IR=1,250)
      IF(IFORM.EQ.2) READ(5,103) (CORE(IT,IR),IR=1,250)
      WRITE(6,3) IDENT,(NAME(IT,I),I=1,10),NEL(IT),R1(IT)
      READ(5,104) NAMEA,IDENT,NZ(IT),ION(IT),DUMMY,IFORM
      IF(NZ(IT).GT.ION(IT)) GO TO 35
      NZTEMP=NZ(IT)
      NZ(IT)=ION(IT)
      ION(IT)=NZTEMP
   35 WRITE(6,11) IDENT,NAMEA,NZ(IT),ION(IT),DUMMY
C
C     CHECK THAT CORE AND ATOMIC CHARGE DENSITIES CORRESPOND
C
      DO 22 I=1,10
   22 IF(NAME(IT,I).NE.NAMEA(I)) GOTO 3000
      IF(IDENT.NE.'AMTC') GOTO 2000
      IF(IFORM.EQ.0) READ(5,7) (AMTC(IT,IR),IR=1,250)
      IF(IFORM.EQ.1) READ(5,5) (AMTC(IT,IR),IR=1,250)
      IF(IFORM.EQ.2) READ(5,103) (AMTC(IT,IR),IR=1,250)
   21 CONTINUE
C
C     LAST CARD OF CHARGE DENSITY DATA MUST CONTAIN 'EOD'
C
      READ(5,6) IDENT
      IF(IDENT.NE.'EOD') GOTO 5000
      IER=1
C
C     GENERATE R-MESH
C
      DO 20 IT=1,NT
      RT=R1(IT)
      DO 20 IR=1,250
   20 RI(IT,IR)=RT*EXP(0.05*(IR-1))
      ITP=1
      ITM=MT(1)
      DO 23 IT=2,NT
      IF(MT(IT).LE.ITM) GO TO 23
      ITM=MT(IT)
      ITP=IT
   23 CONTINUE
      NELTOT=0
      DO 27 IT=1,NT
   27 NELTOT=NELTOT+NEL(IT)*MT(IT)
C
C     REDEFINE <NEL> AS THE NUMBER OF CONDUCTION ELECTRONS FOR THE
C     NEUTRAL ATOM
C
      DO 34 IT=1,NT
   34 NEL(IT)=NEL(IT)+ION(IT)
C
C     NORMAL RETURN
```

```
C
      RETURN
 2000 WRITE(6,2001) IDENT
      RETURN
 3000 WRITE(6,3001) IT,(NAME(IT,I),I=1,10),NAMEA
      RETURN
 5000 WRITE(6,5001) IDENT
      RETURN
      END
      SUBROUTINE RPOT(IS,INO)
C     ****************************************************************
C     *                                                              *
C     *      READ AND CONVERT POTENTIAL PARAMETERS FROM ORIGINAL BAND *
C     *      CALCULATION                                             *
C     *                                                              *
C     ****************************************************************
      REAL*4 LNN,LNJ,LJJ
      COMMON/PARAMT/OMML1T(4,4,2),SPHI2T(4,4,2),PHIRT(4,4,2),
     1FAVT(4,4,2),DNYT(4,4,2),ENYT(4,4,2),SWST,SWT(4),EMINT(2),DET(2),
     2NE(2),SPDNOS(1001,4,4,2),SPDDOS(1001,4,4,2)
      COMMON/NET/RI(2,250),SWP(2),SWS,R1(2),QTR(2),AMDL(2),XWS(2),BLX,
     1CESUM,VOL,EXCHF,CRIT,ICRIT,IXCH,JRIS(2),NQ,NL,NT,NS,NW,NTLFX(6,4),
     2ITP,MT(2),ITXT(4)
      COMMON/POTENT/BGX(250,2,2),POTS(2,2),WS(2),DEO(4),ITER,ICTRM
      INTEGER TXTS(15),TXTC(15),TXTP(20,4),ITIS(4,2,3)
      DATA ITIS/1,2,3,4,1,2,3,4,1,2,3,4,2,1,3,4,1,2,3,4,3,4,1,2/
    1 FORMAT(1H0,'RPOT:',5X,15A4,/,11X,15A4,/)
    2 FORMAT(1H ,10X,20A4)
C
      READ(IS) TXTS,TXTC,SWST,NL,NQ,NLM,NLMQ,NT
      READ(IS) NPX,NPY,NPZ,LAT,IDUM1,BOA,COA,DUM1
C
C     SET ICTRM TO 1 IF THE CORRECTION TO THE ASA WAS INCLUDED
C     IN THE ORIGINAL BAND CALCULATION
C
      ICTRM=0
      IF(TXTC(1).NE.'NO C') ICTRM=1
      WRITE(6,1) TXTS,TXTC
      DO 23 ITT=1,NT
      IT=ITIS(ITT,IS,INO)
      READ(IS) (TXTP(I,IT),I=1,20),SWT(IT),ITXT(IT)
      WRITE(6,2) (TXTP(I,IT),I=1,20)
      SW=SWT(IT)
      SWSQ=SW*SW
      DO 23 IL=1,NL
      TWOL=2*IL-1
      READ(IS) ENYT(IT,IL,IS),OMN,OMJ,FINSQ,FIJSQ,LNN,LNJ,LJJ
      OMML1T(IT,IL,IS)=OMN/SWSQ
      SPHI2T(IT,IL,IS)=FINSQ*SW
      PHIRT(IT,IL,IS)=TWOL*SW*SWSQ*FINSQ/(OMN-OMJ)
      IF(ABS(OMJ).LT.1.E-03) GO TO 21
      BM=(LJJ-1.)/OMJ/OMJ
      GO TO 22
   21 BM=(LNN-1.)/OMN/OMN
   22 FAVT(IT,IL,IS)=SWSQ*SQRT(BM)
      DNYT(IT,IL,IS)=TWOL/(OMN/PHIRT(IT,IL,IS)/OMJ-1.)
   23 CONTINUE
      RETURN
      END
      SUBROUTINE RENORM
C     ****************************************************************
C     *                                                              *
C     *    <RENORM> RENORMALISES CORE AND ATOMIC ELECTRON DENSITIES  *
C     *                                                              *
C     ****************************************************************
      COMMON/CHARA/GAR(2,250),GARS(250,2,2),QTI(2),CHAR(2,250),
     1CHARS(250,2,2),WRKTLS(2,4,2),AMAG(2,4)
      COMMON/POT/COR(2,250),SPLIT(2),NZ(2),NEL(2)
```

```
      COMMON/POTENT/BGX(250,2,2),POTS(2,2),WS(2),DEO(4),ITER,ICTRM
      COMMON/NET/RI(2,250),SWP(2),SWS,R1(2),QTR(2),AMDL(2),XWS(2),BLX,
     1CESUM,VOL,EXCHF,CRIT,ICRIT,IXCH,JRIS(2),NQ,NL,NT,NS,NW,NTLFX(6,4),
     2ITP,MT(2),ITXT(4)
      COMMON/ATM/AMTC(2,250),CORE(2,250)
    1 FORMAT(1H ,'RENORM: C=',E15.8,'   CWS=',E15.8,'    Z=',E15.8,
     1'    ZWS=',E15.8)
    2 FORMAT(1H ,'RENORM: **WARNING** : CORE CONTAINS WRONG NUMBER OF EL
     1ECTRONS')
C
      DO 20 IT=1,NT
      JRI=JRIS(IT)
      P=XWS(IT)-(JRI-1)
      ISTART=1
      IF(MOD(JRI,2).EQ.0) ISTART=2
      Z=AMTC(IT,ISTART)*RI(IT,ISTART)
      C=CORE(IT,ISTART)*RI(IT,ISTART)
      I=ISTART+1
   21 Z=Z+4.*AMTC(IT,I)*RI(IT,I)+2.*AMTC(IT,I+1)*RI(IT,I+1)
      C=C+4.*CORE(IT,I)*RI(IT,I)+2.*CORE(IT,I+1)*RI(IT,I+1)
      I=I+2
      IF(I.LT.JRI) GO TO 21
      Z=.05/3.*(Z-AMTC(IT,JRI)*RI(IT,JRI))
      C=.05/3.*(C-CORE(IT,JRI)*RI(IT,JRI))
      Z0=AMTC(IT,JRI-2)*RI(IT,JRI-2)
      C0=CORE(IT,JRI-2)*RI(IT,JRI-2)
      Z1=AMTC(IT,JRI-1)*RI(IT,JRI-1)
      C1=CORE(IT,JRI-1)*RI(IT,JRI-1)
      Z2=AMTC(IT,JRI)*RI(IT,JRI)
      C2=CORE(IT,JRI)*RI(IT,JRI)
      ZWS=Z+.05*(Z0*(P**3/6.-P*P/4.+1./12.)
     %          +Z1*(P-P**3/3.-.6666666666)+Z2*(P**3/6.+P*P/4.-5./12.))
      CWS=C+.05*(C0*(P**3/6.-P*P/4.+1./12.)
     %          +C1*(P-P**3/3.-.6666666666)+C2*(P**3/6.+P*P/4.-5./12.))
      WRITE(6,1) C,CWS,Z,ZWS
      DC=NZ(IT)-NEL(IT)-CWS
      DZ=NZ(IT)-ZWS+QTR(IT)
      IF(ABS(DC).GT.0.5) WRITE(6,2)
      FACC=DC*3./WS(IT)**3
      FACZ=DZ*3./WS(IT)**3
      DO 22 IR=1,JRI
      RSQ=RI(IT,IR)**2
      COR(IT,IR)=CORE(IT,IR)+FACC*RSQ
   22 GAR(IT,IR)=AMTC(IT,IR)+FACZ*RSQ
   20 CONTINUE
      RETURN
      END
      SUBROUTINE CBLX(I,N,A1,A2)
C     ********************************************************************
C     *                                                                  *
C     *      <CBLX> SELECTS MIXING PARAMETER FUNCTION                     *
C     *                                                                  *
C     ********************************************************************
      COMMON/NET/RI(2,250),SWP(2),SWS,R1(2),QTR(2),AMDL(2),XWS(2),BLX,
     1CESUM,VOL,EXCHF,CRIT,ICRIT,IXCH,JRIS(2),NQ,NL,NT,NS,NW,NTLFX(6,4),
     2ITP,MT(2),ITXT(4)
      COMMON/ABLX/BLXA,BLXB,DBLX,IBLX,BNDBLX
      COMMON/PARAM/OMMLM1(2,4,2),SPHISQ(2,4,2),PHIRAT(2,4,2),CC(2,4,2),
     1FAV(2,4,2),MY(2,4,2),DNY(2,4,2),ENY(2,4,2),VL(2,4,2),TL(2,4,2),
     2BOT(2,4,2),TOP(2,4,2),D1(2,4,2),D2(2,4,2),D3(2,4,2),XMR(2,4,2),
     1AR(2,4,2),BR(2,4,2)
      COMMON/PARAMT/OMML1T(4,4,2),SPHI2T(4,4,2),PHIRT(4,4,2),
     1FAVT(4,4,2),DNYT(4,4,2),ENYT(4,4,2),SWST,SWT(4),EMINT(2),DET(2),
     2NE(2),SPDNOS(1001,4,4,2),SPDDOS(1001,4,4,2)
      DIMENSION ENYQ(2),OMQ(2),SFIQ(2)
    1 FORMAT(1H ,///,3X,'MIXED PARAMETERS FOR USE IN BAND CALCULATION',
     1'  BNDBLX=',F10.5,/)
    2 FORMAT('   ENY,OM,SFISQ =',6F13.6)
```

```
C
      IF(I.NE.1) GO TO 30
C
C     MIX POTENTIAL PARAMETERS AND WRITE IF CONVERGED
C
      WRITE(6,1) BNDBLX
      BLXM=1.-BNDBLX
      DO 32 IT=1,NT
      DO 32 IL=1,NL
      DO 31 IS=1,NS
      ENYQ(IS)=ENYT(IT,IL,IS)*BNDBLX+ENY(IT,IL,IS)*BLXM
      OMQ(IS)=OMML1T(IT,IL,IS)*BNDBLX+OMMLM1(IT,IL,IS)*BLXM
   31 SFIQ(IS)=SPHI2T(IT,IL,IS)*BNDBLX+SPHISQ(IT,IL,IS)*BLXM
      WRITE(6,2) (ENYQ(IS),OMQ(IS),SFIQ(IS),IS=1,NS)
   32 CONTINUE
      RETURN
   30 IF(IBLX-1) 21,22,23
C
C     MIXING OF CHARGE DENSITIES INDEPENDENT OF <ITER>
C
   21 BLX=BLXB
      RETURN
C
C     INCREASE MIXING WHEN THE MOMENTS HAVE SETTLED
C
   22 BLX=AMAX1(BLXB,BLXA-N*DBLX)
      RETURN
   23 RETURN
      END
      SUBROUTINE PRTCD(ICD)
C     ****************************************************************
C     *                                                              *
C     *     PRINT AND PUNCH CHARGE DENSITIES, POTENTIALS,  ETC.      *
C     *                                                              *
C     ****************************************************************
      COMMON/FUNC/F(250,2,4,2),F1(250,2,4,2),F2(250,2,4,2),
     1P(250),Q(250),PS(250),QS(250),PE(250,3)
      COMMON/POTENT/BGX(250,2,2),POTS(2,2),WS(2),DEO(4),ITER,ICTRM
      COMMON/CHARA/GAR(2,250),GARS(250,2,2),QTI(2),CHAR(2,250),
     1CHARS(250,2,2),WRKTLS(2,4,2),AMAG(2,4)
      COMMON/POT/COR(2,250),SPLIT(2),NZ(2),NEL(2)
      COMMON/NET/RI(2,250),SWP(2),SWS,R1(2),QTR(2),AMDL(2),XWS(2),BLX,
     1CESUM,VOL,EXCHF,CRIT,ICRIT,IXCH,JRIS(2),NQ,NL,NT,NS,NW,NTLFX(6,4),
     2ITP,MT(2),ITXT(4)
    1 FORMAT(1H ,6E15.7)
    2 FORMAT(1H1,'CHARGE DENSITIES')
    3 FORMAT(5E16.8)
    4 FORMAT(' POTENTIAL FOR',A4)
    5 FORMAT(1H1,10X,'IS =',I5,'  TYPE =',A4,//)
    6 FORMAT(1H ,5E15.6)
C
      GO TO (20,21,22),ICD
C
C     PRINT CHARGE DENSITIES
C
   20 WRITE(6,2)
      DO 30 IT=1,NT
      JRI=JRIS(IT)
      DO 31 IR=1,JRI
      RRI=RI(IT,IR)
      FACR=12.566371*RRI*RRI
   31 CHAR(IT,IR)=(CHAR(IT,IR)+COR(IT,IR))/FACR
   30 WRITE(6,1) (RI(IT,IR),CHAR(IT,IR),IR=1,JRI)
      RETURN
C
C     PUNCH POTENTIALS
C
```

```
   21 DO 32 IT=1,NT
      WRITE(7,4) ITXT(IT)
      JRI=JRIS(IT)
      DO 33 IR=1,JRI
      RSQ=RI(IT,IR)**2
      DO 33 IS=1,NS
   33 BGX(IR,IT,IS)=BGX(IR,IT,IS)/RSQ
      DO 32 IS=1,NS
   32 WRITE(7,3) (BGX(IR,IT,IS),IR=1,JRI)
      RETURN
C
C     PRINT WAVE FUNCTIONS
C
   22 DO 34 IS=1,NS
      DO 34 IT=1,NT
      WRITE(6,5) IS,ITXT(IT)
      JRI=JRIS(IT)
      DO 34 IR=1,JRI
   34 WRITE(6,6) RI(IT,IR),(F(IR,IT,IL,IS),IL=1,NL)
      RETURN
      END
      SUBROUTINE SXCPOT
C     *********************************************************************
C     *                                                                   *
C     *  SET CONSTANTS FOR <XCPOT>                                        *
C     *                                                                   *
C     *********************************************************************
      COMMON/NET/RI(2,250),SWP(2),SWS,R1(2),QTR(2),AMDL(2),XWS(2),BLX,
     1CESUM,VOL,EXCHF,CRIT,ICRIT,IXCH,JRIS(2),NQ,NL,NT,NS,NW,NTLFX(6,4),
     2ITP,MT(2),ITXT(4)
      COMMON/CXCPOT/ XCCP,XCCF,XCRP,XCRF,XALPHA,OTH,FTH,AA,BB,IXCH1
      IXCH1=IXCH+1
C
      GO TO (31,32,33,34),IXCH1
C
C     BARTH-HEDIN J. PHYS. C5,1629(1972)
C
   31 XCCP=0.0504
      XCCF=0.0254
      XCRP=30.
      XCRF=75.
      OTH=1./3.
      FTH=4./3.
      AA=1./2.**OTH
      BB=1.-AA
      RETURN
C
C     SLATER X-ALPA
C
   32 OTH=1./3.
      XALPHA=6.*EXCHF*(3./16./3.141592654**2)**OTH
      RETURN
C
C     BARTH-HEDIN-JANAK PHYS. REV. B12,1257(1975)
C
   33 XCCP=0.045
      XCCF=0.0225
      XCRP=21.
      XCRF=53.
      OTH=1./3.
      FTH=4./3.
      AA=1./2.**OTH
      BB=1.-AA
      RETURN
C
C     VOSKO-WILK-NUSAIR CAN. J. PHYS. 58,1200(1900)
C
```

232

```
   34 OTH=1./3.
      FTH=4./3.
      AA=2.**FTH-2.
      RETURN
      END
      SUBROUTINE XCPOT(RHO1,RHO2,RHO,V1,V2,EXC)
C     **********************************************************************
C     *                                                                    *
C     *    CALCULATES EXCHANGE CORRELATION POTENTIAL                       *
C     *                                                                    *
C     **********************************************************************
      COMMON/CXCPOT/ XCCP,XCCF,XCRP,XCRF,XALPHA,OTH,FTH,AA,BB,IXCH1
      DATA AP,XPO,BP,CP,QP,CP1,CP2,CP3/0.0621814,-0.10498,3.72744,
     A12.9352,6.1519908,1.2117833,1.1435257,-0.031167608/
      DATA AF,XFO,BF,CF,QF,CF1,CF2,CF3/0.0310907,-0.32500,7.060428,
     A18.0578,4.7309269,2.9847935,2.7100059,-0.1446006/
C
      RS1=(RHO/3.)**OTH
      RS=1./RS1
      GO TO (21,22,21,23),IXCH1
C
C     BARTH-HEDIN EXCHANGE CORRELATION
C     J. PHYS. C5,1629(1972)
C              --
C
   21 RSF=RS/XCRF
      RSF2=RSF*RSF
      RSF3=RSF2*RSF
      RSP=RS/XCRP
      RSP2=RSP*RSP
      RSP3=RSP2*RSP
      FCF=(1.+RSF3)*ALOG(1.+1./RSF)+.5*RSF-RSF2-OTH
      FCP=(1.+RSP3)*ALOG(1.+1./RSP)+.5*RSP-RSP2-OTH
      EPSCP=-XCCP*FCP
      EPSCF=-XCCF*FCF
      EPSXP=-.91633059/RS
      CNY=5.1297628*(EPSCF-EPSCP)
      X=RHO1/RHO
      FX=(X**FTH+(1.-X)**FTH-AA)/BB
      EXC=EPSXP+EPSCP+FX*(CNY+FTH*EPSXP)/5.1297628
      ARS=-1.22177412/RS+CNY
      BRS=-XCCP*ALOG(1.+XCRP/RS)-CNY
      TRX1=(2.*X)**OTH
      V1=ARS*TRX1+BRS
      TRX2=(2.*RHO2/RHO)**OTH
      V2=ARS*TRX2+BRS
      RETURN
C
C     SLATER EXCHANGE POTENTIAL
C
   22 EXC=-0.75*XALPHA*(0.5*RHO)**OTH
      V1=-XALPHA*(RHO1)**OTH
      V2=-XALPHA*(RHO2)**OTH
      RETURN
C
C     VOSKO-WILK-NUSAIR EXCHANGE CORRELATION
C     CAN. J. PHYS. 58,1200(1980)
C                --
C
   23 X=SQRT(RS)
      XPX=X*X+BP*X+CP
      XFX=X*X+BF*X+CF
      S=(RHO2-RHO1)/RHO
      SP=1.+S
      SM=1.-S
      S4=S**4-1.
      FS=(SP**FTH+SM**FTH-2.)/AA
```

```
      BETA=1./(2.74208+3.182*X+0.09873*X*X+0.18268*X**3)
      DFS=FTH*(SP**OTH-SM**OTH)/AA
      DBETA=-(0.27402*X+0.09873+1.591/X)*BETA**2
      ATNP=ATAN(QP/(2.*X+BP))
      ATNF=ATAN(QF/(2.*X+BF))
      ECP=AP*(ALOG(X*X/XPX)+CP1*ATNP-CP3*(ALOG((X-XP0)**2/XPX)+CP2*ATNP
     1))
      ECF=AF*(ALOG(X*X/XFX)+CF1*ATNF-CF3*(ALOG((X-XF0)**2/XFX)+CF2*ATNF
     1))
      EC=ECP+FS*(ECF-ECP)*(1.+S4*BETA)
      TP1=(X*X+BP*X)/XPX
      TF1=(X*X+BF*X)/XFX
      UCP=ECP-AP/3.*(1.-TP1-CP3*(X/(X-XP0)-TP1-XP0*X/XPX))
      UCF=ECF-AF/3.*(1.-TF1-CF3*(X/(X-XF0)-TF1-XF0*X/XFX))
      UC0=UCP+(UCF-UCP)*FS
      UC20=UC0+(ECF-ECP)*SM*DFS
      UC10=UC0-(ECF-ECP)*SP*DFS
      DUC=(UCF-UCP)*BETA*S4*FS+(ECF-ECP)*(-RS/3.)*DBETA*S4*FS
      DUC2=DUC+(ECF-ECP)*BETA*SM*(4.*S**3*FS+S4*DFS)
      DUC1=DUC-(ECF-ECP)*BETA*SP*(4.*S**3*FS+S4*DFS)
      UC1=UC10+DUC1
      UC2=UC20+DUC2
      EPX=-0.91633059/RS*(1.+FTH*FS/5.1297628)
      AMYX2=-1.22177412/RS*SP**OTH
      AMYX1=-1.22177412/RS*SM**OTH
      EXC=EC+EPX
      V1=UC1+AMYX1
      V2=UC2+AMYX2
      RETURN
      END
      SUBROUTINE POISON(YSQ,NZ,W)
C     ****************************************************************
C     *                                                              *
C     *     SOLVES POISSON'S EQUATION BASED UPON LOUCKS:             *
C     *     "AUGMENTED PLANE WAVE METHOD", BENJAMIN,                 *
C     *     NEW YORK (1967) P. 98                                    *
C     *                                                              *
C     *     INPUT  : YSQ = 4.*PI*RADIUS**2*CHARGE DENSITY            *
C     *              NZ  = ATOMIC NUMBER                             *
C     *     OUTPUT : W   = ELECTROSTATIC POTENTIAL FROM NUCLEUS      *
C     *                    ,ELECTRON CHARGE DENSITY AND MADELUNG     *
C     *                    CONTRIBUTION. THE TWO FORMER GIVE         *
C     *                    (TOTAL CHARGE INSIDE WS)/WS AT THE WS-RADIUS *
C     *                                                              *
C     ****************************************************************
      COMMON/POTENT/BGX(250,2,2),POTS(2,2),WS(2),DEO(4),ITER,ICTRM
      COMMON/NET/RI(2,250),SWP(2),SWS,R1(2),QTR(2),AMDL(2),XWS(2),BLX,
     1CESUM,VOL,EXCHF,CRIT,ICRIT,IXCH,JRIS(2),NQ,NL,NT,NS,NW,NTLFX(6,4),
     2ITP,MT(2),ITXT(4)
      DIMENSION YSQ(2,250),W(2,250),E(250),F(250),NZ(2)
C
      A=1.-.0025/48.
      B=-2.-.025/48.
      EDL=EXP(.025)
      C=.0025/6.
      C2=-B/A
      E(1)=.0
      F(1)=EXP(.025)
      ZM=ABS(QTR(ITP))
C
      DO 20 IT=1,NT
      JRI=JRIS(IT)
      P=XWS(IT)-(JRI-1)
C
C     FIND CHARGE INSIDE R(JRI)
C
```

234

```fortran
      ISTART=1
      IF(MOD(JRI,2).EQ.0) ISTART=2
      Z=YSQ(IT,ISTART)*RI(IT,ISTART)
      I=ISTART+1
   24 Z=Z+4.*YSQ(IT,I)*RI(IT,I)+2.*YSQ(IT,I+1)*RI(IT,I+1)
      I=I+2
      IF(I.LT.JRI) GOTO 24
      QJRI=.05/3.*(Z-YSQ(IT,JRI)*RI(IT,JRI))
      W0=YSQ(IT,JRI-2)*RI(IT,JRI-2)
      W1=YSQ(IT,JRI-1)*RI(IT,JRI-1)
      W2=YSQ(IT,JRI)*RI(IT,JRI)
      QWS=QJRI+.05*(W0*(P**3/6.-P*P/4.+1./12.)
     %          +W1*(P-P**3/3.-.6666666666)+W2*(P**3/6.+P*P/4.-5./12.))
      QQ=QWS-NZ(IT)
C
C     SOLVE POISSON
C
      ITOP=JRI-1
      DO 21 J=2,ITOP
      D=C*SQRT(RI(IT,J))*(EDL*YSQ(IT,J+1)+10.*YSQ(IT,J)+YSQ(IT,J-1)/EDL)
      F(J)=C2-1./F(J-1)
   21 E(J)=(D/A+E(J-1))/F(J)
      W(IT,JRI)=2.*QJRI/SQRT(RI(IT,JRI))
      DO 22 J=1,ITOP
      JV=JRI-J
   22 W(IT,JV)=E(JV)+W(IT,JV+1)/F(JV)
C
C     IMPOSE PROPER BOUNDARY CONDITION
C
      W0=W(IT,JRI-2)/SQRT(RI(IT,JRI-2))
      W1=W(IT,JRI-1)/SQRT(RI(IT,JRI-1))
      W2=W(IT,JRI)/SQRT(RI(IT,JRI))
      VS=.5*P*(P-1.)*W0+(1.-P*P)*W1+.5*P*(P+1.)*W2
      IF(ABS(QTR(IT)).GE.1.E-06) GO TO 25
      POTM=0.D0
      GO TO 26
   25 POTM=SIGN(ZM,QTR(IT))*AMDL(IT)
   26 DV=2.*(NZ(IT)+QTR(IT))/WS(IT)-POTM-VS
      TWOZ=2.*NZ(IT)
      DO 23 IR=1,JRI
      R=RI(IT,IR)
   23 W(IT,IR)=W(IT,IR)/SQRT(R)-TWOZ/R+DV
   20 CONTINUE
      RETURN
      END
      SUBROUTINE WAVEFC
C     ****************************************************************
C     *                                                            *
C     *    SOLVES THE DIRAC EQUATIONS WITHOUT SPIN-ORBIT COUPLING AND *
C     *    EVALUATES POTENTIAL PARAMETERS. ORIGINAL VERSION BY U.K. *
C     *    POULSEN BASED PARTIALLY ON LOUCKS' <FOVG>                *
C     *                                                            *
C     ****************************************************************
      REAL LP1,LDCSQ,MY,M0,M00
      COMMON/FUNC/F(250,2,4,2),F1(250,2,4,2),F2(250,2,4,2),
     1P(250),Q(250),PS(250),QS(250),PE(250,3)
      COMMON/POTENT/V(250,2,2),POTS(2,2),WS(2),DEO(4),ITER,ICTRM
      COMMON/NET/RI(2,250),SWP(2),SWS,R1(2),QTR(2),AMDL(2),XWS(2),BLX,
     1CESUM,VOL,EXCHF,CRIT,ICRIT,IXCH,JRIS(2),NQ,NL,NT,NS,NW,NTLFX(6,4),
     2ITP,MT(2),ITXT(4)
      COMMON/PARAM/OMMLM1(2,4,2),SPHISQ(2,4,2),PHIRAT(2,4,2),CC(2,4,2),
     1FAV(2,4,2),MY(2,4,2),DNY(2,4,2),ENY(2,4,2),VL(2,4,2),TL(2,4,2),
     2BOT(2,4,2),TOP(2,4,2),D1(2,4,2),D2(2,4,2),D3(2,4,2),XMR(2,4,2),
     1AR(2,4,2),BR(2,4,2)
      COMMON/PARAMT/OMML1T(4,4,2),SPHI2T(4,4,2),PHIRT(4,4,2),
     1FAVT(4,4,2),DNYT(4,4,2),ENYT(4,4,2),SWST,SWT(4),EMINT(2),DET(2),
```

```
      2NE(2),SPDNOS(1001,4,4,2),SPDDOS(1001,4,4,2)
       DIMENSION PEL(3),PEH(3),PSEL(3),PSEH(3),PWS(3),PSWS(3),D(3),
      1VM(250,2,2),RN(3),BWPCT(4)
       DATA BWPCT/0.01,0.01,0.01,0.013/
    9 FORMAT(' HARD TEST IN WAVEFCT : IT=',I1,'  IL=',I1,'  IS=',I1,
      1'  IE=',I1,'  N=',I3,'  PREL=',1PE12.5,'  QREL=',1PE12.5,'  ****')
C
       ITST=1
       TEST=1E+04
       IF(ITER.GT.3) TEST=1E+05
       NRK=6
       C=274.07446
       DX=.05
       CSQ=C*C
       EXPDXH=EXP(DX/2.)
       DXD4=DX/4.
       DXD8=DX/8.
       A1=DX*3.3
       A2=-DX*4.2
       A3=DX*7.8
       A4=DX*14./45.
       A5=DX*64./45.
       A6=DX*24./45.
       FAC3=7./8.*EXPDXH**2
       FAC4=.5*EXPDXH**2
C
       DO 21 IT=1,NT
       S=WS(IT)
       S2=S*S
       SQS=SQRT(S)
       XS=XWS(IT)
       JRI=JRIS(IT)
       CALL DIFF(V,DX,JRI,VM,IT)
C
       DO 20 IS=1,NS
C
C      V=BGX=RADIUS**2*POTENTIAL
C
       TWOZ=-V(1,IT,IS)/RI(IT,1)
C
       DO 19 IL=1,NL
C
       L=IL-1
       LP1=L+1.
       LDCSQ=L/CSQ
       DE=DEO(IL)
       IF(ITER.GT.2) DE=SPHISQ(IT,IL,IS)*(11.+L*7)*BWPCT(IL)
       IF(DE.GT.0.05) DE=DEO(IL)
       FAC1=.5/DE
       FAC2=1./DE/DE
C
C      SOLVE THE DIRAC EQUATIONS FOR THREE ENERGIES IN ORDER TO
C      EVALUATE ENERGY DERIVATIVES
C
       DO 15 IE=1,3
C
       E=ENY(IT,IL,IS)+(IE-2)*DE
C
       R=RI(IT,1)
       VN=V(1,IT,IS)
       VMN=VM(1,IT,IS)
       P(1)=1.
C
C      FREE-ELECTRON CASE,  I.E.  V(IR,IT,IS)=0.
C
       Q(1)=-E/(2.*L+3.)*R*P(1)
       BETA=LP1
       IF(TWOZ.EQ.0.) GOTO 1
```
236

```
C
C      WITH POTENTIAL :
C
       BETA=SQRT(LP1*L+1.-(TWOZ/C)**2)
       SB0=(BETA-LP1)*CSQ/TWOZ
       SA1=(BETA-1.-(TWOZ/C)**2)/(2.*BETA+1.)
       SB1=-SB0+CSQ/TWOZ*(BETA-L)*SA1
       SA2=(2.*L*SA1-3.*L+TWOZ/CSQ*(BETA+L+3.)*SB1)/4./(BETA+1.)
       SB2=-SB1+CSQ/TWOZ*(BETA-L+1.)*SA2
       DELTA=R*CSQ/TWOZ
       Q(1)=(SB0+DELTA*(SB1+DELTA*SB2))/(1.+DELTA*(SA1+DELTA*SA2))*P(1)
     1 CONTINUE
       C1=VN/R**2-E
       C2=1.-C1/CSQ
       C3=(VMN-2.*VN)/C2/C2*LDCSQ
       PS(1)=R*C2*Q(1)+LP1*P(1)
       QS(1)=-LP1*Q(1)+(R*C1-C3/R**3)*P(1)
C
       N=1
     2 PC=P(N)
       QC=Q(N)
       DP1=DX*(R*C2*QC+LP1*PC)
       DQ1=DX*(-LP1*QC+(R*C1-C3/R**3)*PC)
       PC=PC+.5*DP1
       QC=QC+.5*DQ1
       R=R*EXPDXH
       VNP1=V(N+1,IT,IS)
       VMNP1=VM(N+1,IT,IS)
       VH=(VN+VNP1)*.5+(VMN-VMNP1)*DXD8
       VMH=1.5*(VNP1-VN)/DX-(VMN+VMNP1)/4.
       C1=VH/R/R-E
       C2=1.-C1/CSQ
       C3=(VMH-2.*VH)/C2/C2*LDCSQ
       DP2=DX*(R*C2*QC+LP1*PC)
       DQ2=DX*(-LP1*QC+(R*C1-C3/R**3)*PC)
       PC=PC+.5*(DP2-DP1)
       QC=QC+.5*(DQ2-DQ1)
       DP3=DX*(R*C2*QC+LP1*PC)
       DQ3=DX*(-LP1*QC+(R*C1-C3/R**3)*PC)
       PC=PC+DP3-.5*DP2
       QC=QC+DQ3-.5*DQ2
       N=N+1
       R=RI(IT,N)
       C1=VNP1/R/R-E
       C2=1.-C1/CSQ
       C3=(VMNP1-2.*VNP1)/C2/C2*LDCSQ
       DP4=DX*(R*C2*QC+LP1*PC)
       DQ4=DX*(-LP1*QC+(R*C1-C3/R**3)*PC)
       P(N)=P(N-1)+(DP1+2.*(DP2+DP3)+DP4)/6.
       Q(N)=Q(N-1)+(DQ1+2.*(DQ2+DQ3)+DQ4)/6.
       PS(N)=R*C2*Q(N)+LP1*P(N)
       QS(N)=-LP1*Q(N)+(R*C1-C3/R**3)*P(N)
       VN=VNP1
       VMN=VMNP1
C
       IF(N-NRK) 2,3,3
C
     3 IF(N.GE.JRI) GOTO 12
       PSN  =PS(NRK)
       QSN  =QS(NRK)
       PSNM1=PS(NRK-1)
       QSNM1=QS(NRK-1)
       PSNM2=PS(NRK-2)
       QSNM2=QS(NRK-2)
       PSNM3=PS(NRK-3)
       QSNM3=QS(NRK-3)
       PSNM4=PS(NRK-4)
       QSNM4=QS(NRK-4)
```

```
     4 R=RI(IT,N+1)
       C1=V(N+1,IT,IS)/R/R-E
       C2=1.-C1/CSQ
       C3=(VM(N+1,IT,IS)-2.*V(N+1,IT,IS))/C2/C2*LDCSQ
       PP=P(N-5)+A1*(PSN+PSNM4)+A2*(PSNM1+PSNM3)+A3*PSNM2
       QP=Q(N-5)+A1*(QSN+QSNM4)+A2*(QSNM1+QSNM3)+A3*QSNM2
       NIT=0
     5 PSNP1=R*C2*QP+LP1*PP
       QSNP1=-LP1*QP+(R*C1-C3/R**3)*PP
       PC=P(N-3)+A4*(PSNP1+PSNM3)+A5*(PSN+PSNM2)+A6*PSNM1
       QC=Q(N-3)+A4*(QSNP1+QSNM3)+A5*(QSN+QSNM2)+A6*QSNM1
       IF(ABS(TEST*(PC-PP))-ABS(PC)) 6,6,7
     6 IF(ABS(TEST*(QC-QP))-ABS(QC)) 11,11,7
     7 IF(NIT-6) 10,8,10
     8 PREL=(PC-PP)/PC
       QREL=(QC-QP)/QC
       WRITE(6,9) IT,IL,IS,IE,N,PREL,QREL
       GOTO 11
    10 NIT=NIT+1
       PP=PC
       QP=QC
       GOTO 5
    11 N=N+1
       P(N)=PC
       Q(N)=QC
       PS(N)=PSNP1
       QS(N)=QSNP1
       PSNM4=PSNM3
       PSNM3=PSNM2
       PSNM2=PSNM1
       PSNM1=PSN
       PSN =PSNP1
       QSNM4=QSNM3
       QSNM3=QSNM2
       QSNM2=QSNM1
       QSNM1=QSN
       QSN =QSNP1
       IF(N-JRI) 4,12,12
C
    12 JM=JRI-1
       X=DX*(XS-JM)
       CALL INTPOL(0.,DX,P(JM),P(JRI),PS(JM),PS(JRI),X,PU,PSU)
       CALL INTPOL(0.,DX,Q(JM),Q(JRI),QS(JM),QS(JRI),X,QU,DUMMY)
       CALL INTPOL(0.,DX,V(JM,IT,IS),V(JRI,IT,IS),VM(JM,IT,IS),
      %            VM(JRI,IT,IS),X,VU,DUMMY)
       IF(IE.EQ.2) POTS(IT,IS)=VU/S2
       D(IE)=S*(1.-(VU/S2-E)/CSQ)*QU/PU+L
C
       SUM=0.
       I=2
    13 SUM=SUM+RI(IT,I)*(P(I)**2+P(I+1)**2*FAC3)
       I=I+2
       IF(I.LT.JRI) GOTO 13
       H=ALOG(S/RI(IT,I-1))
       SUM=16.*SUM + ( 7.*(RI(IT,1)-P(I-1)**2*RI(IT,I-1))
      %                +DX*(RI(IT,1)*(1.+2.*PS(1))
      %                -P(I-1)*RI(IT,I-1)*(P(I-1)+2.*PS(I-1))) )
       SUM=SUM*DX/15. + ( RI(IT,1)/(1.+2.*BETA)
      %          + PU*PU*S*(H/2.-H*H/12.*(1.+2.*PSU/PU))
      %          +P(I-1)**2*RI(IT,I-1)*(H/2.+H*H/12.*
      %          (1.+2.*PS(I-1)/P(I-1))))
       RNIE=1./SQRT(SUM)
       RN(IE)=RNIE
       PWS(IE)=PU*RNIE
       PSWS(IE)=PSU*RNIE
       DO 14 I=1,JRI
```

238

```
   14 PE(I,IE)=P(I)*RNIE
      PSEL(IE)=PS(JM)*RNIE
      PSEH(IE)=PS(JRI)*RNIE
   15 CONTINUE
C
C     END OF IE-LOOP
C
      RNM1=RN(1)
      RNO=RN(2)
      RNP1=RN(3)
C
C     EVALUATE POTENTIAL PARAMETERS
C
      MO=1./PWS(2)**2/S
      AO=-PWS(2)/2./DE*(PWS(3)-PWS(1))/S
      BO=-PWS(2)*(PWS(3)-2.*PWS(2)+PWS(1))/S/3./DE/DE/S2
      IF(FAVT(IT,IL,IS).GT.1.E-06) GO TO 23
      BO=0.DO
      GO TO 17
   23 IF(BO.GT.1.D-06) GO TO 17
      WRITE(6,22) BO,AO,MO,DE
   22 FORMAT(1H ,'**WARNING FROM <WAVEFC> : B=',F10.6,' A=',F10.6,' M=',
     1F10.6,' DE=',F10.6)
      BO=0.DO
C
   17 DNY(IT,IL,IS)=D(2)
      OM=1./MO/(AO+1./(D(2)+LP1))
      OMO=1./MO/(AO+1./(D(2)))
      OML=1./MO/(AO+1./(D(2)-L))
      OMX=1./MO/AO
      OMMLM1(IT,IL,IS)=OM/S2
      BM=BO*MO
      PHIRAT(IT,IL,IS)=1.-(2.*LP1-1.)/(LP1+D(2)+1./AO)
      SPHISQ(IT,IL,IS)=1./MO/S2/(1.+AO*(D(2)+LP1))**2
      SFISQL=1./MO/S2/(1.+AO*(D(2)-L))**2
      FAV(IT,IL,IS)=S2*SQRT(BM)
      CC(IT,IL,IS)=ENY(IT,IL,IS)+OM/S2/(1.+BM*OM**2)
      BOT(IT,IL,IS)=ENY(IT,IL,IS)+OMO/S2/(1.+BM*OMO**2)
      TOP(IT,IL,IS)=ENY(IT,IL,IS)+OMX/S2/(1.+BM*OMX**2)
      VL(IT,IL,IS)=ENY(IT,IL,IS)+OML/S2/(1.+BM*OML**2)
      MY(IT,IL,IS)=2./S2/SPHISQ(IT,IL,IS)*(1.+BM*OM**2)**2/(1.-BM*OM**2)
      TL(IT,IL,IS)=(2*L+3)/S2/SFISQL*(1.+BM*OML**2)**2/(1.-BM*OML**2)
C
      D1(IT,IL,IS)=DNY(IT,IL,IS)+1.DO/(AO+DSQRT(BO/MO))
      D2(IT,IL,IS)=DNY(IT,IL,IS)+1.DO/AO
      D3(IT,IL,IS)=DNY(IT,IL,IS)+1.DO/(AO-DSQRT(BO/MO))
C
      DO 16 IR=1,JRI
      PNM1=PE(IR,1)
      PNO=PE(IR,2)
      PNP1=PE(IR,3)
      F(IR,IT,IL,IS)= PNO
      F1(IR,IT,IL,IS)=(PNP1-PNM1)*FAC1
   16 F2(IR,IT,IL,IS)=(PNP1-2.*PNO+PNM1)*FAC2
C
      MOO=-(D(3)-D(1))/2./DE/S2
      AOO=1./((PWS(3)*D(3)-PWS(1)*D(1)) / (PWS(3)-PWS(1)) - D(2))
      IF(ITST.EQ.0) GO TO 19
C
      I=2
      SUM1=0.
      SUM2=0.
C
   18 R=RI(IT,I)
      RX=RI(IT,I+1)
      PEI2=PE(I,2)
      PEI2X=PE(I+1,2)
```

```fortran
      SUM1=SUM1+2.*(PEI2-PE(I,1))**2*R+(PEI2X-PE(I+1,1))**2*RX
      SUM2=SUM2+2.*(PEI2-PE(I,3))**2*R+(PEI2X-PE(I+1,3))**2*RX
      I=I+2
      IF(I.LT.JRI) GOTO 18
C
      SUM1=DX/3.*(2.*SUM1+(PE(1,2)-PE(1,1))**2*RI(IT,1)
     %           -(PE(I-1,2)-PE(I-1,1))**2*RI(IT,I-1))
     2      +ALOG(S/RI(IT,I-1))*.5*( (PWS(2)-PWS(1))**2*S
     3       +(PE(I-1,2)-PE(I-1,1))**2*RI(IT,I-1) )
     4        +(PE(1,2)-PE(1,1))**2*RI(IT,1)/(1.+2.*BETA)
      SUM2=DX/3.*(2.*SUM2+(PE(1,2)-PE(1,3))**2*RI(IT,1)
     A           -(PE(I-1,2)-PE(I-1,3))**2*RI(IT,I-1))
     B      +ALOG(S/RI(IT,I-1))*.5*( (PWS(2)-PWS(3))**2*S
     C       +(PE(I-1,2)-PE(I-1,3))**2*RI(IT,I-1) )
     D        +(PE(1,2)-PE(1,3))**2*RI(IT,1)/(1.+2.*BETA)
      B00=PWS(2)**2/S**3*FAC2*(SUM1+SUM2)/2.
C
      XMR(IT,IL,IS)=1.-M00/M0
      AR(IT,IL,IS)=1.-A00/A0
      IF(B0.EQ.0.D0) GO TO 19
      BR(IT,IL,IS)=1.-B00/B0
   19 CONTINUE
C
C     END IL-LOOP
C
   20 CONTINUE
C
C     END IS-LOOP
C
   21 CONTINUE
C
C     END IT-LOOP
C
      RETURN
      END
      SUBROUTINE SWSC(NT,NQ,MT,SC,SW,SWS)
C     ******************************************************************
C     *                                                                *
C     *    SCALE ATOMIC RADII ACCORDING TO SC(IT) UNDER VOLUME         *
C     *    CONSERVATION                                                *
C     *                                                                *
C     ******************************************************************
      DIMENSION F(10),SC(NT),SW(NT),MT(NT)
C
      OM=NQ*SWS**3
      SC1=SC(1)
      SUM=0.D0
      DO 20 I=1,NT
      F(I)=SC(I)/SC1
   20 SUM=SUM+MT(I)*F(I)**3
      R1=(OM/SUM)**(1.D0/3.D0)
      DO 21 I=1,NT
   21 SW(I)=F(I)*R1
      RETURN
      END
      SUBROUTINE INTPOL(A,B,FA,FB,FMA,FMB,X,FX,FMX)
C     ******************************************************************
C     *                                                                *
C     *    INTERPOLATE FROM THE VALUES AND THE DERIVATIVES AT A AND B  *
C     *                                                                *
C     ******************************************************************
C
      DX=B-A
      D=(X-A)/DX
      IF(D*(1.-D).LT.0.) WRITE(6,1)
    1 FORMAT(' WRONG USE OF INTPOL  ****************************')
      C2=3.*(FB-FA)-(FMB+2.*FMA)*DX
      C3=2.*(FA-FB)+(FMA+FMB)*DX
```

240

```
      FX = FA+D*(DX*FMA+D*(C2+D*C3))
      FMX=FMA+D*(2.*C2+3.*C3*D)/DX
      RETURN
      END
      SUBROUTINE DIFF(V,DX,N,VM,IT)
C     ******************************************************************
C     *                                                                *
C     *    DIFFERENTIATE THE POTENTIAL FOR USE IN <WAVEFC>             *
C     *                                                                *
C     ******************************************************************
      DIMENSION V(250,2,2),VM(250,2,2)
C
      NM2=N-2
      DO 2 IS=1,2
      VM(1,IT,IS)=((6.*V(2,IT,IS)+6.6666666667*V(4,IT,IS)+
     %          1.2*V(6,IT,IS))-(2.45*V(1,IT,IS)+7.5*V(3,IT,IS)+
     %          3.75*V(5,IT,IS)+.16666666667*V(7,IT,IS)))/DX
      VM(2,IT,IS)=((6.*V(3,IT,IS)+6.6666666667*V(5,IT,IS)+
     %          1.2*V(7,IT,IS))-(2.45*V(2,IT,IS)+7.5*V(4,IT,IS)+
     %          3.75*V(6,IT,IS)+.16666666667*V(8,IT,IS)))/DX
      DO 1 I=3,NM2
    1 VM(I,IT,IS)=((V(I-2,IT,IS)+8.*V(I+1,IT,IS))-
     %          (8.*V(I-1,IT,IS)+V(I+2,IT,IS)))/12./DX
      VM(N-1,IT,IS)=(V(N,IT,IS)-V(N-2,IT,IS))/2./DX
      VM(N,IT,IS)=(V(N-2,IT,IS)/2.-2.*V(N-1,IT,IS)+3./2.*V(N,IT,IS))/DX
    2 CONTINUE
      RETURN
      END
      SUBROUTINE FORWRD
C     ******************************************************************
C     *                                                                *
C     *    <FORWRD> PREPARES INPUT FOR THE NEXT ITERATION BASED ON THE *
C     *    ELECTRON DENSITY OF THE PRESENT AND THE PREVIOUS ITERATION  *
C     *                                                                *
C     ******************************************************************
      REAL MY
      COMMON/MOM/EMOM(4,2,4,2),PINDEX(2,4,2)
      COMMON/POTENT/BGX(250,2,2),POTS(2,2),WS(2),DEO(4),ITER,ICTRM
      COMMON/POT/COR(2,250),SPLIT(2),NZ(2),NEL(2)
      COMMON/CHARA/GAR(2,250),GARS(250,2,2),QTI(2),CHAR(2,250),
     1CHARS(250,2,2),WRKTLS(2,4,2),AMAG(2,4)
      COMMON/FERMI/EF,SHIFTE,NELTOT
      COMMON/PARAM/OMMLM1(2,4,2),SPHISQ(2,4,2),PHIRAT(2,4,2),CC(2,4,2),
     1FAV(2,4,2),MY(2,4,2),DNY(2,4,2),ENY(2,4,2),VL(2,4,2),TL(2,4,2),
     2BOT(2,4,2),TOP(2,4,2),D1(2,4,2),D2(2,4,2),D3(2,4,2),XMR(2,4,2),
     1AR(2,4,2),BR(2,4,2)
      COMMON/NET/RI(2,250),SWP(2),SWS,R1(2),QTR(2),AMDL(2),XWS(2),BLX,
     1CESUM,VOL,EXCHF,CRIT,ICRIT,IXCH,JRIS(2),NQ,NL,NT,NS,NW,NTLFX(6,4),
     2ITP,MT(2),ITXT(4)
      DIMENSION W(2,250)
    1 FORMAT(' WARNING FROM FORWRD: IR=',I3,'  IT=',I1,'   CHARS(',
     1I1,') =',1PE15.8,'  IS NEGATIVE.  ***************')
C
      DO 21 IT=1,NT
      JRI=JRIS(IT)
      QTR(IT)=BLX*QTR(IT)+(1.-BLX)*QTI(IT)
      DO 20 IS=1,NS
      DO 20 IR=1,JRI
   20 GARS(IR,IT,IS)=BLX*GARS(IR,IT,IS)+
     %          (1.-BLX)*(CHARS(IR,IT,IS)+0.5*COR(IT,IR))
      DO 21 IR=1,JRI
   21 GAR(IT,IR)=BLX*GAR(IT,IR)+(1.-BLX)*(CHAR(IT,IR)+COR(IT,IR))
C
      CALL POISON(GAR,NZ,W)
C
      DO 22 IT=1,NT
      JRI=JRIS(IT)
```

```
           DO 22 IR=1,JRI
           R=RI(IT,IR)
           RCE=R*R
C
           DO 23 IS=1,NS
           IF(GARS(IR,IT,IS).GE.0.) GO TO 23
           WRITE(6,1) IR,IT,IS,GARS(IR,IT,IS)
           GO TO 24
C
C          EXCHANGE AND CORRELATION
C
      23   RHO=GAR(IT,IR)/RCE
           RHO1=GARS(IR,IT,1)/RCE
           RHO2=GARS(IR,IT,2)/RCE
           CALL XCPOT(RHO1,RHO2,RHO,VXC1,VXC2,EXC)
      24   CONTINUE
           BGX(IR,IT,1)=RCE*(W(IT,IR)+VXC1)
           IF(NS.EQ.1) GO TO 22
           BGX(IR,IT,2)=RCE*(W(IT,IR)+VXC2)
      22   CONTINUE
C
C          FIND ENY
C
           DO 25 IS=1,NS
           DO 25 IL=1,NL
           DO 25 IT=1,NT
           IF(NTLFX(IT,IL).EQ.1) GO TO 26
           IF(NTLFX(IT,IL).EQ.2) GO TO 32
           ENY(IT,IL,IS)=ENY(IT,IL,IS)+EMOM(2,IT,IL,IS)/EMOM(1,IT,IL,IS)
           GO TO 25
      26   ENY(IT,IL,IS)=ENY(IT,IL,IS)
           GO TO 25
      32   ENY(IT,IL,IS)=CC(IT,IL,IS)
      25   CONTINUE
           RETURN
           END
           SUBROUTINE NWCHAR(IEXIT,ITERX)
C     *****************************************************************
C     *                                                               *
C     *     <NEWCHAR>                                                  *
C     *     1) CALCULATES THE NEW CHARGE DENSITY FROM THE WAVE FUNCTIONS*
C     *        AND THE POTENTIAL PARAMETERS AS OBTAINED IN <WAVEFCT>   *
C     *     2) SERVES AS OUTPUT ROUTINE                                *
C     *                                                               *
C     *****************************************************************
           REAL MY
           COMMON/FUNC/F(250,2,4,2),F1(250,2,4,2),F2(250,2,4,2),
          1P(250),Q(250),PS(250),QS(250),FG(250,3)
           COMMON/POT/COR(2,250),SPLIT(2),NZ(2),NEL(2)
           COMMON/DENS/DOS(2,4,2),TOTDOS
           COMMON/FERMI/EF,SHIFTE,NELTOT
           COMMON/POTENT/BGX(250,2,2),POTS(2,2),WS(2),DEO(4),ITER,ICTRM
           COMMON/NET/RI(2,250),SWP(2),SWS,R1(2),QTR(2),AMDL(2),XWS(2),BLX,
          1CESUM,VOL,EXCHF,CRIT,ICRIT,IXCH,JRIS(2),NQ,NL,NT,NS,NW,NTLFX(6,4),
          2ITP,MT(2),ITXT(4)
           COMMON/PARAMT/OMML1T(4,4,2),SPHI2T(4,4,2),PHIRT(4,4,2),
          1FAVT(4,4,2),DNYT(4,4,2),ENYT(4,4,2),SWST,SWT(4),EMINT(2),DET(2),
          2NE(2),SPDNOS(1001,4,4,2),SPDDOS(1001,4,4,2)
           COMMON/MOM/EMOM(4,2,4,2),PINDEX(2,4,2)
           COMMON/PARAM/OMMLM1(2,4,2),SPHISQ(2,4,2),PHIRAT(2,4,2),CC(2,4,2),
          1FAV(2,4,2),MY(2,4,2),DNY(2,4,2),ENY(2,4,2),VL(2,4,2),TL(2,4,2),
          2BOT(2,4,2),TOP(2,4,2),D1(2,4,2),D2(2,4,2),D3(2,4,2),XMR(2,4,2),
          1AR(2,4,2),BR(2,4,2)
           COMMON/CHARA/GAR(2,250),GARS(250,2,2),QTI(2),CHAR(2,250),
          1CHARS(250,2,2),WRKTLS(2,4,2),AMAG(2,4)
           DIMENSION HP(4,2),CDIFF(4,2),STO1(2,4),VOA(2,4,2)
           DATA VOA/16*-1./
```

```
C 395 FORMAT(1H )
  396 FORMAT('    V             =',8F13.6)
  398 FORMAT('    A             =',8F13.6)
  397 FORMAT('    TAU           =',8F13.6)
  399 FORMAT('0  OMEGA(-)       =',8F13.6)
  400 FORMAT('    C             =',8F13.6)
  401 FORMAT('    10S*PHISQ(-) =',8F13.6)
  402 FORMAT('    PHI(-)/PHI(+)=',8F13.6)
  403 FORMAT('    1/SQ(BM)/S2  =',8F13.6)
  404 FORMAT('0  MY            =',8F13.6)
  405 FORMAT('    CPS*S2,CDS*S2=',13X,3F13.6,13X,3F13.6)
  406 FORMAT('0  ATOM:',6X,A4)
  407 FORMAT('0  D1            =',8F13.6)
  408 FORMAT('    DNYD          =',8F13.6)
  409 FORMAT('    MAGNETISATION=',8F13.6)
  410 FORMAT('    STONER-I      =',8F13.6)
  411 FORMAT('    D3            =',8F13.6)
  412 FORMAT('    B             =',8F13.6)
  413 FORMAT('    AR            =',8F13.6)
  414 FORMAT('    BR            =',8F13.6)
  415 FORMAT('    MR            =',8F13.6)
  416 FORMAT('0  SUSCEP        =',3PE13.2,4X,' (EMU/MOLE)',2X,' M/DE   ='
     1,3PE13.2,4X,' DE       =',3PE13.2)
  500 FORMAT(1H0,19X,'L=0,S=-1/2',3X,'L=1,S=-1/2',3X,'L=2,S=-1/2',3X,'L=
     13,S=-1/2',3X,'L=0,S=+1/2',3X,'L=1,S=+1/2',3X,'L=2,S=+1/2',3X,
     2'L=3,S=+1/2')
  501 FORMAT('    D-NY          =',8F13.6)
  502 FORMAT('    E-NY          =',8F13.6)
  510 FORMAT('0  ATOM:',6X,A4,6X,'** S =',F13.6,8X,'SWS =',F13.6,' **',
     1/)
  540 FORMAT('    FERMI ENERGY =',F13.6,' TMAG =',F13.6,' QTI =',F13.6,
     12F13.6)
  541 FORMAT('    N.O.S.-EF     =',8F13.6)
  545 FORMAT('    D.O.S.-EF     =',8F13.6)
  550 FORMAT('    N*<EPS>**',I1,'    =',8F13.6)
  570 FORMAT('    V(S) UP,DOWN =',F13.6,39X,F13.6)
C
C     SHIFT THE ENERGY BANDS FOR SUSCEPTIBILITY CALCULATION
C
      IF(NS.EQ.1) GO TO 47
      DO 21 IT=1,NT
      DO 21 IL=1,NL
      OMMLM1(IT,IL,1)=SHIFTE+OMMLM1(IT,IL,1)
      OMMLM1(IT,IL,2)=-SHIFTE+OMMLM1(IT,IL,2)
      CC(IT,IL,1)=SHIFTE+CC(IT,IL,1)
   21 CC(IT,IL,2)=-SHIFTE+CC(IT,IL,2)
C
C     TEST FOR CONVERGENCE
C
   47 SM=0.
      DO 44 IT=1,NT
      DO 44 IL=1,NL
      DO 44 IS=1,NS
      IF(NTLFX(IT,IL).EQ.1) GO TO 44
      SM=SM+ABS(EMOM(2,IT,IL,IS))
C
   44 CONTINUE
C     THE CONVERGENCE CRITERIA IS EITHER THAT THE SUM OF THE MOMENTS
C     IS SMALLER THAN <CRIT> OR THAT <ITERX> ITERATIONS HAVE BEEN
C     PERFORMED, WHICHEVER HAPPENS FIRST, PROVIDED THAT AT LEAST
C     <ICRIT> ITERATIONS  HAVE BEEN PERFORMED
C
      IF(SM.LE.CRIT.AND.ITER.GT.ICRIT) IEXIT=1
      IF(ITER.EQ.ITERX) IEXIT=1
      IF(ITER.LE.3.OR.IEXIT.EQ.1) WRITE(6,500)
C
C     WRITE POTENTIAL PARAMETERS
C
```

```
      IF(ITER.GT.3.AND.IEXIT.EQ.0) GO TO 45
      DO 20 IT=1,NT
      WRITE(6,510) ITXT(IT),WS(IT),SWS
      WRITE(6,502) ((ENY(IT,IL,IS),IL=1,4),IS=1,NS)
      WRITE(6,501) ((DNY(IT,IL,IS),IL=1,4),IS=1,NS)
      WRITE(6,407) ((D1(IT,IL,IS),IL=1,4),IS=1,NS)
      WRITE(6,408) ((D2(IT,IL,IS),IL=1,4),IS=1,NS)
      WRITE(6,411) ((D3(IT,IL,IS),IL=1,4),IS=1,NS)
      WRITE(6,413) ((AR(IT,IL,IS),IL=1,4),IS=1,NS)
      WRITE(6,414) ((BR(IT,IL,IS),IL=1,4),IS=1,NS)
      WRITE(6,415) ((XMR(IT,IL,IS),IL=1,4),IS=1,NS)
      WRITE(6,399) ((OMMLM1(IT,IL,IS),IL=1,4),IS=1,NS)
      DO 22 IL=1,NL
      DO 22 IS=1,NS
      IF(FAV(IT,IL,IS).EQ.0.) GO TO 22
      VOA(IT,IL,IS)=1./FAV(IT,IL,IS)
   22 HP(IL,IS)=10.*SPHISQ(IT,IL,IS)
      WRITE(6,401) ((HP(IL,IS),IL=1,4),IS=1,NS)
      WRITE(6,402) ((PHIRAT(IT,IL,IS),IL=1,4),IS=1,NS)
      WRITE(6,403) ((VOA(IT,IL,IS),IL=1,4),IS=1,NS)
      WRITE(6,404) (( MY(IT,IL,IS),IL=1,4),IS=1,NS)
      WRITE(6,397) ((TL(IT,IL,IS),IL=1,4),IS=1,NS)
      WRITE(6,398) ((TOP(IT,IL,IS),IL=1,4),IS=1,NS)
      WRITE(6,400) ((CC(IT,IL,IS),IL=1,4),IS=1,NS)
      WRITE(6,412) ((BOT(IT,IL,IS),IL=1,4),IS=1,NS)
      WRITE(6,396) ((VL(IT,IL,IS),IL=1,4),IS=1,NS)
      DO 23 IL=2,NL
      DO 23 IS=1,NS
   23 CDIFF(IL,IS)=(CC(IT,IL,IS)-CC(IT,1,IS))*WS(IT)**2
      WRITE(6,405) ((CDIFF(IL,IS),IL=2,4),IS=1,NS)
      WRITE(6,570) (POTS(IT,IS),IS=1,NS)
   48 WRITE(6,395)
   20 CONTINUE
C
   45 CALL EFERMI
C
C     DETERMINE IONICITY
C
      DO 32 IT=1,NT
      QQ=0.
      DO 33 IS=1,NS
      DO 33 IL=1,NL
   33 QQ=QQ+EMOM(1,IT,IL,IS)
      QQ=QQ/MT(IT)
      QTI(IT)=QQ-NEL(IT)
   32 CONTINUE
C
      IF(ITER.NE.1.OR.NS.EQ.1) GO TO 27
C
C     START SPLITTING
C
      DO 41 IT=1,NT
      IF(SPLIT(IT).LT.0.) ISPLIT=-1
      ISPLIT=1
      DNEF1=EMOM(1,IT,NL,1)
      DNEF2=EMOM(1,IT,NL,2)
      DNEF1=DNEF1-SPLIT(IT)/2.
      EMOM(1,IT,NL,1)=DNEF1
      I=PINDEX(IT,NL,1)+3.*ISPLIT
   42 I=I-ISPLIT
      IF(DNEF1.LT.SPDNOS(I,IT,NL,1)) GO TO 42
      PINDEX(IT,NL,1)=I+(DNEF1-SPDNOS(I,IT,NL,1))/(SPDNOS(I+1,IT,NL,1)-
     1SPDNOS(I,IT,NL,1))
      DNEF2=DNEF2+SPLIT(IT)/2.
      EMOM(1,IT,NL,2)=DNEF2
      J=PINDEX(IT,NL,2)-3.*ISPLIT
   43 J=J+ISPLIT
      IF(DNEF2.GT.SPDNOS(J,IT,NL,2)) GO TO 43
```

```
      PINDEX(IT,NL,2)=J+(DNEF2-SPDNOS(J,IT,NL,2))/(SPDNOS(J+1,IT,NL,2)-
     1SPDNOS(J,IT,NL,2))
   41 CONTINUE
C
   27 CALL MOMENT
C
      TMAG=0.
      IF(NS.EQ.1) GO TO 37
      DO 38 IT=1,NT
      DO 38 IL=1,NL
      AMAGT=EMOM(1,IT,IL,2)-EMOM(1,IT,IL,1)
      AMAG(IT,IL)=AMAGT
      TMAG=TMAG+AMAGT
      IF(ABS(AMAGT).GT.1.E-05) GO TO 40
      STOI(IT,IL)=0.
      GO TO 38
C
C     EVALUATE STONER I
C
   40 STOI(IT,IL)=(CC(IT,IL,1)-CC(IT,IL,2)-2.*SHIFTE)/AMAGT
   38 CONTINUE
      IF(SHIFTE.EQ.0.) GO TO 37
      DEFF=TMAG/SHIFTE
      SUSCEP=2.3762D-06*DEFF
C
C     WRITE NOS AND DOS AT THE FERMI LEVEL
C
   37 WRITE(6,540) EF,TMAG,(QTI(IT),IT=1,NT)
      IF(ITER.GT.3.AND.IEXIT.EQ.0) GO TO 46
      DO 24 IT=1,NT
      WRITE(6,406) ITXT(IT)
      WRITE(6,541) ((EMOM(1,IT,IL,IS),IL=1,4),IS=1,NS)
      WRITE(6,545) ((DOS(IT,IL,IS),IL=1,4),IS=1,NS)
      DO 39 IM=2,3
      IMM=IM-1
   39 WRITE(6,550) IMM,((EMOM(IM,IT,IL,IS),IL=1,4),IS=1,NS)
      IF(NS.EQ.1) GO TO 24
      WRITE(6,409) (AMAG(IT,IL),IL=1,4),TMAG
      WRITE(6,410) (STOI(IT,IL),IL=1,4)
   24 CONTINUE
      IF(SHIFTE.EQ.0.) GO TO 46
      WRITE(6,416) SUSCEP,DEFF,SHIFTE
C
C     OBTAIN ELECTRON DENSITY FROM THE ENERGY MOMENTS OF THE STATE
C     DENSITY
C
C
   46 DO 25 IT=1,NT
      JRI=JRIS(IT)
      XMT=MT(IT)
      DO 25 IR=1,JRI
      DO 25 IS=1,NS
      CHD=0.
      DO 26 IL=1,NL
      FRTLS=F(IR,IT,IL,IS)
      FRTLS1=F1(IR,IT,IL,IS)
   26 CHD=CHD+FRTLS*FRTLS*EMOM(1,IT,IL,IS)+2.*FRTLS*FRTLS1*EMOM(2,IT,IL,
     1IS)+(FRTLS1*FRTLS1+FRTLS*F2(IR,IT,IL,IS))*EMOM(3,IT,IL,IS)
      CHARS(IR,IT,IS)=CHD/XMT
   25 CONTINUE
C
      IF(NS.EQ.2) GO TO 28
      DO 29 IT=1,NT
      JRI=JRIS(IT)
      DO 29 IR=1,JRI
      CHAT=CHARS(IR,IT,1)
      CHARS(IR,IT,1)=.5*CHAT
```

```
   29 CHAR(IT,IR)=CHAT
      GO TO 30
   28 DO 31 IT=1,NT
      JRI=JRIS(IT)
      DO 31 IR=1,JRI
   31 CHAR(IT,IR)=CHARS(IR,IT,1)+CHARS(IR,IT,2)
C
C     RENORMALISE CHARGE DENSITY
C
   30 DO 35 IT=1,NT
      JRI=JRIS(IT)
      PP=XWS(IT)-JRI+1
      I=1
      IF(MOD(JRI,2).EQ.0) I=2
      Z=CHAR(IT,I)*RI(IT,I)
      I=I+1
   36 Z=Z+4.*CHAR(IT,I)*RI(IT,I)+2.*CHAR(IT,I+1)*RI(IT,I+1)
      I=I+2
      IF(I.LT.JRI) GO TO 36
      QJRI=0.05/3.*(Z-CHAR(IT,JRI)*RI(IT,JRI))
      WO=CHAR(IT,JRI-2)*RI(IT,JRI-2)
      W1=CHAR(IT,JRI-1)*RI(IT,JRI-1)
      W2=CHAR(IT,JRI)*RI(IT,JRI)
      QWS=QJRI+.05*(WO*(PP**3/6.-PP*PP/4.+1./12.)
     %    +W1*(PP-PP**3/3.-.6666666666)+W2*(PP**3/6.+PP*PP/4.-5./12.))
      FAC=(NEL(IT)+QTI(IT))/QWS
      DO 34 IR=1,JRI
      CHAR(IT,IR)=FAC*CHAR(IT,IR)
      DO 34 IS=1,NS
   34 CHARS(IR,IT,IS)=FAC*CHARS(IR,IT,IS)
   35 CONTINUE
      RETURN
      END
      SUBROUTINE MOMENT
C     ************************************************************
C     *                                                          *
C     *    <MOMENT> CALCULATES THE ENERGY MOMENTS OF THE S-,P-,AND D- *
C     *    STATE DENSITIES BASED ON CANONICAL SCALING            *
C     *                                                          *
C     ************************************************************
      REAL MY,NOSO,NOSNM1,NOSNP1
      COMMON/PARAMT/OMML1T(4,4,2),SPHI2T(4,4,2),PHIRT(4,4,2),
     1FAVT(4,4,2),DNYT(4,4,2),ENYT(4,4,2),SWST,SWT(4),EMINT(2),DET(2),
     2NE(2),SPDNOS(1001,4,4,2),SPDDOS(1001,4,4,2)
      COMMON/MOM/EMOM(4,2,4,2),PINDEX(2,4,2)
      COMMON/POTENT/BGX(250,2,2),POTS(2,2),WS(2),DEO(4),ITER,ICTRM
      COMMON/PARAM/OMMLM1(2,4,2),SPHISQ(2,4,2),PHIRAT(2,4,2),CC(2,4,2),
     1FAV(2,4,2),MY(2,4,2),DNY(2,4,2),ENY(2,4,2),VL(2,4,2),TL(2,4,2),
     2BOT(2,4,2),TOP(2,4,2),D1(2,4,2),D2(2,4,2),D3(2,4,2),XMR(2,4,2),
     1AR(2,4,2),BR(2,4,2)
      COMMON/NET/RI(2,250),SWP(2),SWS,R1(2),QTR(2),AMDL(2),XWS(2),BLX,
     1CESUM,VOL,EXCHF,CRIT,ICRIT,IXCH,JRIS(2),NQ,NL,NT,NS,NW,NTLFX(6,4),
     2ITP,MT(2),ITXT(4)
      DIMENSION NOUT(2,4,2)
    1 FORMAT(' WARNING FROM MOMENT:',16I5,'   ***************')
C
      DMDNYF(EPST)=-1./SFISQT/(1./EPST+FI1FIT-EPST*FI12MT)
      SOM(DMDNY)=OML1+SPHI2L/(SRATL/(1.+TWOLP1/(DNYMLT+DMDNY))-PHIR)
C
      N=0
      DO 20 IT=1,NT
      SRAT=SWT(IT)/WS(IT)/SWST*SWS
      DO 20 IL=1,NL
      TWOLP1=2.*IL-1.
      DO 20 IS=1,NS
C
```

```
      DNYMLT=DNYT(IT,IL,IS)
      SFISQT=SPHI2T(IT,IL,IS)*(1.+OMML1T(IT,IL,IS)*(1.-PHIRT(IT,IL,IS))/
     %         TWOLP1/SPHI2T(IT,IL,IS))**2
      FI1FIT=1./(TWOLP1*SPHI2T(IT,IL,IS)/(PHIRT(IT,IL,IS)-1.)
     %            -OMML1T(IT,IL,IS))
      FI12MT=FAVT(IT,IL,IS)**2
      NOUT(IT,IL,IS)=0
      OML1=OMMLM1(IT,IL,IS)
      SPHI2L=SPHISQ(IT,IL,IS)*TWOLP1
      FAVLS=FAV(IT,IL,IS)
      PHIR=PHIRAT(IT,IL,IS)
      SRATL=SRAT**TWOLP1
C
      EM1T=EMINT(IS)-2.*DET(IS)
      NIP1=PINDEX(IT,IL,IS)+1.
      STORE=SPDNOS(NIP1,IT,IL,IS)
      SPDNOS(NIP1,IT,IL,IS)=EMOM(1,IT,IL,IS)
      NOSNM1=SPDNOS(1,IT,IL,IS)
      NOSO  =SPDNOS(2,IT,IL,IS)
C
      EPST=EMINT(IS)-ENYT(IT,IL,IS)
      DMDNY=DMDNYF(EPST)
      DA=DMDNY+DNYMLT+IL-1
      SOMG=SOM(DMDNY)
      EPS=SOMG/(1.+(SOMG*FAVLS)**2)
      DNOS=NOSO-NOSNM1
      EMOM1=DNOS*EPS
      EMOM2=DNOS*EPS**2
C
      DO 21 IE=3,NIP1
      NOSNP1=SPDNOS(IE,IT,IL,IS)
      EPST=(EM1T-ENYT(IT,IL,IS)+IE*DET(IS))+1E-44
      DMDNY=DMDNYF(EPST)
      DA=DMDNY+DNYMLT+IL-1
      SOMG=SOM(DMDNY)
      IF(ABS(SOMG*FAVLS).LT.1.) GO TO 22
      NOUT(IT,IL,IS)=NOUT(IT,IL,IS)+1
      GO TO 23
   22 EPS=SOMG/(1.+(SOMG*FAVLS)**2)
      DNOS=NOSNP1-NOSNM1
      EMOM1=EMOM1+DNOS*EPS
      EMOM2=EMOM2+DNOS*EPS**2
   23 NOSNM1=NOSO
      NOSO  =NOSNP1
   21 CONTINUE
C
      EPSEFT=EMINT(IS)-ENYT(IT,IL,IS)+(PINDEX(IT,IL,IS)-1.)*DET(IS)
      DMDNY=DMDNYF(EPSEFT)
      SOMG=SOM(DMDNY)
      EPS=SOMG/(1.+(SOMG*FAVLS)**2)
      EMOM(2,IT,IL,IS)=.5*(EMOM1+(EMOM(1,IT,IL,IS)-NOSNM1)*EPS)
      EMOM(3,IT,IL,IS)=.5*(EMOM2+(EMOM(1,IT,IL,IS)-NOSNM1)*EPS**2)
      SPDNOS(NIP1,IT,IL,IS)=STORE
      N=N+NOUT(IT,IL,IS)
   20 CONTINUE
C
      IF(N.NE.0) WRITE(6,1) NOUT
      RETURN
      END
      FUNCTION TOTNOS(E)
C     ********************************************************************
C     *                                                                  *
C     *    FUNCTION <TOTNOS> CALCULATES THE TOTAL NUMBER OF              *
C     *    ELECTRONS AT ENERGY <E> BASED ON CANONICAL SCALING.          *
C     *    ON RETURN THE PARTIAL CONTRIBUTIONS TO <TOTNOS> ARE          *
C     *    STORED IN <EMOM(1,IT,IL,IS)>                                 *
C     *                                                                  *
C     ********************************************************************
```

247

```
      REAL MY
      COMMON/MOM/EMOM(4,2,4,2),PINDEX(2,4,2)
      COMMON/POTENT/BGX(250,2,2),POTS(2,2),WS(2),DEO(4),ITER,ICTRM
      COMMON/PARAM/OMMLM1(2,4,2),SPHISQ(2,4,2),PHIRAT(2,4,2),CC(2,4,2),
     1FAV(2,4,2),MY(2,4,2),DNY(2,4,2),ENY(2,4,2),VL(2,4,2),TL(2,4,2),
     2BOT(2,4,2),TOP(2,4,2),D1(2,4,2),D2(2,4,2),D3(2,4,2),XMR(2,4,2),
     1AR(2,4,2),BR(2,4,2)
      COMMON/PARAMT/OMML1T(4,4,2),SPHI2T(4,4,2),PHIRT(4,4,2),
     1FAVT(4,4,2),DNYT(4,4,2),ENYT(4,4,2),SWST,SWT(4),EMINT(2),DET(2),
     2NE(2),SPDNOS(1001,4,4,2),SPDDOS(1001,4,4,2)
      COMMON/DENS/DOS(2,4,2),TOTDOS
      COMMON/NET/RI(2,250),SWP(2),SWS,R1(2),QTR(2),AMDL(2),XWS(2),BLX,
     1CESUM,VOL,EXCHF,CRIT,ICRIT,IXCH,JRIS(2),NQ,NL,NT,NS,NW,NTLFX(6,4),
     2ITP,MT(2),ITXT(4)
      DIMENSION RNSX(4)
      DATA RNSX/2.,6.,10.,14./
    1 FORMAT(' WARNING FROM TOTNOS: OUT OF RANGE  N=',I4,' AND NE(',I1,
     %        ')=',I4,'    IT=',I1,'    L=',I1,'  ****************')
    2 FORMAT(1H ,'E,EPS,ET=',3E15.8)
C
C
      TOTNOS=0.
      TOTDOS=0.
      DO 21 IT=1,NT
      SRAT=WS(IT)/SWT(IT)/SWS*SWST
      DO 21 IS=1,NS
      DO 21 IL=1,NL
C
C
      ENERGY SCALING
C
      TWOLP1=2.*IL-1.
      EPS=E-ENY(IT,IL,IS)
      SFISQ=SPHISQ(IT,IL,IS)*(1.+OMMLM1(IT,IL,IS)*(1.-PHIRAT(IT,IL,IS))/
     %      TWOLP1/SPHISQ(IT,IL,IS))**2
      FI1FI=1./(TWOLP1*SPHISQ(IT,IL,IS)/(PHIRAT(IT,IL,IS)-1.)
     %      -OMMLM1(IT,IL,IS))
      FI12M=FAV(IT,IL,IS)**2
      DNYML=DNY(IT,IL,IS)-(IL-1)
      SRATL=SRAT**TWOLP1
      SOMT=OMML1T(IT,IL,IS)+TWOLP1*SPHI2T(IT,IL,IS)/(SRATL/(1.+TWOLP1/
     %(DNYML-EPS/SFISQ/(1.+EPS*FI1FI-EPS*EPS*FI12M)))-PHIRT(IT,IL,IS))
      FT=(SOMT*FAVT(IT,IL,IS))**2
      IF(FT.LE.1.) GO TO 20
      FT=EXP((1.-FT)*3.)
   20 ET=ENYT(IT,IL,IS)+SOMT/(1.+FT)
      SRATL=1./SRATL
      SOMG=OMMLM1(IT,IL,IS)+SPHISQ(IT,IL,IS)/(1./TWOLP1*(PHIRT(IT,IL,IS)
     1*SRATL-PHIRAT(IT,IL,IS))+SRATL*SPHI2T(IT,IL,IS)/
     2(SOMT-OMML1T(IT,IL,IS)))
      F=(SOMG*FAV(IT,IL,IS))**2
      IF(F.GE.1.) F=0.
      PN=(ET-EMINT(IS))/DET(IS)+1.
      PINDEX(IT,IL,IS)=PN
      N=PN
      IF(N.GE.0.AND.N+2.LE.NE(IS)) GO TO 27
      L=IL-1
      WRITE(6,1) N,IS,NE(IS),IT,L
      WRITE(6,2) E,EPS,ET
      DOS(IT,IL,IS)=SPDDOS(NE(IS),IT,IL,IS)
      EMOM(1,IT,IL,IS)=SPDNOS(NE(IS),IT,IL,IS)
      IF(EMOM(1,IT,IL,IS).GE.RNSX(IL)/NS) GO TO 26
      DOS(IT,IL,IS)=(SPDNOS(NE(IS),IT,IL,IS)-SPDNOS(NE(IS)-50,IT,IL,IS))
     %   /50./DET(IS)
      EMOM(1,IT,IL,IS)=SPDNOS(NE(IS),IT,IL,IS)+DOS(IT,IL,IS)*
     %      (PN-NE(IS))*DET(IS)
      IF(EMOM(1,IT,IL,IS).LT.RNSX(IL)/NS) GO TO 26
      EMOM(1,IT,IL,IS)=RNSX(IL)/NS
```

248

```fortran
      DOS(IT,IL,IS)=0.
      GO TO 26
   27 CONTINUE
      P=PN-N
      IF(N.LT.2) GO TO 23
C
C     <ET> INSIDE RANGE OF CALCULATED STATE DENSITY
C
      EMOM(1,IT,IL,IS)=(P*P-1.)*(SPDNOS(N,IT,IL,IS)*(P/2.-1.)
     1                 +SPDNOS(N+2,IT,IL,IS)*P/6.)
     2       +P*(P-2.)*((1.-P)/6.*SPDNOS(N-1,IT,IL,IS)
     3                 -(1.+P)/2.*SPDNOS(N+1,IT,IL,IS))
      DOS(IT,IL,IS)=(P*P-1.)*(SPDDOS(N,IT,IL,IS)*(P/2.-1.)
     1                 +SPDDOS(N+2,IT,IL,IS)*P/6.)
     2       +P*(P-2.)*((1.-P)/6.*SPDDOS(N-1,IT,IL,IS)
     3                 -(1.+P)/2.*SPDDOS(N+1,IT,IL,IS))
   26 DOS(IT,IL,IS)=DOS(IT,IL,IS)*SPHISQ(IT,IL,IS)/SPHI2T(IT,IL,IS)
     A      *(SOMT-OMML1T(IT,IL,IS))**2/(SOMG-OMMLM1(IT,IL,IS))**2
     B      *(1.-FT)/(1.-F) * ((1.+F)/(1.+FT) )**2
      GO TO 24
   23 EMOM(1,IT,IL,IS)=0.
      DOS(IT,IL,IS)=0.
   24 TOTNOS=TOTNOS+EMOM(1,IT,IL,IS)
      TOTDOS=TOTDOS+DOS(IT,IL,IS)
   21 CONTINUE
      RETURN
      END
      SUBROUTINE EFERMI
C     ******************************************************************
C     *                                                                *
C     *    FINDS THE FERMI LEVEL                                       *
C     *                                                                *
C     ******************************************************************
      COMMON/POTENT/BGX(250,2,2),POTS(2,2),WS(2),DEO(4),ITER,ICTRM
      COMMON/FERMI/EF,SHIFTE,NELTOT
C
    3 FORMAT(1H ,'***TOO MANY ITERATIONS IN <FERMI>***  EF =',E16.8,
     1' RES=',E15.8,' NELTOT=',I5)
      N=1
      DEE=0.04/ITER**1.25
      DE=AMAX1(0.001,DEE)
      TEST=1E04
      RES=TOTNOS(EF)-NELTOT
C
    1 EF=EF-SIGN(DE,RES)
      RESL=RES
      RES=TOTNOS(EF)-NELTOT
      N=N+1
      IF(IFIX(TEST*RES).EQ.0) RETURN
      IF(N.GT.40) GO TO 2
      IF(RES*RESL.GT.0.) GOTO 1
      DE=-DE/2.
      GOTO 1
    2 WRITE(6,3)EF,RES,NELTOT
      STOP
      END
      SUBROUTINE WORK
C     ******************************************************************
C     *                                                                *
C     *    <WORK> CALCULATES THE PRESSURE                              *
C     *                                                                *
C     ******************************************************************
      REAL MY
      COMMON/PARAM/OMMLM1(2,4,2),SPHISQ(2,4,2),PHIRAT(2,4,2),CC(2,4,2),
     1FAV(2,4,2),MY(2,4,2),DNY(2,4,2),ENY(2,4,2),VL(2,4,2),TL(2,4,2),
     2BOT(2,4,2),TOP(2,4,2),D1(2,4,2),D2(2,4,2),D3(2,4,2),XMR(2,4,2),
     1AR(2,4,2),BR(2,4,2)
```

```
      COMMON/MOM/EMOM(4,2,4,2),PINDEX(2,4,2)
      COMMON/CHARA/GAR(2,250),GARS(250,2,2),QTI(2),CHAR(2,250),
     1CHARS(250,2,2),WRKTLS(2,4,2),AMAG(2,4)
      COMMON/NET/RI(2,250),SWP(2),SWS,R1(2),QTR(2),AMDL(2),XWS(2),BLX,
     1CESUM,VOL,EXCHF,CRIT,ICRIT,IXCH,JRIS(2),NQ,NL,NT,NS,NW,NTLFX(6,4),
     2ITP,MT(2),ITXT(4)
      COMMON/POT/COR(2,250),SPLIT(2),NZ(2),NEL(2)
      COMMON/POTENT/BGX(250,2,2),POTS(2,2),WS(2),DEO(4),ITER,ICTRM
      DIMENSION PT(2)
    1 FORMAT('    EPS-XC(S)    =',F13.6,7X,' 4*PI*S**2*RHO(S) =',F13.6)
    2 FORMAT('0  PV (',A4,')    =',8F13.6)
    3 FORMAT('0  PRESS (MBAR) =',F13.6,3(3X,A4,F13.6))
C
      PTOT=0.
      ZM=ABS(QTI(ITP))
      DO 20 IT=1,NT
      JRI=JRIS(IT)
      P=XWS(IT)-(JRI-1)
      S=WS(IT)
      S2=S*S
      PIT=0.
C
C     CHARGE DENSITY AT S
C
      CHART=.5*P*(P-1.)*(CHAR(IT,JRI-2)+COR(IT,JRI-2))
     A    +(1.-P*P)*(CHAR(IT,JRI-1)+COR(IT,JRI-1))
     B    +.5*P*(P+1.)*(CHAR(IT,JRI)+COR(IT,JRI))
      CHARU=CHART/S2
      RHO=CHARU
      RHO1=(.5*P*(P-1.)*(CHARS(JRI-2,IT,1)+.5*COR(IT,JRI-2))
     1       +(1.-P*P)*(CHARS(JRI-1,IT,1)+.5*COR(IT,JRI-1))
     2    +.5*P*(P+1.)*(CHARS(JRI,IT,1)+.5*COR(IT,JRI)))/S2
      RHO2=RHO-RHO1
      CALL XCPOT(RHO1,RHO2,RHO,V1,V2,EPSXC)
C
      IF(ABS(QTI(IT)).GE.1.E-06) GO TO 22
      POTM=0.
      GO TO 25
   22 POTM=SIGN(ZM,QTI(IT))*AMDL(IT)
   25 ECOUL=2.*QTI(IT)/S-POTM
      DO 21 IL=1,NL
      TWOLP1=2*IL-1
      L=IL-1
      DO 21 IS=1,NS
      PHIM=DSQRT(SPHISQ(IT,IL,IS)/S)
      FINY=PHIM*(1.+OMMLM1(IT,IL,IS)/TWOLP1/SPHISQ(IT,IL,IS)*
     1(1.-PHIRAT(IT,IL,IS)))
      FID=(PHIRAT(IT,IL,IS)-1.)/TWOLP1/S/PHIM
      FIDD=-3.*FINY*FAV(IT,IL,IS)**2
      DL=DNY(IT,IL,IS)
      SFISQ=S*FINY*FINY
      AA=(DL+L+1.)*(DL-L)+(ENY(IT,IL,IS)-ECOUL-EPSXC)*S2
      BB=2.*DL+1.
      PVP=SFISQ*AA*EMOM(1,IT,IL,IS)+
     1(2.*S*FINY*FID*AA-BB+S2*SFISQ)*EMOM(2,IT,IL,IS)+
     2((FID*FID+FINY*FIDD)*AA-S-FID/FINY*BB+1./SFISQ+2.*S2*S*FINY*FID)*
     3EMOM(3,IT,IL,IS)
      PVP=PVP/3.
      WRKTLS(IT,IL,IS)=PVP
      PIT=PIT+PVP/VOL
      AA=ENY(IT,IL,IS)-EPSXC
   21 CONTINUE
      PIT=PIT-0.5*QTI(IT)*POTM*MT(IT)/3./VOL
      PT(IT)=PIT*147.18
      PTOT=PTOT+PIT
      WRITE(6,2) ITXT(IT),((WRKTLS(IT,IL,IS),IL=1,4),IS=1,NS)
      WRITE(6,1) EPSXC,CHART
   20 CONTINUE
```

250

```
C
      PRESS=147.18*PTOT
      WRITE(6,3) PRESS,(ITXT(IT),PT(IT),IT=1,NT)
      RETURN
      END
      SUBROUTINE PRMW
C     ********************************************************************
C     *                                                                  *
C     *   <PRMW> READS AND UPDATES POTENTIAL PARAMETERS                   *
C     *                                                                  *
C     ********************************************************************
      COMMON/ABLX/BLXA,BLXB,DBLX,IBLX,BNDBLX
      COMMON/NET/RI(2,250),SWP(2),SWS,R1(2),QTR(2),AMDL(2),XWS(2),BLX,
     1CESUM,VOL,EXCHF,CRIT,ICRIT,IXCH,JRIS(2),NQ,NL,NT,NS,NW,NTLFX(6,4),
     2ITP,MT(2),ITXT(4)
      COMMON/POTENT/BGX(250,2,2),POTS(2,2),WS(2),DEO(4),ITER,ICTRM
      COMMON/PARAM/OMMLM1(2,4,2),SPHISQ(2,4,2),PHIRAT(2,4,2),CC(2,4,2),
     1FAV(2,4,2),MY(2,4,2),DNY(2,4,2),ENY(2,4,2),VL(2,4,2),TL(2,4,2),
     2BOT(2,4,2),TOP(2,4,2),D1(2,4,2),D2(2,4,2),D3(2,4,2),XMR(2,4,2),
     1AR(2,4,2),BR(2,4,2)
      COMMON/PARAMT/OMML1T(4,4,2),SPHI2T(4,4,2),PHIRT(4,4,2),
     1FAVT(4,4,2),DNYT(4,4,2),ENYT(4,4,2),SWST,SWT(4),EMINT(2),DET(2),
     2NE(2),SPDNOS(1001,4,4,2),SPDDOS(1001,4,4,2)
      INTEGER TXTA(20),TXTP(20,6),TTXT(6)
      DIMENSION SWSP(6)
    1 FORMAT(20A4)
    2 FORMAT(F10.6,A4)
    3 FORMAT(5F10.6)
    4 FORMAT(1H0,'  ***POTENTIAL PARAMETERS HAVE BEEN WRITTEN ON FILE 4'
     1,/,7X,'VS =',F10.6,/)
C
      REWIND 4
      READ(4,1) TXTA
      DO 20 IT=1,NT
      MTT=MT(IT)
      DO 20 LT=1,MTT
      READ(4,1) (TXTP(I,IT),I=1,20)
      READ(4,2) SWSP(IT),TTXT(IT)
      DO 20 IL=1,NL
   20 READ(4,3) DUMMY
C
C     UPDATE ONLY IF THE RADIUS READ AGREES WITH CURRENT RADIUS
C
      IF(DABS(SWSP(1)-WS(1)).GT.1.E-05) RETURN
C
C     CALCUATE MUFFIN-TIN ZERO IF CORRECTION TO THE ASA WAS
C     ORIGINALLY INCLUDED
C
      VS=0.
      IF(ICTRM.EQ.0) GO TO 23
      SUM=0.
      DO 22 IS=1,NS
      DO 22 IT=1,NT
      MTT=MT(IT)
      DO 22 LT=1,MTT
      SUM=SUM+1.D0
   22 VS=VS+POTS(IT,IS)
      VS=VS/SUM
C
   23 REWIND 4
C
C     MIX POTENTIAL PARAMETERS IF BNDBLX .GT. 0 AND WRITE
C     ON FILE 4
C
      BLXM=1.-BNDBLX
      WRITE(4,1) TXTA
      DO 21 IS=1,NS
```

251

```
      DO 21 IT=1,NT
      MTT=MT(IT)
      DO 21 LT=1,MTT
      WRITE(4,1) (TXTP(I,IT),I=1,20)
      WRITE(4,2) SWSP(IT),TTXT(IT)
      DO 21 IL=1,NL
      ENYQ=ENYT(IT,IL,IS)*BNDBLX+(ENY(IT,IL,IS)-VS)*BLXM
      OMQ=OMML1T(IT,IL,IS)*BNDBLX+OMMLM1(IT,IL,IS)*BLXM
      SFIQ=SPHI2T(IT,IL,IS)*BNDBLX+SPHISQ(IT,IL,IS)*BLXM
      WRITE(4,3) ENYQ,OMQ,SFIQ,PHIRAT(IT,IL,IS),FAV(IT,IL,IS)
   21 CONTINUE
      WRITE(6,4) VS
      RETURN
      END
```

9.6.3 Execution of SCFC

The execution of SCFC requires a data set with total and projected state
densities and number-of-states functions as generated by DDNS and stored
in DOS/YY/B. An example of the additional input, which consists of control
variables and the atomic charge densities, is shown in Table 9.7.

Table 9.7. Input for SCFC

```
      2    1   50    0   00 0001
0.            0.025     0.025     0.005     0.001          R
0.9           0.8       0.01          0     0.0       0.001          20
2.684         2.70
      1
0.
0.
0.4
1.0
0.            0.
0.            0.        0.        0.
CR 3D5 4S1 BARTH-HEDIN                    CORE 6            6.930357569E-051
 6.792870901E-04 7.500376726E-04 8.281039091E-04 9.142364228E-04 1.009263414E-03
 1.114099335E-03 1.229753132E-03. 1.357337613E-03 1.498080235E-03 1.653333767E-03
 1.824589986E-03 2.013493648E-03 2.221857918E-03 2.451681636E-03 2.705168175E-03
 2.984746237E-03 3.293092741E-03 3.633158003E-03 4.008193459E-03 4.421782147E-03
 4.877872251E-03 5.380813991E-03 5.935400186E-03 6.546910853E-03 7.221162236E-03
 7.964560679E-03 8.784161828E-03 9.687735665E-03 1.068383793E-02 1.178188854E-02
 1.299225767E-02 1.432636022E-02 1.579675941E-02 1.741728050E-02 1.920313529E-02
 2.117105877E-02 2.333945862E-02 2.572857910E-02 2.836068033E-02 3.126023449E-02
 3.445414043E-02 3.797195817E-02 4.184616518E-02 4.611243611E-02 5.080994817E-02
 5.598171402E-02 6.167494463E-02 6.794144444E-02 7.483804134E-02 8.242705415E-02
 9.077680052E-02 9.996214814E-02 1.100651123E-01 1.211755033E-01 1.333916263E-01
 1.468210386E-01 1.615813652E-01 1.778011789E-01 1.956209469E-01 2.151940466E-01
 2.366878561E-01 2.602849195E-01 2.861841923E-01 3.146023673E-01 3.457752837E-01
 3.799594196E-01 4.174334685E-01 4.584999987E-01 5.034871923E-01 5.527506612E-01
 6.066753328E-01 6.656773975E-01 7.302063070E-01 8.007468097E-01 8.778210037E-01
 9.619903877E-01 1.053857881E+00 1.154069780E+00 1.263317615E+00 1.382339861E+00
 1.511923444E+00 1.652904988E+00 1.806171731E+00 1.972662022E+00 2.153365309E+00
 2.349321521E+00 2.561619706E+00 2.791395826E+00 3.039829532E+00 3.308139780E+00
 3.597579120E+00 3.909426460E+00 4.244970120E+00 4.605536992E+00 4.992399516E+00
 5.406840409E+00 5.850094774E+00 6.323337506E+00 6.827659757E+00 7.364042331E+00
 7.933325877E+00 8.536177835E+00 9.173056128E+00 9.844169702E+00 1.054943612E+01
 1.128843651E+01 1.206036845E+01 1.286399726E+01 1.369760686E+01 1.455895094E+01
```

Table 9.7 (cont.)

```
1.544520614E+01 1.635292859E+01 1.727801584E+01 1.821567637E+01 1.916040909E+01
2.010599558E+01 2.104550800E+01 2.197133586E+01 2.287523461E+01 2.374839941E+01
2.458156682E+01 2.536514706E+01 2.608938879E+01 2.674457750E+01 2.732126761E+01
2.781054699E+01 2.820433067E+01 2.849567915E+01 2.867913384E+01 2.875106031E+01
2.870998730E+01 2.855692708E+01 2.829566049E+01 2.793296798E+01 2.747878684E+01
2.694627395E+01 2.635175443E+01 2.571453763E+01 2.505658587E+01 2.440202561E+01
2.377649735E+01 2.320634870E+01 2.271768402E+01 2.233529555E+01 2.208150653E+01
2.197497939E+01 2.202953618E+01 2.225305795E+01 2.264653016E+01 2.320330433E+01
2.390864364E+01 2.473961359E+01 2.566536702E+01 2.664785620E+01 2.764298365E+01
2.860217842E+01 2.947435726E+01 3.020820180E+01 3.075465555E+01 3.106951993E+01
3.111600876E+01 3.086710804E+01 3.030758284E+01 2.943547869E+01 2.826298048E+01
2.681651896E+01 2.513605278E+01 2.327350219E+01 2.129036698E+01 1.925462306E+01
1.723705599E+01 1.530724852E+01 1.352949053E+01 1.195890506E+01 1.063809557E+01
9.594595774E+00 8.839342797E+00 8.366316453E+00 8.153384991E+00 8.164293015E+00
8.351629401E+00 8.660529796E+00 9.032804575E+00 9.411442241E+00 9.743047064E+00
9.984143766E+00 1.010066086E+01 1.007086048E+01 9.885406659E+00 9.546720439E+00
9.067482335E+00 8.468509882E+00 7.776272586E+00 7.020307434E+00 6.230769753E+00
5.436305936E+00 4.662369314E+00 3.930053769E+00 3.255439748E+00 2.649420802E+00
2.117935767E+00 1.662512462E+00 1.281020137E+00 9.685298968E-01 7.181927305E-01
5.220613758E-01 3.718023655E-01 2.592655243E-01 1.768988539E-01 1.180144739E-01
7.691903008E-02 4.893770244E-02 3.036849476E-02 1.836841252E-02 1.081928870E-02
6.202921696E-03 3.456404173E-03 1.868985080E-03 9.790922430E-04 4.960446935E-04
2.426044395E-04 1.143183385E-04 5.179487564E-05 2.251545837E-05 9.369598319E-06
3.723770310E-06 1.409916174E-06 5.072583353E-07 1.729480227E-07 5.572166097E-08
1.691542842E-08 4.823296986E-09 1.287577963E-09 3.209064358E-10 7.437950020E-11
1.592897917E-11 3.180538011E-12 5.916130651E-13 9.841963641E-14 1.415593109E-14
1.843625130E-15 2.163004964E-16 2.273901091E-17 0.              0.
0.              0.              0.              0.              0.
CR 3D5 4S1 BARTH-HEDIN                     AMTC24                              1
6.794932114E-04 7.502652620E-04 8.283551865E-04 9.145138356E-04 1.009569661E-03
1.114437392E-03 1.230126283E-03 1.357749477E-03 1.498534804E-03 1.653835444E-03
1.825143628E-03 2.014104608E-03 2.222532101E-03 2.452425554E-03 2.705989006E-03
2.985651898E-03 3.294091960E-03 3.634260406E-03 4.009409654E-03 4.423123831E-03
4.879352320E-03 5.382446658E-03 5.937201120E-03 6.548897326E-03 7.223353281E-03
7.966977273E-03 8.786827091E-03 9.690675072E-03 1.068707955E-02 1.178546330E-02
1.299619965E-02 1.433070694E-02 1.580155223E-02 1.742256495E-02 1.920896153E-02
2.117748202E-02 2.334653969E-02 2.573638494E-02 2.836928463E-02 3.126971837E-02
3.446459316E-02 3.798347799E-02 4.185886016E-02 4.612642516E-02 5.082536204E-02
5.599869652E-02 6.169365387E-02 6.796205424E-02 7.486074271E-02 8.245205701E-02
9.080433547E-02 9.999246845E-02 1.100984961E-01 1.212122558E-01 1.334320827E-01
1.468655665E-01 1.616303677E-01 1.778550982E-01 1.956802675E-01 2.152592996E-01
2.367596230E-01 2.603638370E-01 2.862709572E-01 3.146977418E-01 3.458801013E-01
3.800745910E-01 4.175599886E-01 4.586389532E-01 5.036397661E-01 5.529181461E-01
6.068591361E-01 6.658790516E-01 7.304274792E-01 8.009893123E-01 8.780868055E-01
9.622816251E-01 1.054176870E+00 1.154419030E+00 1.263699840E+00 1.382757993E+00
1.512380648E+00 1.653404676E+00 1.806717576E+00 1.973257970E+00 2.154015595E+00
2.350030679E+00 2.562392585E+00 2.792237598E+00 3.040745700E+00 3.309136192E+00
3.598661968E+00 3.910602283E+00 4.246253807E+00 4.606919757E+00 4.993896917E+00
5.408460303E+00 5.851845306E+00 6.325227068E+00 6.829696949E+00 7.366235901E+00
7.935684661E+00 8.538710673E+00 9.175771770E+00 9.847076696E+00 1.055254268E+01
1.129175039E+01 1.206389673E+01 1.286774622E+01 1.370158171E+01 1.456315564E+01
1.544964313E+01 1.635759851E+01 1.728291729E+01 1.822080558E+01 1.916575962E+01
2.011155805E+01 2.105126978E+01 2.197728080E+01 2.288134280E+01 2.375464700E+01
2.458792590E+01 2.537158561E+01 2.609587078E+01 2.675106310E+01 2.732771360E+01
2.781690730E+01 2.821055730E+01 2.850172320E+01 2.868494697E+01 2.875659650E+01
2.871520502E+01 2.856179187E+01 2.830014797E+01 2.793706737E+01 2.748250492E+01
2.694963959E+01 2.635482355E+01 2.571739868E+01 2.505936580E+01 2.440489625E+01
2.377968230E+01 2.321013057E+01 2.272241214E+01 2.234139408E+01 2.208948311E+01
2.198543435E+01 2.204317257E+01 2.227069252E+01 2.266910576E+01 2.323190402E+01
2.394450711E+01 2.478415653E+01 2.572020397E+01 2.671482782E+01 2.772418884E+01
2.870001205E+01 2.959155367E+01 3.034788425E+01 3.092039096E+01 3.126537785E+01
```

Table 9.7 (cont.)

```
3.134662260E+01 3.113773565E+01 3.062416245E+01 2.980467517E+01 2.869221583E+01
2.731398000E+01 2.571066847E+01 2.393488287E+01 2.204869829E+01 2.012050860E+01
1.822130472E+01 1.642060575E+01 1.478231471E+01 1.336079641E+01 1.219748636E+01
1.131831576E+01 1.073217713E+01 1.043057716E+01 1.038852139E+01 1.056657048E+01
1.091390780E+01 1.137217155E+01 1.187973587E+01 1.237608239E+01 1.280589296E+01
1.312252383E+01 1.329058921E+01 1.328748132E+01 1.310377042E+01 1.274254518E+01
1.221785530E+01 1.155249105E+01 1.077537210E+01 9.918820506E+00 9.015964042E+00
8.098467116E+00 7.194718869E+00 6.328559225E+00 5.518541147E+00 4.777698404E+00
4.113744398E+00 3.529604208E+00 3.024176037E+00 2.593217078E+00 2.230260018E+00
1.927482739E+00 1.676474021E+00 1.468860035E+00 1.296770370E+00 1.153167876E+00
1.032019529E+00 9.282841948E-01 8.380826267E-01 7.584855262E-01 6.871214882E-01
6.223252540E-01 5.623617991E-01 5.062188055E-01 4.530269688E-01 4.025091807E-01
3.545751894E-01 3.092761074E-01 2.667936142E-01 2.273511924E-01 1.911753324E-01
1.584563422E-01 1.293201483E-01 1.038097845E-01 8.187668368E-02 6.338128281E-02
4.810183742E-02 3.574974708E-02 2.598923053E-02 1.845895554E-02 1.279326947E-02
8.641023547E-03 5.680482001E-03 3.629416024E-03 2.250492008E-03 1.352241815E-03
7.857969503E-04 4.404272699E-04 2.379012181E-04 1.235719783E-04 5.961587235E-05
2.819056009E-05 1.136589799E-05 4.311767030E-06 1.561812925E-06 5.390935220E-07
EOD
```

In the first line NW = 2 indicates that we wish to calculate the potential parameters and the pressure at two volumes to be given by the Wigner-Seitz radii, 2.70 a.u. and 2.684 a.u., in Line 4. NS = 1 indicates calculations without spin-polarisation, and ITRX = 50 indicates that we want at most 50 scaling iterations. IXCH = 0 indicates the Barth-Hedin exchange correlation [9.7], ICD = 0 means no action of the subroutine PRTCD which is used to print and punch potentials, etc., and INO = 0 is the standard option for the use of the state-density data on FILE1 and FILE2. NTLFX = 0 for IL = 1 through 4 means that all E_ν's are free to adjust to the centre of gravity of the occupied part of the $t\ell$ band. If one of these four numbers had been different from zero, the corresponding E_ν would be fixed through all the iterations to the value given in Line 11.

In the second line EXCHF is the Slater $X\alpha$ factor, but since IXCH = 0 this is not used. The next four numbers are the energy increments in Rydbergs used in the first two iterations to evaluate the energy derivatives. As mentioned in the paragraph about Step 3 in Sect.9.6, later iterations use a fraction of the relevant bandwidth. Further, RKEY = ' R' indicates that the entries in Line 4 are the radii in atomic units rather than a percentage of the radius used in the original band calculation.

In the third line IBLX = 0 means that the charge density for the N + 1'th iteration will be BLXB = 0.8 times that of iteration N - 1 plus 0.2 times that of iteration N. If IBLX = 1, the mixing starts at BLXA = 0.9 and decreases in steps of DBLX = 0.01. In most applications a mixing of 0.8 will work, but in difficult cases it should be as high as 0.95 - 0.99. The symptoms of too small a mixing are that the Fermi level, the magnetic moment

and charge transfer printed at each iteration oscillate. In very difficult
cases one must use IBLX = 1 and BLXA = 1.0. Also, BNDBLX = 0 indicates that
the subroutine PRMW which updates the potential parameter data set writes
the potential parameters of the last iteration without mixing. Otherwise,
they are mixed with the original parameters, which may be necessary for compounds with appreciable charge transfer. Furthermore, CRIT = 0.001 means that
when the sum of the absolute values of the first-order moments of the state
densities is less than 0.001, convergence is assumed to have occurred, and
ICRIT = 20 indicates that at least 20 iterations are performed.

The fourth line contains the NW Wigner-Seitz radii in atomic units since
RKEY = ' R'. If RKEY is different from ' R', Line 4 should contain the desired
percentage increase in the current Wigner-Seitz radius over that of the original radius.

In the fifth line, MT(IT) = 1 indicates that there is only one atom of
type IT in the unit cell. For an alloy such as Pt_3Sn this line should have
contained the numbers 3 and 1.

The sixth line contains the Madelung constant in terms of the Wigner-
Seitz radius. In the present case it is not used, and therefore set equal
to zero. In the case of the NaCl structure the data should be the number
2.168199 typed twice.

The seventh line contains the guess at the charge transfer, QTRO(IT),
and the eighth line the guess at the spin polarisation in Bohr magnetons to
be used in the first iteration. In the present case there is neither charge
transfer nor a magnetic moment.

The ninth line contains the numbers used to scale the individual atomic
radii as described by (9.10) in Sect.9.3.2.

The tenth line contains the guess at the Fermi level E_F and the small
energy shift ESHIFT used to calculate susceptibilities. The guess at E_F is
not critical, zero is usually adequate, and since we want no susceptibility
calculation, ESHIFT = 0.

The next NT lines contain the guess at $E_{\nu t\ell}$ to be used in the first
iteration. If one or more of the NTLX's in Line 1 is different from zero,
the guesses at $E_{\nu t\ell}$ will be used through all iterations for those $t\ell$ values
which have non-zero NTLFX.

The remainder of the data for SCFC is the standard output of the atomic
programme RHFS, and consists essentially of the core- and the total atomic
charge densities. The first line has 40 characters of text describing the
atomic calculation followed by IDENT which must be 'CORE', the number of
conduction electrons NEL, here 6, the first r value R1 on the radial mesh,
and IFORM which determines the format to be used when reading the following

250 numbers of the core charge density. After that follows one line with 40 characters of text which must be equal to the above 40 characters, an identifier which this time must be 'AMTC', the atomic number NZ, the ionicity ION which here is zero, and the format specification IFORM. Then the total atomic charge density follows, ended by IDENT = 'EOD'.

9.6.4 Sample Output from SCFC

```
                  INPUT DATA FOR SELF-CONSISTENCY PROGRAMME


RPOT:     STRUCTURE CONSTANTS FOR BCC LATTICE
          CORRECTION MATRIX FOR BCC STRUCTURE

          POTENTIAL PARAMETERS FOR NON-MAGNETIC CR          100381

READT:    EB= 0.200000 EF=  1.100000 DE=  0.001000 NE=  901 NB1=    1 N


READT:    NW=  2 NS= 1 ITRX= 50 IXCH= 0 ICD= 0 INO= 1 IFRM=  0
READT:    BNDBLX=     0.0000  AMDL=  0.000000
READT:    CRIT=  0.001000  ICRIT=  20
READT:    SWP=  1.000000
READT:    SPLIT=  0.400000
READT:    QTR=  0.000000
READT:    MT=     1
READT:    EF=  0.000000 SHIFTE=  0.000000
READT:    NTLFX 0001


          CORE: CR 3D5 4S1 BARTH-HEDIN
          NEL=     6 R1=   6.930358E-05


          AMTC: CR 3D5 4S1 BARTH-HEDIN
          NZ=   24ION=    0 CESUM=   0.

MAIN:  AMDL=  0.000000  SW=  2.684000  XWS=212.286446  JRI=  213  VOL=
RENORM: C=  .17991391E+02  CWS=  .17987868E+02  Z=  .22751452E+02   ZWS
                  ORIGINAL POTENTIAL PARAMETERS


             ENY              DNY-L            OM(-)           S*FI**2


     CR      0.461561        -0.508563         0.208272        0.396336
     CR      0.620718        -0.682886         1.080268        0.361979
     CR      0.639058        -2.798975         0.151915        0.040556



                       **START OF CASE NO :  1 **

                       BARTH-HEDIN EXCHANGE

                       S(  0.0 %)  S(  0.0 %)

                AVERAGE:  2.684000    2.684000
                     CR :  2.684000    2.684000

ITER=  1  N=  1  BLX=     0.800000

          L=0,S=-1/2   L=1,S=-1/2    L=2,S=-1/2    L=3,S=-1/2    L
```

```
ATOM:        CR       ** S =     2.684000        SWS =     2.684000 **

E-NY      =      0.000000     0.000000     0.000000     0.000000
D-NY      =     -1.020638     0.233747    -0.207738     0.000000

D1        =      3.671378     4.911190     1.805744     0.000000
DNYD      =      6.223945     8.561112     2.859014     0.000000
D3        =     14.867429    38.141287     6.222979     0.000000
AR        =      0.000071     0.000058    -0.000337     0.000000
BR        =      0.009526     0.008425     0.002252     0.000000
MR        =     -0.000178    -0.000107     0.000263     0.000000

OMEGA(-)  =     -0.008285     1.038303     0.329231     0.000000
10S*PHISQ(-) =   4.026065     3.665119     0.617162     0.000000
PHI(-)/PHI(+)=   0.861571     0.715939     0.146614     0.000000
1/SQ(BM)/S2  =   5.330874     6.291072     1.320618    -1.000000

MY        =      0.689584     0.821702     5.411349     0.000000
TAU       =      0.786347     0.989027    -3.303566     0.000000
A         =      2.237814     3.051189     0.542403     0.000000
C         =     -0.008285     1.010770     0.309967     0.000000
B         =     -0.471823     0.133973    -0.050123     0.000000
V         =     -0.471823    -0.494400    -0.632407     0.000000
CPS*S2,CDS*S2=                7.341129     2.292641     0.000000
V(S) UP,DOWN =  -0.724461

FERMI ENERGY =     0.300273 TMAG =    0.000000 QTI =     0.000041

ATOM:        CR
N.O.S.-EF =      0.658669     0.919101     4.422271     0.000000
D.O.S.-EF =      0.274258     0.678137     4.628240     0.000000
N*<EPS>**1 =    -0.125764    -0.014606     0.475269     0.000000
N*<EPS>**2 =     0.039295     0.018245     0.097703     0.000000

  <PRINTOUT SHORTENED>

ITER= 19  N= 15  BLX=     0.800000
FERMI ENERGY =    -0.003676 TMAG =    0.000000 QTI =    -0.000078

ITER= 20  N= 16  BLX=     0.800000
FERMI ENERGY =    -0.003832 TMAG =    0.000000 QTI =    -0.000029

ITER= 21  N= 17  BLX=     0.800000

              L=0,S=-1/2   L=1,S=-1/2   L=2,S=-1/2   L=3,S=-1/2   L

ATOM:        CR       ** S =     2.684000        SWS =     2.684000 **

E-NY      =     -0.349565    -0.194289    -0.166372     0.000000
D-NY      =     -0.513426     0.320458    -0.818361     0.000000

D1        =      3.982912     4.942974     1.487308     0.000000
DNYD      =      6.582336     8.379156     2.331419     0.000000
D3        =     16.305977    31.720947     4.150549     0.000000
AR        =      0.000039     0.000071     0.000022     0.000000
BR        =      0.008007     0.009607     0.001304     0.000000
MR        =     -0.000316    -0.000120    -0.000944     0.000000

OMEGA(-)  =      0.206201     1.082903     0.151001     0.000000
10S*PHISQ(-) =   3.965862     3.623422     0.408915     0.000000
PHI(-)/PHI(+)=   0.868115     0.710959     0.062163     0.000000
1/SQ(BM)/S2  =   5.558100     6.515989     1.007943    -1.000000

MY        =      0.702943     0.832099     7.260529     0.000000
TAU       =      0.796198     0.981920    -1.207781     0.000000
A         =      2.058766     2.925492     0.159025     0.000000
C         =     -0.143648     0.859509    -0.018686     0.000000
```

257

```
    B         =      -0.599692    -0.009192    -0.293796      0.000000
    V         =      -0.599692    -0.638263    -0.459847      0.000000
    CPS*S2,CDS*S2=                 7.226597     0.900203      0.000000
    V(S) UP,DOWN =   -0.814899

    FERMI ENERGY =   -0.003941 TMAG =      0.000000 QTI =    -0.000017

    ATOM:         CR
    N.O.S.-EF   =     0.618827     0.797849     4.583307      0.000000
    D.O.S.-EF   =     0.064456     0.928394     8.440850      0.000000
    N*<EPS>**1  =    -0.000058    -0.000077    -0.000509      0.000000
    N*<EPS>**2  =     0.007470     0.009985     0.031452      0.000000

    PV ( CR)    =     0.167112     0.252593    -0.492271      0.000000
    EPS-XC(S)   =    -0.632026     4*PI*S**2*RHO(S) =         3.580257

    PRESS (MBAR) =   -0.131870     CR    -0.131870



    MIXED PARAMETERS FOR USE IN BAND CALCULATION  BNDBLX=    0.00000

    ENY,OM,SFISQ =   -0.349565     0.206201     0.396586
    ENY,OM,SFISQ =   -0.194289     1.082903     0.362342
    ENY,OM,SFISQ =   -0.166372     0.151001     0.040892

    ***POTENTIAL PARAMETERS HAVE BEEN WRITTEN ON FILE 4
        VS = -0.814899


MAIN: AMDL=  0.000000  SW=  2.700000  XWS=212.405317  JRI=  213  VOL=
RENORM: C=  .17991391E+02  CWS=  .17988532E+02  Z=  .24356816E+02   ZWS
                 ORIGINAL POTENTIAL PARAMETERS

                ENY            DNY-L          OM(-)           S*FI**2

    CR        0.461561       -0.508563       0.208272        0.396336
    CR        0.620718       -0.682886       1.080268        0.361979
    CR        0.639058       -2.798975       0.151915        0.040556



                      **START OF CASE NO :  2 **

                       BARTH-HEDIN EXCHANGE

                       S(  0.0 %)  S(  0.6 %)

            AVERAGE:   2.684000     2.700000
                CR :   2.684000     2.700000

ITER=  1  N=  1  BLX=      0.800000

               L=0,S=-1/2   L=1,S=-1/2   L=2,S=-1/2   L=3,S=-1/2   L

ATOM:     CR     ** S =     2.700000       SWS =     2.700000 **

E-NY     =     -0.349565    -0.194289    -0.166372     0.000000
D-NY     =     -0.541153     0.297355    -0.923650     0.000000

D1       =      3.962762     4.910753     1.449763     0.000000
DNYD     =      6.523677     8.285783     2.281773     0.000000
D3       =     15.835280    30.057354     4.011975     0.000000
AR       =      0.000061     0.000076    -0.000119     0.000000
BR       =     -0.017815     0.004967     0.000388     0.000000
MR       =     -0.000156    -0.000127    -0.000135     0.000000
```

```
OMEGA(-)       =      0.190449      1.052612      0.134199      0.000000
10S*PHISQ(-) =      3.897457      3.558475      0.392242      0.000000
PHI(-)/PHI(+)=      0.867086      0.708335      0.053348      0.000000
1/SQ(BM)/S2  =      5.492053      6.441997      0.973809     -1.000000

MY             =      0.706460      0.834983      7.403142      0.000000
TAU            =      0.799161      0.981604     -1.153560      0.000000
A              =      2.010264      2.875534      0.137641      0.000000
C              =     -0.159345      0.830950     -0.034674      0.000000
B              =     -0.608030     -0.025275     -0.301830      0.000000
V              =     -0.608030     -0.646540     -0.415285      0.000000
CPS*S2,CDS*S2=                     7.219254      0.908850      0.000000
V(S) UP,DOWN =     -0.809697

FERMI ENERGY =     -0.020386 TMAG =      0.000000 QTI =      0.000096

ATOM:          CR
N.O.S.-EF      =      0.618932      0.797574      4.583589      0.000000
D.O.S.-EF      =      0.065289      0.942220      8.717285      0.000000
N*<EPS>**1   =     -0.007647     -0.011044     -0.053778      0.000000
N*<EPS>**2   =      0.007336      0.009877      0.030314      0.000000


<PRINTOUT SHORTENED>


ITER= 20  N= 19  BLX=      0.800000
FERMI ENERGY =     -0.017100 TMAG =      0.000000 QTI =      0.000021

ITER= 21  N= 20  BLX=      0.800000

                   L=0,S=-1/2   L=1,S=-1/2   L=2,S=-1/2   L=3,S=-1/2   L

ATOM:          CR      ** S =      2.700000          SWS =      2.700000 **

E-NY           =     -0.360699     -0.206541     -0.175303      0.000000
D-NY           =     -0.513478      0.319962     -0.816365      0.000000

D1             =      3.966340      4.931699      1.485589      0.000000
DNYD           =      6.551839      8.325668      2.323112      0.000000
D3             =     16.195082     30.637902      4.118613      0.000000
AR             =      0.000038      0.000073      0.000058      0.000000
BR             =      0.004209      0.004917      0.000709      0.000000
MR             =     -0.000317     -0.000123     -0.000978      0.000000

OMEGA(-)       =      0.202828      1.067321      0.147286      0.000000
10S*PHISQ(-) =      3.900361      3.566938      0.397806      0.000000
PHI(-)/PHI(+)=      0.867582      0.709462      0.060700      0.000000
1/SQ(BM)/S2  =      5.454997      6.454890      0.986836     -1.000000

MY             =      0.706315      0.834594      7.371408      0.000000
TAU            =      0.799051      0.981480     -1.200514      0.000000
A              =      2.000962      2.874928      0.141767      0.000000
C              =     -0.158151      0.832375     -0.031227      0.000000
B              =     -0.606934     -0.024124     -0.299445      0.000000
V              =     -0.606934     -0.645471     -0.458338      0.000000
CPS*S2,CDS*S2=                     7.220934      0.925273      0.000000
V(S) UP,DOWN =     -0.809558

FERMI ENERGY =     -0.017100 TMAG =      0.000000 QTI =      0.000026

ATOM:          CR
N.O.S.-EF      =      0.619068      0.799573      4.581385      0.000000
D.O.S.-EF      =      0.065522      0.932377      8.628612      0.000000
N*<EPS>**1   =      0.000001      0.000001      0.000001      0.000000
N*<EPS>**2   =      0.007261      0.009798      0.029870      0.000000
```

```
PV ( CR)    =     0.156330     0.236493   -0.490184      0.000000
EPS-XC(S)   =    -0.627576          4*PI*S**2*RHO(S) =   3.535799

PRESS (MBAR) =    -0.173803     CR    -0.173803


MIXED PARAMETERS FOR USE IN BAND CALCULATION  BNDBLX=   0.00000

ENY,OM,SFISQ =   -0.360699     0.202828     0.390036
ENY,OM,SFISQ =   -0.206541     1.067321     0.356694
ENY,OM,SFISQ =   -0.175303     0.147286     0.039781
```

10. Self-Consistent Potential Parameters for 61 Metals

In this chapter we list the four standard potential parameters (4.1) for 61
metals as obtained in self-consistent LMTO calculations using the exchange-
correlation potential given by *von Barth* and *Hedin* [10.1]. Table 10.1 containing parameters for d-transition metals was prepared by O.K. Andersen
and D. Glötzel, whom we wish to thank for permission to quote these results.
Table 10.2 was prepared by the author.

The entries in the two tables plus the programmes STR, (COR), and LMTO
allow the reader to reproduce the self-consistent energy bands for 61 metals
at the observed equilibrium radius. The tables may also be used to estimate
the gross features of the electronic structure of these metals by means of
(2.28,29) and (4.2,3,9,10,16).

Table 10.1. Potential parameters at experimental equilibrium atomic volume

Element	S=exp [a.u.]	E_ν s [mRy]	E_ν p [mRy]	E_ν d [mRy]	E_ν f [mRy]	$\omega(-)$ s [mRy]	$\omega(-)$ p [mRy]	$\omega(-)$ d [mRy]
K	4.862	-270	-247	-233	-214	31	347	468
Rb	5.197	-260	-241	-226	-163	21	334	392
Cs	5.656	-243	-229	-214	-212	18	308	266
Ca	4.122	-340	-286	-257	-251	16	418	306
Sr	4.494	-320	-282	-250	-220	2	400	300
Ba	4.652	-271	-246	-222	-219	27	473	236
Sc	3.427	-351	-264	-222	-251	74	642	238
Y	3.761	-337	-260	-229	-217	50	607	276
Lu	3.624	-377	-271	-230	-223	10	645	337
Ti	3.052	-351	-233	-207	-184	186	837	215
Zr	3.347	-337	-226	-219	-190	106	804	266
Hf	3.301	-397	-242	-224	-193	45	804	319
V	2.818	-350	-233	-198	-177	188	1011	189
Nb	3.071	-329	-208	-211	-184	161	1005	251
Ta	3.069	-399	-223	-209	-180	80	982	302
Cr	2.684	-374	-227	-215	-185	215	1099	157
Mo	2.922	-352	-209	-240	-197	189	1092	214
W	2.945	-433	-225	-242	-193	95	1052	264
Mn	2.699	-438	-290	-248	-223	184	1038	104
Tc	2.840	-381	-235	-245	-216	200	1140	155
Re	2.872	-464	-246	-242	-206	98	1094	197
Fe	2.662	-474	-325	-263	-256	182	1053	71
Ru	2.791	-424	-277	-277	-242	205	1160	107
Os	2.825	-504	-285	-269	-220	99	1118	143
Co	2.621	-497	-346	-272	-264	174	1067	50
Rh	2.809	-485	-348	-330	-302	192	1116	60
Ir	2.835	-565	-354	-320	-274	92	1086	84
Ni	2.602	-521	-370	-282	-281	161	1063	30
Pd	2.873	-537	-422	-387	-375	171	1047	23
Pt	2.897	-620	-435	-383	-357	76	1020	34
Cu	2.669	-545	-404	-357	-359	105	961	11
Ag	3.005	-545	-450	-509	-496	89	878	0
Au	3.002	-638	-491	-471	-465	27	911	2

Table 10.1 (cont.)

		K / Rb / Cs	Ca / Sr / Ba	Sc / Y / Lu	Ti / Zr / Hf	V / Nb / Ta	Cr / Mo / W	Mn / Tc / Re	Fe / Ru / Os	Co / Rh / Ir	Ni / Pd / Pt	Cu / Ag / Au
$10\,s\,\phi^2(-)$ [m Ry]	f	1160 / 920 / 771	1747 / 1181 / 989	2422 / 1634 / 1913	3011 / 2012 / 2225	3552 / 2378 / 2533	3733 / 2656 / 2707	3999 / 2814 / 2827	3946 / 2967 / 2908	4367 / 3112 / 2925	4494 / 3061 / 2991	4415 / 2998 / 2899
	s	1031 / 910 / 797	1468 / 1256 / 1323	2318 / 1981 / 2009	2036 / 2629 / 2544	3611 / 3207 / 3040	3958 / 3542 / 3302	3743 / 3701 / 3433	3744 / 3755 / 3480	3772 / 3561 / 3317	3719 / 3223 / 2988	3325 / 2702 / 2562
	p	1045 / 970 / 868	1423 / 1286 / 1386	2171 / 1957 / 2099	2805 / 2567 / 2617	3320 / 3129 / 3117	3624 / 3425 / 3371	3446 / 3566 / 3507	3466 / 3615 / 3569	3507 / 3447 / 3441	3482 / 3165 / 3167	3162 / 2705 / 2796
	d	621 / 544 / 386	411 / 469 / 449	394 / 541 / 646	414 / 614 / 715	418 / 667 / 783	397 / 651 / 773	319 / 601 / 728	280 / 539 / 672	256 / 440 / 569	230 / 337 / 448	174 / 220 / 325
	f	831 / 640 / 549	1630 / 851 / 829	2300 / 1278 / 1603	2867 / 1649 / 1923	3366 / 1980 / 2220	3292 / 2254 / 2363	3740 / 2325 / 2451	3416 / 2429 / 2498	4046 / 2587 / 2443	4146 / 2385 / 2523	4022 / 2221 / 2280
$10^3\,\phi(-)/\phi(+)$	s	846 / 847 / 852	846 / 848 / 965	860 / 863 / 854	866 / 871 / 861	869 / 875 / 866	868 / 876 / 867	863 / 874 / 866	860 / 872 / 864	857 / 868 / 860	853 / 862 / 854	845 / 850 / 843
	p	694 / 711 / 727	685 / 706 / 758	702 / 728 / 722	711 / 746 / 736	715 / 758 / 749	711 / 753 / 746	697 / 744 / 740	691 / 733 / 734	687 / 720 / 723	682 / 709 / 712	671 / 693 / 700
	d	469 / 472 / 431	283 / 354 / 351	173 / 285 / 314	128 / 251 / 285	94 / 225 / 265	67 / 195 / 240	33 / 159 / 207	10 / 129 / 179	-6 / 97 / 146	-21 / 69 / 113	-29 / 56 / 88
	f	402 / 331 / 405	561 / 360 / 448	553 / 405 / 466	550 / 424 / 467	550 / 434 / 469	508 / 449 / 464	556 / 441 / 461	514 / 444 / 458	562 / 471 / 454	565 / 455 / 483	571 / 458 / 471

Table 10.1 (cont.)

$\langle \dot{\phi}_\nu^2 \rangle^{-1/2}$ [π Ry]

		K / Rb / Cs	Ca / Sr / Ba	Sc / Y / Lu	Ti / Zr / Hf	V / Nb / Ta	Cr / Mo / W	Mn / Tc / Re	Fe / Ru / Os	Co / Rh / Ir	Ni / Pd / Pt	Cu / Ag / Au
s		1300 / 1150 / 1040	1840 / 1590 / 1840	3100 / 2710 / 2640	4200 / 3750 / 3490	5070 / 4690 / 4280	5560 / 5200 / 4680	5120 / 5420 / 4850	5060 / 5460 / 4990	5030 / 5080 / 4580	5870 / 4490 / 4030	4190 / 3560 / 3300
p		2000 / 1880 / 1680	2670 / 2470 / 2280	4000 / 3500 / 3960	5020 / 4250 / 4700	5900 / 4890 / 5320	6550 / 5560 / 5880	6470 / 6130 / 6330	6570 / 6580 / 6690	6660 / 6640 / 6730	6600 / 6280 / 6360	5920 / 5330 / 5620
d		1190 / 1120 / 810	810 / 960 / 960	850 / 1140 / 1360	950 / 1320 / 1530	1000 / 1460 / 1700	1000 / 1460 / 1710	860 / 1390 / 1640	800 / 1290 / 1540	770 / 1110 / 1360	730 / 930 / 1140	610 / 710 / 910
f		∞	∞	∞	∞	∞	∞	∞	∞	∞	∞	∞

Table 10.2

	Cs/Fr	Ba/Ra	La/Ac	Ce/Th	Pr/Pa	Nd/U	Pm/Np	Sm/Pu	Eu/Am	Gd/Cm	Tb/Bk	Dy/Cf	Ho/Es	Er/Fm	Tm/Md	Yb/No	Lu/Lw
S_{exp} [a.u.]	5.621 / (5.9)	4.652 / 4.790	3.920 / (3.9)	(3.8) / 3.756	3.818 / 3.430	3.804 / 3.221	3.783 / 3.140	3.768 / 3.181	4.263 / 3.614	3.764 / 3.641	3.720 / (3.55)	3.704 / –	3.687 / –	3.668 / –	3.649 / –	4.052 / –	3.62 / (3.5)
E_ν [mRy] s	-241 / -252	-266 / -276	-235 / -212	-220 / -257	-250 / -161	-265 / -96	-276 / -54	-288 / -110	-310 / -305	-312 / -312	-315 / -291	-324 /	-333 /	-341 /	-348 /	-340 /	-367 / -358
p	-227 / -241	-238 / 200	-182 / 500	-164 / 400	200 / 900	100 / 1300	0 / 1500	-100 / 1200	-271 / 300	-235 / 200	-234 / 250	-239 /	-244 /	-247 /	-250 /	-289 /	-259 / 0
d	-211 / -226	-217 / -226	-181 / -152	-164 / -184	-184 / -129	-191 / -77	-195 / -56	-200 / -91	-239 / -228	-207 / -230	-203 / -209	-206 /	-208 /	-209 /	-210 /	-245 /	-217 / -187
f	– / -225	– / 0	– / -130	– / -143	– / -55	– / 18	– / 44	– / 0	– / -171	– / -175	– / -156						– / 800
$\omega(-)$ [mRy] s	17 / 3	25 / -5	87 / 59	95 / 64	81 / 113	73 / 171	67 / 204	60 / 164	9 / 41	43 / 28	40 / 30	34 /	28 /	22 /	16 /	-15 /	4 / -32
p	312 / 316	467 / 25	754 / 210	815 / 325	331 / 212	409 / 107	497 / 99	583 / 235	501 / 389	676 / 445	684 / 482	676 /	668 /	661 /	656 /	515 /	637 / 603
d	265 / 283	234 / 270	244 / 327	261 / 308	258 / 382	261 / 444	265 / 503	269 / 497	295 / 388	277 / 398	288 / 433	294 /	300 /	308 /	315 /	386 /	330 / 401
f	– / 827	– / 530	– / 755	– / 391	– / 230	– / 175	– / 133	– / 91	– / 34	– / 24	– / 22						– / 1063
$10S\phi^2(-)$ [mRy] s	815 / 684	1324 / 1205	2055 / 2114	2201 / 2257	2128 / 2851	2111 / 3254	2106 / 3516	2091 / 3348	1461 / 2281	2029 / 2197	2065 / 2337	2056 /	2048 /	2044 /	2040 /	1572 /	2 / 2166
p	884 / 830	1384 / 1197	2213 / 1951	2396 / 3138	1959 / 2554	1986 / 2837	2031 / 3040	2072 / 2970	1534 / 2265	2088 / 2234	2125 / 2372	2118 /	2113 /	2112 /	2113 /	1619 /	2097 / 2385

Table 10.2 (cont.)

	Cs Fr	Ba Ra	La Ac	Ce Th	Pr Pa	Nd U	Pm Np	Sm Pu	Eu Am	Gd Cm	Tb Bk	Dy Cf	Ho Es	Er Fm	Tm Md	Yb No	Lu Lw
$10^4 \phi(-)/\phi(+)$ d	390 / 405	450 / 515	613 / 828	660 / 825	627 / 1067	618 / 1248	616 / 1369	612 / 1288	508 / 848	602 / 833	619 / 905	622 / –	625 / –	631 / –	637 / –	615 / –	646 / 849
f	– / 729	– / 497	– / 578	– / 245	– / 170	– / 159	– / 145	– / 109	– / 43	– / 36	– / 37	–	–	–	–	–	– / 1689
s	8528 / 8429	8646 / 8580	8793 / 8799	8802 / 8789	8766 / 8873	8742 / 8927	8723 / 8950	8702 / 8897	8544 / 8698	8655 / 8666	8645 / 8677	8627	8608	8590	8573	8434	8532 / 8520
p	7287 / 7409	7571 / 6711	8287 / 7530	8396 / 7620	7292 / 7611	7353 / 7571	7432 / 7609	7501 / 7680	7270 / 7429	7520 / 7453	7496 / 7533	7438	7387	7343	7302	7115	7218 / 7395
d	4282 / 4673	3501 / 4024	3240 / 3973	3256 / 3762	3185 / 3922	3149 / 4034	3125 / 4116	3102 / 4064	3451 / 3791	3079 / 3809	3090 / 3880	3086	3085	3089	3097	3698	3114 / 3602
f	– / 5415	– / 3091	– / 2792	– / 1100	66	–349	–604	–802	–942	–965	–1028						– / 4532
$\langle \phi^2_\zeta \rangle^{-1/2}$ [meV] s	1074 / 865	1825 / 1620	3062 / 3152	3270 / 3391	3137 / 4348	3080 / 5026	3045 / 5427	2994 / 5178	1925 / 3284	2830 / 3138	2864 / 3314	2823	2787	2770	2725	1879	2634 / 2829
p	1695 / 1579	2278 / 1989	∞ / 2932	∞ / 3226	3215 / 3787	3279 / 4194	3340 / 4455	3393 / 4310	2909 / 3673	3503 / 3643	3601 / 3738	3688	3761	3825	3880	3156	3945 / 4120
d	819 / 882	956 / 1122	1308 / 1796	1404 / 1795	1338 / 2326	1320 / 2723	1315 / 3001	1305 / 2823	1074 / 1858	1281 / 1823	1314 / 1975	1317	1323	1333	1344	1271	1358 / 1817
f	– / 2124	– / 1115	– / 1386	– / 779	– / 682	– / 684	– / 669	– / 569	– / 315	– / 286	– / 297						– / 4040

11. List of Symbols

Below are listed those symbols used most frequently throughout the book. The numbers in the second column refer to that section or, if in brackets, to that equation where the symbols in the first column are defined.

Symbol	Definition in text	Legend
$a_{\ell m}^{jk}$	(2.5)	Expansion coefficients for muffin-tin orbitals, ASA
$a_{n\ell m}^{jk}$	(1.18)	Expansion coefficients for atomic orbitals
A_ℓ	(2.22)	Top of the ℓ band
A_L^{jk}	(6.30)	
$b_{\ell m}^{jk}$	(1.20)	Expansion coefficients for partial waves
B_ℓ	(2.22)	Bottom of the ℓ band
B_L^{jk}	(6.30)	
$B_{\ell'm';\ell m}^k(\kappa)$	(6.10)	KKR structure constants
$c^{\ell''}(\ell'm';\ell m)$	(5.15)	Coefficients tabulated by *Condon* and *Shortley* [11.1]
C_ℓ	(2.22)	Centre of the ℓ band
$C_{LL'L''}$	(5.15)	Gaunt coefficients
c_ℓ^{jk}	(6.33)	ℓ character
$D_\ell(E)$	(2.3)	Logarithmic-derivative function
$D_\ell\{\phi\}$	Sect.3.2	$\dfrac{S}{\phi_\ell(S)} \left.\dfrac{\partial\phi_\ell(r)}{\partial r}\right|_S$
D_ℓ		Logarithmic derivative

Symbol	Definition in text	Legend
$D_{\nu\ell}$	(3.6)	$D_\ell(E_\nu)$
$D_{\nu\ell}^\bullet$	(3.6)	$D_\ell\{\dot{\phi}_\ell\}$
E_j	(1.1)	One-electron energy
E_F	(1.15)	Fermi level
$E_{\nu\ell}$	Sect.4.2	Fixed but arbitrary energy
$E_j(\mathbf{k})$	(1.12)	Energy-band number j
$E^{j\mathbf{k}}$		$E_j(\mathbf{k})$
$E_{n\ell i}(\mathbf{k})$	(2.12)	Unhybridised nℓ band
$E_\ell(D)$	Sect.3.5	Energy function inverse to the logarithmic-derivative function
$\mathbf{G}$	(1.9)	Reciprocal-lattice vector
$\underline{\underline{H}}$	(6.17)	Hamiltonian matrix
$H_{L'L}^k$	(5.46)	LMTO Hamiltonian matrix
$j_\ell(\kappa r)$	Sect.5.1	Spherical Bessel function
$J_\ell(r)$	(6.12)	Augmented spherical Bessel function, ASA
$J_\ell(\kappa r)$	Sect.5.4	Augmented spherical Bessel function
$\mathbf{k}$	(1.5)	Bloch vector
L	(5.5)	ℓm
n_ℓ	(6.42)	Number of ℓ electrons in the atomic sphere
$n(\mathbf{r})$	(1.2)	Electron density
$n(r)$	(2.38) Sect.6.8	Spherically averaged electron density
$n_\ell(\kappa r)$	Sect.5.1	Spherical Neumann function
$n(E)$	(1.14)	Number-of-states function
$n_\ell(E)$	(2.37) (1.14)	ℓ-projected number-of-states function
$N_\ell(r)$	(6.13)	Augmented spherical Neumann function, ASA
$N_\ell(\kappa r)$	Sect.5.4	Augmented spherical Neumann function
$N(E)$	(1.13)	State-density function

Symbol	Definition in text	Legend
$N_\ell(E)$	(6.38)	ℓ-projected state-density function
$o(\varepsilon^n)$		Terms of order higher than ε^n
$\underline{\underline{0}}$	(6.17)	Overlap matrix
$0^k_{L'L}$	(5.47)	LMTO overlap matrix
$p_\ell(E)$	(2.2)	Potential function
$\mathbf{P}(E)$	Sect.2.5	Potential-function vector: $\{P_s, P_p,...\}$
$P_\ell(E)$	(2.9)	Potential function
$\mathbf{q}$	Sect.8.2.1	Position of atom q inside the primitive cell
$\mathbf{Q}$	(5.52), Sect.8.1.1	Position of atom Q
$\mathbf{R}$	(1.6)	Lattice vector in real space
S	(1.8)	Atomic Wigner-Seitz radius
$S^k_{\ell i}$	Sect.2.2	Canonical bands
$S^k_{L'L}$	Sect.6.4	Canonical structure constants
$v(\mathbf{r})$	Sect.1.2	One-electron potential
$v_s(\mathbf{r})$	Sect.7.2,3	Effective one-electron potential
$v_{MT}(r)$	(5.1)	Spherically symmetric potential inside muffin-tin well
V_ℓ	(2.21)	Square-well pseudopotential
V_{MTZ}	Sect.5.1	Muffin-tin zero
w_ℓ	Sect.2.3	ℓ bandwidth on structure-constant scale
W_ℓ	Sect.3.4	ℓ bandwidth on energy scale
$Y_{\ell m}(\hat{r})$	Sect.5.1	Spherical harmonic with phase given by *Condon* and *Shortley* [1.1]
$\alpha^{jk}_{\ell m}$	(5.27)	Expansion coefficient for muffin-tin orbitals, general κ^2
γ_ℓ	(4.16)	Distortion parameter
ε_ℓ	Sect.3.3	$E-E_{\nu\ell}$
η_ℓ	(5.11)	Phase shift

Symbol	Definition in text	Legend
θ		Step function
κ^2	(5.3)	Kinetic energy in the interstitial region
μ_ℓ	(3.39)	Band mass at C_ℓ
σ	Sect.1.2	Spin
τ_ℓ	(3.40)	Band mass at V_ℓ
$\phi_\ell(E,r)$	Sect.3.1	Partial wave normalised to unity over the atomic sphere
$\phi_{\nu\ell}(r)$	(3.4)	$\phi_\ell(E_{\nu\ell},r)$
$\phi_{\nu\ell}$		$\phi_{\nu\ell}(S)$
$\overset{u}{\phi}$	Sect.3.2	$\partial^u\phi/dE^u$
$\dot{\phi}_{\nu\ell}(r)$	(3.5)	$[\partial\phi_\ell(E,r)/\partial E]_{E_{\nu\ell}}$
$<\dot{\phi}_{\nu\ell}^2>$		$\int_0^S \dot{\phi}^2(r)r^2\,dr$
$\Phi_{\ell m}(D,\mathbf{r})$	(3.47)	Trial function with logarithmic derivative D
$\Phi_\ell(D,r)$	(3.44)	Radial trial function
$\Phi_\ell(D)$	(3.46)	$\Phi(D,S)$
$\Phi_\ell(-)$	(4.1)	$\Phi_\ell(-\ell-1)$
$\Phi_\ell(+)$	(4.1)	$\Phi_\ell(\ell)$
$\Phi_\ell\{j\}$		$\Phi_\ell(D\{j\},S)$
$\Phi_\ell\{n\}$		$\Phi_\ell(D\{n\},S)$
$\chi_L(E,\kappa,\mathbf{r})$	Sect.5.2	Muffin-tin orbital
$\chi_{\ell m}(\mathbf{r})$	(6.11)	Muffin-tin orbital, ASA
$\chi_{n\ell m}(\mathbf{r})$	(1.18)	Atomic orbital
$\psi_\ell(E,r)$	Sect.5.1	Partial wave
$\overset{u}{\psi}$	Sect.3.2	$\partial^u\psi/\partial E^u$
$\dot{\psi}_\ell(E,r)$	(1.22)	$\partial\psi_\ell(E,r)/\partial E$
$\omega_\ell(D)$	(3.45)	Omega function

Symbol	Definition of text	Legend
$\omega_\ell(-)$		$\omega_\ell(-\ell-1)$
$\omega_\ell(+)$		$\omega_\ell(\ell)$
$\omega_\ell\{j\}$		$\omega_\ell(D\{j\})$
$\omega_\ell\{n\}$		$\omega_\ell(D\{n\})$
Ω	(1.7)	Volume of primitive cell

References

Chapter 1

1.1 M. Born, J.R. Oppenheimer: Ann. Phys. *87*, 457 (1927)
1.2 D.R. Hartree: Proc. Cambridge Philos. Soc. *24*, 89 (1928)
1.3 V. Fock: Z. Phys. *61*, 126 (1930); *62*, 795 (1930)
1.4 J.C. Slater: "The Self-Consistent Field for Molecules and Solids", in *Quantum Theory of Molecules and Solids*, Vol.4 (McGraw-Hill, New York 1974)
1.5 P. Hohenberg, W. Kohn: Phys. Rev. *136*, B864 (1964)
1.6 W. Kohn, L.J. Sham: Phys. Rev. *140*, A1133 (1965)
1.7 J.F. Cornwell: *Group Theory and Electronic Energy Bands in Solids* (North-Holland, Amsterdam 1969);
J. Zak: *The Irreducible Representations of Space Groups* (Benjamin, New York 1969)
1.8 E.U. Condon, G.H. Shortley: *The Theory of Atomic Spectra* (University Press, Cambridge 1951)
1.9 C. Herring: Phys. Rev. *57*, 1169 (1940)
1.10 J.C. Phillips. L. Kleinman: Phys. Rev. *116*, 287 (1959); *116*, 880 (1959); see also *Solid State Physics*, Vol.24 (Academic, New York 1969)
1.11 A. Zunger, M.L. Cohen: Phys. Rev. B*18*, 5449 (1978); B*20*, 4082 (1979)
1.12 F. Bloch: Z. Phys. *52*, 555 (1928);
N.F. Mott, H. Jones: *The Theory of Metals and Alloys* (Oxford University Press, London 1936)
1.13 V. Heine: "Electronic Structure from the Point of View of the Local Atomic Environment", p.1, and
D.W. Bullet: "The Renaissance and Quantitative Development of the Tight Binding Method", in *Solid State Physics*, Vol.35 (Academic, New York 1980) p. 129
1.14 E.P. Wigner, F. Seitz: Phys. Rev. *43*, 804 (1933); *46*, 509 (1934)
1.15 L. Fritsche, M. Rafat mehr, R. Glocker, J. Noffke: Z. Phys. B*33*, 1 (1979)
1.16 J.C. Slater: Phys. Rev. *51*, 846 (1937)
1.17 J. Korringa: Physica *13*, 392 (1947)
1.18 W. Kohn, N. Rostoker: Phys. Rev. *94*, 1111 (1954)
1.19 O.K. Andersen: Phys. Rev. B*12*, 3060 (1975)
1.20 A.R. Williams, J. Kübler, C.D. Gelatt, Jr.: Phys. Rev. B*19*, 1990 (1979)
1.21 O.K. Andersen: "Comments on the KKR Wavefunctions, Extension of the Spherical Wave Expansion Beyond the Muffin-Tins", in *Computational Methods in Band Theory*, ed. by P.M. Markus, J.F. Janak, A.R. Williams (Plenum, New York 1971) p.178
1.22 O.K. Andersen, R.V. Kasowski: Phys. Rev. B*4*, 1064 (1971)
1.23 R.V. Kasowski, I.K. Andersen: Solid State Commun. *11*, 799 (1972)
1.24 O.K. Andersen: "The Electronic Structure of Transition Metals", in *Winter College on Electrons in Crystalline Solids* (IAEA, Vienna 1972) unpublished
1.25 O.K. Andersen, R.G. Woolley: Mol. Phys. *26*, 905 (1973)

1.26 O.K. Andersen: Solid State Commun. *13*, 133 (1973)
1.27 O.K. Andersen: "Band Structure of Transition Metals", in *Mont Tremblant International Summer School 1973*, unpublished
1.28 R.V. Kasowski: Phys. Rev. B*8*, 1378 (1973)
1.29 R.V. Kasowski: Solid State Commun. *14*, 103 (1974)
1.30 R.V. Kasowksi: Phys. Rev. Lett. *33*, 83 (1974)
1.31 R.V. Kasowski: Solid State Commun. *17*, 179 (1975)
1.32 O. Jepsen: Phys. Rev. B*12*, 2988 (1975)
1.33 O. Jepsen, O.K. Andersen, A.R. Mackintosh: Phys. Rev. B*12*, 3084 (1975)
1.34 D.D. Koelling, G. Arbman: J. Phys. F*5*, 2041 (1975)
1.35 P.M. Holtham, J.-P. Jan, H.L. Skriver: J. Phys. F*7*, 635 (1977)
1.36 H.L. Skriver: Phys. Rev. B*14*, 5187 (1976)
1.37 H.L. Skriver: Phys. Rev. B*15*, 1894 (1977)
1.38 J.-P. Jan, H.L. Skriver: J. Phys. F*7*, 957 (1977)
1.39 J.-P. Jan, H.L. Skriver: J. Phys. F*7*, 1719 (1977)
1.40 H.L. Skriver, H.P. Lengkeek: Phys. Rev. B*19*, 900 (1979)
1.41 A.E. Dunsworth, J.-P. Jan, H.L. Skriver: J. Phys. F*8*, 1427 (1978); F*9*, 261 (1979)
1.42 A.E. Dunsworth, J.-P. Jan, H.L. Skriver: J. Phys. F*9*, 1077 (1979)
1.43 R.M. Boulet, A.E. Dunsworth, J.-P. Jan, H.L. Skriver: J. Phys. F*10*, 2197 (1980)
1.44 T. Jarlborg, G. Arbman: J. Phys. F*6*, 189 (1976); F*7*, 1635 (1977)
1.45 N.E. Christensen: J. Phys. F*8*, L51 (1978)
1.46 O.K. Andersen, W. Klose, H. Nohl: Phys. Rev. B*17*, 1209 (1978)
1.47 D.G. Pettifor: J. Phys. F*7*, 613 (1977)
1.48 D.G. Pettifor: J. Phys. F*7*, 1009 (1977)
1.49 D.G. Pettifor: J. Phys. F*8*, 219 (1978)
1.50 J.C. Duthie, D.G. Pettifor: Phys. Rev. Lett. *38*, 564 (1977)
1.51 L. Hedin, B.I. Lundqvist: J. Phys. C*4*, 2064 (1971)
1.52 U. von Barth, L. Hedin: J. Phys. C*5*, 1629 (1972)
1.53 J. Madsen, O.K. Andersen, U.K. Poulsen, O. Jepsen: "Canonical Band Theory of the Volume and Structure Dependence of the Iron Magnetic Moment", in *Magnetism and Magnetic Materials 1975* (AIP, New York 1976) p.328
1.54 U.K. Poulsen, J. Kollár, O.K. Andersen: J. Phys. F*9*, L241 (1976)
1.55 O.K. Andersen, O. Jepsen: Physica B*91*, 317 (1977)
1.56 O.K. Andersen, J. Madsen, U.K. Poulsen, O. Jepsen, J. Kollár: Physica B*86-88*, 249 (1977)
1.57 O. Jepsen, D. Glötzel, A.R. Mackintosh: Phys. Rev. B*23*, 2684 (1981)
1.58 D. Glötzel: J. Phys. F*8*, L163 (1978)
1.59 H.L. Skriver, O.K. Andersen, B. Johansson: Phys. Rev. Lett. *41*, 42 (1978)
1.60 H.L. Skriver, O.K. Andersen, B. Johansson: Phys. Rev. Lett. *44*, 1230 (1980)
1.61 O.K. Andersen, H.L. Skriver, H. Nohl, B. Johansson: Pure Appl. Chem. *52*, 93 (1980)
1.62 H.L. Skriver, O.K. Andersen, B. Johansson: J. Mag. Mag. Mat. *15-18*, 861 (1980)
1.63 H.L. Skriver, O.K. Andersen: "Self-Consistent Calculations of Ground-State Properties for Ordered Transition Metal Alloys", in *Transition Metals 1977*, ed. by M.J.G. Lee, J.M. Perz, E. Fawcett, Institute of Physics Conf. Ser. No.39, 100 (1978)
1.64 J. Kübler: J. Phys. F*8*, 2301 (1978)
1.65 T. Jarlborg: J. Phys. F*9*, 283 (1979)
1.66 T. Jarlborg: J. Phys. F*9*, 1065 (1979)
1.67 A.J. Freeman, T. Jarlborg: J. Appl. Phys. *50*, 1876 (1979)
1.68 T. Jarlborg, A.J. Freeman: Phys. Rev. Lett. *44*, 178 (1980)
1.69 P.J. Kelly, M.S.S. Brooks: "Cohesive Properties of Uranium Dioxides", in *The Physics of Actinides and Related 4f Materials*, ed. by P. Wachter (North-Holland, Amsterdam 1980) p.81

1.70 D. Glötzel, B. Segall, O.K. Andersen: Solid State Commun. *36*, 403 (1980)
1.71 T. Jarlborg, A.J. Freeman: Phys. Lett. A*74*, 349 (1979)
1.72 D. Glötzel, A.K. McMahan: Phys. Rev. B*20*, 3210 (1979)
1.73 F. Perrot: Phys. Rev. B*21*, 3167 (1980)
1.74 H.L. Skriver, J.-P. Jan: Phys. Rev. B*21*, 1489 (1980)
1.75 A.K. McMahan, H.L. Skriver, B. Johansson: Phys. Rev. B*23*, 5016 (1981)
1.76 A.R. Williams, C.D. Gelatt, V.L. Moruzzi: Phys. Rev. Lett. *44*, 429 (1980)
1.77 O. Jepsen, J. Madsen, O.K. Andersen: Phys. Rev. B*18*, 605 (1978)
1.78 O. Jepsen, J. Madsen, O.K. Andersen: J. Mag. Mag. Mat. *15-18*, 967 (1980)
1.79 R.V. Kasowski: Phys. Rev. B*14*, 3398 (1976)
1.80 M. Posternak, H. Krakauer, A.R. Freeman, D.D. Koelling: Phys. Rev. B*21*, 5601 (1980)
1.81 O. Gunnarsson, J. Harris, R.O. Jones: J. Phys. C*9*, 2739 (1976)
1.82 O. Gunnarsson, J. Harris, R.O. Jones: Phys. Rev. B*15*, 3027 (1977)
1.83 O. Gunnarsson, J. Harris, R.O. Jones: J. Chem. Phys. *67*, 3970 (1977)
1.84 J. Harris, R.O. Jones: J. Chem. Phys. *68*, 1190 (1978)
1.85 J. Harris, R.O. Jones: J. Chem. Phys. *70*, 830 (1979)
1.86 A.R. Mackintosh, O.K. Andersen: "The Electronic Structure of Transition Metals", in *Electrons at the Fermi Surface*, ed. by M. Springford (University Press, Cambridge 1980) p.149
1.87 D. Glötzel, O.K. Andersen, O. Jepsen: Adv. Phys. (To be published)
1.88 O.K. Andersen: "Electronic Structure of Metals" in *Troisieme Cycle de la Physique en Suisee Romande* (Geneve 1978) unpublished
1.89 O.K. Andersen: New methods for the one-electron problem. Europhys. News *12*, 4 (1981);
O.K. Andersen: "Linear Methods in Band Theory", in *The Electronic Structure of Complex Systems*, ed. by P. Phariseau, NATO ASI series (Plenum, New York 1983)

Chapter 2

2.1 O.K. Andersen: Phys. Rev. B*12*, 3060 (1975)
2.2 O.K. Andersen, O. Jepsen: Physica B*91*, 317 (1977)
2.3 A.R. Mackintosh, O.K. Andersen: "The Electronic Structure of Transition Metals", in *Electrons at the Fermi Surface*, ed. by M. Springford (University Press, Cambridge 1980) p.149
2.4 O.K. Andersen: Solid State Commun. *13*, 133 (1973)
2.5 O.K. Andersen: "Electronic Structure of Metals" in *Troisieme Cycle de la Physique en Suisse Romande* (Geneve 1978) unpublished
2.6 P.M. Holtham, J.-P. Jan, H.L. Skriver: J. Phys. F*7*, 957 (1977)
2.7 H.L. Skriver: Phys. Rev. B*14*, 5187 (1976)
2.8 H.L. Skriver: Phys. Rev. B*15*, 1894 (1977)
2.9 J.-P. Jan, H.L. Skriver: J. Phys. F*7*, 1719 (1977)
2.10 A.E. Dunsworth, J.-P. Jan, H.L. Skriver: J. Phys. F*8*, 1427 (1978); F*9*, 261 (1979)
2.11 H.L. Skriver, J.-P. Jan: Phys. Rev. B*21*, 1489 (1980)
2.12 L. Hodges, H. Ehrenreich, R.E. Watson: Phys. Rev. B*5*, 3953 (1972)
2.13 H.L. Skriver: J. Phys. F*11*, 97 (1981)
2.14 O.K. Andersen: New methods for the one-electron problem. Europhys. News *12*, 4 (1981);
O.K. Andersen: "Linear Methods in Band Theory", in *The Electronic Structure of Complex Systems*, ed. by P. Phariseau, NATO ASI series (Plenum, New York 1983)
2.15 O.K. Andersen, W. Klose, H. Nohl: Phys. Rev. B*17*, 1209 (1978)

Chapter 3

3.1 O.K. Andersen: "Band Structure of Transition Metals", in *Mont Tremblant International Summer School 1973*, unpublished
3.2 O.K. Andersen, R.G. Woolley: Mol. Phys. *26*, 905 (1973)
3.3 O.K. Andersen: Phys. Rev. B*12*, 3060 (1975)

Chapter 4

4.1 H.L. Skriver: J. Phys. F*11*, 97 (1981)
4.2 H.A. Antosiewicz: "Bessel Functions of Fractional Order", in *Handbook of Mathematical Functions*, ed. by M. Abramowitz, I.A. Stegun (Dover, New York 1965) p.435

Chapter 5

5.1 O.K. Andersen: "Band Structure of Transition Metals", in *Mont Tremblant International Summer School 1973*, unpublished
5.2 O.K. Andersen, R.G. Woolley: Mol. Phys. *26*, 905 (1973)
5.3 E.U. Condon, G.H. Shortley: *The Theory of Atomic Spectra* (University Press, Cambridge 1951)
5.4 H.A. Antosiewicz: "Bessel Functions of Fractional Order", in *Handbook of Mathematical Functions*, ed. by M. Abramowitz, I.A. Stegun (Dover, New York 1965) p.435
5.5 O.K. Andersen: "Comments on the KKR Wavefunctions, Extension of the Spherical Wave Expansion Beyond the Muffin-Tins", in *Computational Methods in Band Theory*, ed. by P.M. Markus, J.F. Janak, A.R. Williams (Plenum, New York 1971) p.178
5.6 J.D. Talman: *Special Functions* (Benjamin, New York 1968)
5.7 J.H. Wilkinson: *The Algebraic Eigenvalue Problem* (Oxford University Press, London 1965);
R.S. Martin, J.H. Wilkinson: Num. Math. *12*, 377 (1968)
5.8 R.V. Kasowski, O.K. Andersen: Solid State Commun. *11*, 799 (1972)
5.9 O. Gunnarsson, J. Harris, R.O. Jones: J. Phys. C*9*, 2739 (1976)
5.10 O. Gunnarsson, J. Harris, R.O. Jones: Phys. Rev. B*15*, 3027 (1977)
5.11 O. Gunnarsson, J. Harris, R.O. Jones: J. Chem. Phys. *67*, 3970 (1977)
5.12 J. Harris, R.O. Jones: J. Chem. Phys. *68*, 1190 (1978)
5.13 J. Harris, R.O. Jones: J. Chem. Phys. *70*, 830 (1979)

Chapter 6

6.1 O.K. Andersen: "Band Structure of Transition Metals", in *Mont Tremblant International Summer School 1973*, unpublished
6.2 O.K. Andersen: Phys. Rev. B*12*, 3060 (1975)
6.3 B. Segall, F.S. Ham: "The Green's Function Method of Korringa, Kohn and Rostoker for the Calculation of the Electronic Band Structure of Solids", in *Methods of Computational Physics*, Vol.8, ed. by B. Adler, S. Fernbach, M. Rotenburg (Academic, New York 1968)
6.4 G. Lehman: Phys. Status Solidi *38*, 151 (1970)
6.5 O.K. Andersen: Solid State Commun. *13*, 133 (1973)
6.6 O.K. Andersen: New Methods for the one-electron problem.
Europhys. News *12*, 4 (1981);
O.K. Andersen: "Linear Methods in Band Theory", in *The Electronic Structure of Complex Systems*, ed. by P. Phariseau, NATO ASI series (Plenum, New York 1983)
6.7 J.C. Slater, G.F. Koster: Phys. Rev. *94*, 1498 (1954)
6.8 O.K. Andersen, W. Klose, H. Nohl: Phys. Rev. B*17*, 1209 (1978)

6.9 O.K. Andersen, H.L. Skriver, H. Nohl, B. Johansson: Pure Appl. Chem. *52*, 93 (1980)
6.10 H.A. Antosiewicz: "Bessel Functions of Fractional Order", in *Handbook of Mathematical Functions*, ed. by M. Abramowitz, I.A. Stegun (Dover, New York 1965) p.435

Chapter 7

7.1 P. Hohenberg, W. Kohn: Phys. Rev. *136*, B864 (1964)
7.2 W. Kohn, L.J. Sham: Phys. Rev. *140*, A1133 (1965)
7.3 O.K. Andersen: "Electronic Structure of Metals", in *Troisieme Cycle de la Physique en Suisse Romande* (Geneve 1978) unpublished
7.4 A.R. Mackintosh, O.K. Andersen: "The Electronic Structure of Transition Metals", in *Electrons at the Fermi Surface*, ed. by M. Springford (University Press, Cambridge 1980)
7.5 J.F. Janak, V.L. Moruzzi, A.R. Williams: Phys. Rev. B*12*, 1257 (1975)
7.6 H.L. Skriver, O.K. Andersen, B. Johansson: Phys. Rev. Lett. *41*, 42 (1978)
7.7 V.L. Moruzzi, A.R. Williams, J.F. Janak: Phys. Rev. B*15*, 2854 (1977)
7.8 O.K. Andersen, H.L. Skriver, H. Nohl, B. Johansson: Pure Appl. Chem. *52*, 93 (1980)
7.9 H.L. Skriver, O.K. Andersen, B. Johansson: J. Mag. Mag. Mat. *15-18*, 861 (1980)
7.10 L. Hedin, B.I. Lundqvist: J. Phys. C*4*, 2064 (1971)
7.11 U. von Barth, L. Hedin: J. Phys. C*5*, 1629 (1972)
7.12 O. Gunnarsson, B.I. Lundqvist: Phys. Rev. B*13*, 4274 (1976)
7.13 J.C. Duthie, D.G. Pettifor: Phys. Rev. Lett. *38*, 564 (1977)
7.14 D.G. Pettifor: "Theory of the Crystal Structures of Transition Metals at Absolute Zero", in *Metallurgical Chemistry*, ed. by O. Kubaschewsky (HMSO, London 1972); Calphad. *1*, 305 (1977)
7.15 H.L. Skriver: Phys. Rev. Lett. *49*, 1768 (1982)
7.16 D.G. Pettifor: Commun. Phys. *1*, 141 (1976)
7.17 D.A. Liberman: Phys. Rev. B*3*, 2081 (1971)
7.18 R.M. Nieminen, C.H. Hodges: J. Phys. F*6*, 573 (1976)
7.19 H.L. Skriver: J. Phys. F*11*, 97 (1981)

Chapter 8

8.1 O.K. Andersen: Phys. Rev. B*12*, 3060 (1975)
8.2 O.K. Andersen, H.L. Skriver, H. Nohl, B. Johansson: Pure Appl. Chem. *52*, 93 (1980)
8.3 J.C. Slater, G.F. Koster: Phys. Rev. *94*, 1498 (1954)

Chapter 9

9.1 O.K. Andersen: "The Electronic Structure of Transition Metals", in *Winter College on Electrons in Crystalline Solids* (IAEA, Vienna 1972) unpublished
9.2 O.K. Andersen: "Electronic Structure of Metals" in *Troisieme Cycle de la Physique en Suisee Romande* (Geneve 1978) unpublished
9.3 J.H. Wilkinson: *The Algebraic Eigenvalue Problem* (Oxford University Press, London 1965);
R.S. Martin, J.H. Wilkinson: Num. Math. *12*, 377 (1968)
9.4 O. Jepsen, O.K. Andersen: Solid State Commun. *9*, 1763 (1971)
9.5 G. Lehman, M. Taut: Phys. Status Solidi (b) *54*, 469 (1972)
9.6 O. Jepsen: "Electronic Structure of Ytterbium", Thesis, Danmarks tekniske Højskole, Lyngby (1971) unpublished

9.7   U. von Barth, L. Hedin: J. Phys. C5, 1629 (1972)
9.8   D.D. Koelling, B.N. Harmon: J. Phys. C10, 3107 (1977)
9.9   M.E. Rose: *Elementary Theory of Angular Momentum* (Wiley, New York 1957)
9.10 T.L. Loucks: *Augmented Plane Wave Method* (Benjamin, New York 1967)

Chapter 10

10.1 U. von Barth, L. Hedin: J. Phys. C5, 1629 (1972)

Chapter 11

11.1 E.U. Condon, G.H. Shortley: *The Theory of Atomic Spectra* (University
      Press, Cambridge 1951)

Subject Index

Landolt-Börnstein

Numerical Data and Functional Relationships in
Science and Technology
Zahlenwerte und Funktionen aus Naturwissenschaf-
ten und Technik
New Series/Neue Serie

Editor in Chief/Gesamtherausgeber: **K.-H. Hellwege**

Group III/Gruppe III: Crystal and Solid State
Physics/Kristall- und Festkörperphysik

Volume 17/Band 17
Semiconductors/Halbleiter

Editors/Herausgeber: **O. Madelung, M. Schulz,
H. Weiss**

Subvolume a/Teilband a
**Physics of Group IV Elements and III–V
Compounds/Physik der Elemente der IV. Gruppe
und der III–V Verbindungen**

By/Von *D. Bimberg, R. Blachnik, M. Cardona,
P. J. Dean, T. Grave, G. Harbeke, K. Hübner, U. Kauf-
mann, W. Kress, O. Madelung, W. v. Münch,
U. Rössler, J. Schneider, M. Schulz, M. S. Skolnick*
Editor/Herausgeber: **O. Madelung**
1982. 1316 figures. XI, 642 pages
ISBN 3-540-10610-3

Semiconductors/Halbleiter

Editors/Herausgeber: **O. Madelung, M. Schulz,
H. Weiss**

Subvolume b/Teilband b
**Physics of II–VI and I–VII Compounds, Semimag-
netic Semiconductors/Physik der II–VI und
I–II-Verbindungen, semimagnetische Halbleiter**

By/Von *I. Broser, R. Broser, H. Finkenrath,
R. R. Gatazka, H. E. Gumlich, A. Hoffmann, J. Kossut,
E. Mollwo, H. Nelkowski, G. Nimtz,
W. von der Osten, M. Rosenzweig, H. J. Schulz,
D. Theis, D. Tschierse*
Editor/Herausgeber: **O. Madelung**
1982. XI, 543 pages. ISBN 3-540-11308-8

Semiconductors/Halbleiter

Editors/Herausgeber: **O. Madelung, M. Schulz,
M. Weiss**

Subvolume c/Teilband c
**Technology of Si, Ge and SiC/
Technologie von Si, Ge, und SiC**

By/Von *W. Dietze, E. Doering, P. Glasow, W. Lang-
heinrich, A. Ludsteck, H. Mader, A. Mühlbauer,
W. v. Münch, H. Runge, L. Schleicher, M. Schnöller,
M. Schulz, E. Sirtl, E. Uden, W. Zulehner*
Editors: **M. Schulz, H. Weiss**
1984. 740 figures. Approx. 660 pages
ISBN 3-540-11474-2

Semiconductors/Halbleiter

Editors/Herausgeber: **O. Madelung, M. Schulz,
H. Weiss**

Subvolume d/Teilband d
**Technology of III–V, II–VI and Non-tetrahedrally
Bonded Compounds/Technologie der III–V, II–VI
und nicht-tetraedrisch gebundenen Verbindungen**

Editor/Herausgeber: **M. Schulz**
1984. ISBN 3-540-11779-2
In preparation

Semiconductors/Halbleiter

Editors/Herausgeber: **O. Madelung, M. Schulz,
H. Weiss**

Subvolume e/Teilband e
**Physics of Non-Tetrahedrally Bonded Elements and
Binary Compounds I/Physik der nicht-tetraedrisch
gebundenen Elemente und binären Verbindungen I**

By/Von *W. Freyland, A. Goltzené, P. Grosse,
G. Harbeke, H. Lehmann, O. Madelung, W. Richter,
C. Schwab, G. Weiser, H. Werheit, W. Żdanowicz*
Editor/Herausgeber: **O. Madelung**
1983. XIII, 533 pages
ISBN 3-540-11780-6

Semiconductors/Halbleiter

Editors: **O. Madelung, M. Schulz,
H. Weiss**

Subvolume f/Teilband f
**Physics of Non-tetrahedrally Bonded Binary Com-
pounds II/Physik der nicht-tetraedrisch gebundenen
Verbindungen II**
By/Von *R. Clasen, G. Harbeke, A. Krost, F. Lévy,
O. Madelung, K. Maschke, G. Nimtz, B. Schlicht,
F. J. Schmitte, J. Treusch*
Editor/Herausgeber: **O. Madelung**
1983. Approx. 1090 figures. Approx. 575 pages
ISBN 3-540-12160-9

Springer-Verlag
Berlin
Heidelberg
New York
Tokyo

B.C.Eu

Semiclassical Theories of Molecular Scattering

1983. 17 figures. Approx. 240 pages
(Springer Series in Chemical Physics, Volume 26)
ISBN 3-540-12410-1

Contents: Introduction. – Mathematical Preparation and Rules of Tracing. – Scattering Theory of Atoms and Molecules. – Elastic Scattering. – Inelastic Scattering: Coupled-State Approach. – Inelastic Scattering: Time-Dependent Approach.–Curve-Crossing Problems. I. – Curve Crossing Problems. II: Multistate Models. – Curve-Crossing Problems. III: Predissociations. – A Multisurface Scattering Theory.– Scattering of an Ellipsoidal Practicle: The WKB Approximation. – Concluding Remarks.– Appendix 1: Asymptotic Forms for Parabolic Cylinder Functions. – Commonly Used Symbols. – References.– Subject Index.

Superconductivity in Ternary Compounds I

Structural, Electronic, and Lattice Properties

Editors: Ø. Fischer, M. B. Maple

1982. 152 figures. XVI, 283 pages
(Topics in Current Physics, Volume 32)
ISBN 3-540-11670-2

Contents: *Ø. Fischer, M. B. Maple:* Superconducting Terenary Compounds: Prospects and Perspectives. – *R. Chevrel, M. Sergent:* Chemistry and Structure of Ternary Molybdenum Chalcogenides. – *K. Yvon:* Structure and Bonding of Ternary Superconductors. – *R. Flükiger, R. Baillif:* Metallurgy and Structural Transformations in Ternary Molybdenum Chalcogenides. – *J. A. Woollam, S. A. Alterovitz, H.-L. Luo:* Thin-Film Ternary Superconductors. – *H. Nohl, W. Klose, O. K. Andersen:* Band Structures of $M_xMo_6X_8$- and $M_2Mo_6X_6$-Cluster Compounds. – *S. D. Bader, S. K. Sinha, B. P. Schweiss, B. Renker:* Phonons in Ternary Molybdenum Chalcogenide Superconductors. – *F. Pobell, D. Rainer, H. Wühl:* Electron-Phonon Interaction in Chevrel-Phase Compounds. – Subject Index.

The Structure and Properties of Matter

Editor: T. Matsubara

1982. 229 figures. XI, 446 pages
(Springer Series in Solid-State Sciences, Volume 28)
ISBN 3-540-11098-4

Contents: *H. Matsuda:* Atoms as Constitutents of Matter. – *T. Tsuneto:* System of Protons and Electrons. – *T. Tsuneto:* Helium   *T. Tsuneto:* Superfluid Helium 3. – *T. Matsubara:* Metals. – *T. Matsubara:* Non-metals. – *T. Matsubara:* Localized Electron Approximation. – *T. Murao:* Magnetism. – *T. Murao:* Magnetic Properties of Dilute Alloys – the Kondo Effect. – *H. Matsuda:* Random Systems. – *F. Yonezawa:* Coherent Potential Approximation (CPA).– References.– Subject Index.

Springer-Verlag
Berlin
Heidelberg
New York
Tokyo